GW00371922

Inorganic Solids

Inorganic Solids

An Introduction to Concepts in Solid-state Structural Chemistry

D. M. ADAMS

Reader in Physical-Inorganic Chemistry,
University of Leicester

JOHN WILEY & SONS

London · New York · Sydney · Toronto

Library of Congress Cataloging in Publication Data:

Adams, David Michael, 1933–
Inorganic solids; an introduction to concepts in solid-state structural chemistry.

1. Solid-state chemistry. I. Title.

QD478.43 541'.042'1 73–16863
ISBN 0 471 00470 7 Cloth bound
ISBN 0 471 00471 5 Paper bound

Made and printed in Great Britain by
William Clowes & Sons, Limited, London, Beccles and Colchester

To
SUSAN and MARK

'Read not to contradict and confute nor to believe and take for granted, but to weigh and consider.'

Francis Bacon

'Plagiarise, Plagiarise, let no one's work escape your eyes.'

Tom Lehrer

Preface

The establishment of inorganic chemistry in British universities, as in many other countries, was a gradual process which took place largely in the 50s and 60s and even now is incomplete in a few of them. Because the upsurge of interest in inorganic work was spearheaded mainly by new initiatives in synthetic work, especially in organometallic chemistry, the inorganic chemistry presented to undergraduates today tends to be primarily synthetic and mechanistic in nature, although often with a strong physical-methods content. A result of this development seems to be that the study of solids is seriously undervalued in British universities, with a few notable exceptions.

Most chemistry students deal with the solid state in connection with the study of semiconduction and heterogeneous catalysis. This, together with the fact that, by default, we have allowed physicists to appropriate much of this area of science by labelling it 'solid-state physics', tends to give the impression that this is not a field in which inorganic chemists should be involved. On the contrary! Great tracts of inorganic chemistry consist of solids, their structures, properties and reactions. Indeed, a good case can be made for structuring a majority of the inorganic syllabus around a core of solid-state theory, and I hope to do this elsewhere.

This lack of solid-state emphasis is the more serious in that our understanding of the reasons governing adoption of particular solid-state structures has advanced considerably over the same period to the point at which some aspects can be accorded the status of a branch of quantitative science. The extent to which even quite major conceptual advances in solid-state theory have failed to penetrate to the undergraduate textbook literature suggests the need for a broad survey at a level suitable both for lecturers assembling courses dealing with solids and for undergraduates. This is what I have tried to do in this volume.

Throughout the book the principal question I have kept in mind is the classic problem of crystallography—'why does *this* solid have *this* structure?' In seeking to provide answers I have commonly posed an important, and all too rarely asked, subsidiary question—'why is *this* structure adopted and *not* some other one exhibited by other compounds of the same stoichiometry?' Since this book is meant for undergraduates, much basic factual material on structures is included, as are many of the set-pieces of the subject, although I have tried to freshen them up a little. Whenever I was uncertain whether or not to include a description of a particular structure, I omitted it because the predominant aim is to discuss *reasons* for structure adoption: I have assumed that anyone needing further descriptions of structures will consult Wells's *Structural Inorganic Chemistry*, or some other source.

The treatment is non-mathematical throughout mainly because, *ipso facto*, this is likely to make the book more generally acceptable to chemistry students. More seriously, there are many solid-state physics texts which cover the mathematical aspects with varying degrees of sophistication. What I have tried to do is to show the broad picture, the wood rather than the trees; to graft onto this story all the background theory would have lengthened the treatment hopelessly.

A particular emphasis of this book is the attempt to present the subject within a deductive framework: this is especially evident in Chapters 3 and 4. This confers the advantage that the student is caused to think about *reasons* for structure adoption even whilst coping with the basic introductory factual material.

In covering such a wide area, it is probable that I shall, through ignorance, have committed some errors and, perhaps have succeeded only in demonstrating the limitations of my own understanding of this enormous tract of science. Nevertheless, I hope that this book will be judged on its overall approach and philosophy, by the broad view which it seeks to give. The only major area not touched upon is that of alloys: this is mainly because it seems to me to be a specialist subject only marginally relevant in chemistry courses.

Although I started to write this book over six years ago, most of the serious preparation was done during a sabbatical term spent at the University of Waterloo, Ontario, in 1971. I owe an especial debt of gratitude to my colleagues there for their generous hospitality, and for inviting me to deliver a graduate course of lectures on solids, which greatly helped in straightening out some of my ideas. Professor W. B. Pearson also kindly allowed me access to a proof copy of his book on *The Crystal Chemistry and Physics of Metals and Alloys* (Wiley, New York, 1972). The majority of the final manuscript was written in ten weeks in the summer of 1972. I am especially indebted to my inorganic colleagues at Leicester for displaying tolerance well beyond the call of duty in bearing with my maunderings on the subject of solids, whilst I sought to improve my understanding and presentation, and for their many helpful comments.

I am particularly grateful to Professor J. S. Anderson, F.R.S., for a most careful and critical reading of Chapter 9 and for saving me from a number of pitfalls. I also wish to thank Professor J. E. Enderby for his valuable remarks on Chapter 8, and Mr G. Beveridge for reading most of the manuscript and giving me an undergraduate's-eye view of it. I need hardly add that they are in no way responsible for any remaining errors and infelicities. Mrs C. A. Crane drew almost all of the figures, and Miss C. Ainsley somehow turned my ghastly scribblings into a manuscript of notable accuracy; their skilful help greatly eased my task. Finally, I thank my wife for many long hours of proof reading, and for her support in so many ways.

Leicester
April 1973

Contents

CHAPTER 1

Preliminaries

From time immemorial crystals have held the fascination of man. Peking man collected quartz crystals, possibly for use as tools; Australian aborigines use stones such as amethysts in rain-making rites and also attribute to crystals powers of a malevolent nature (Burke, 1966). Even today the attraction of well-cut specimens of an allotrope of carbon is well known.

Down the centuries the symmetry and beauty of crystals have also attracted the attention of scientists. Before the demonstration in 1912 by von Laue, Friedrich and Knipping, that X-rays are diffracted by crystals—an observation which opened the way for the study of their internal structure—only their external forms were susceptible to experimental investigation. Some, with the insight of genius, also made great strides in understanding the internal structure of crystals long before the demonstration of X-ray diffraction, but the crowning glory was the deduction (independently) by E. S. Fedorov (1853–1919), Artur Schoenflies (1853–1928) and W. Barlow (1845–1934), that the internal structure of solids is to be described in terms of just 230 types of array in space. This theoretical understanding underlies all X-ray crystallography; we are the heirs of more than half a century of work by crystallographers who have shown the majority of what is known about the arrangement in space of the atoms of solids.

Our purpose in this book is to look at some of these accumulated facts and, especially, to examine some of the reasons and factors which cause a given solid to have the structure it does.

1.1 WHY STUDY SOLIDS?

The study of solids covers an immense area of science. It involves many chemists, engineers and about half of all graduate physicists. The importance of the subject is due to the tremendous technological importance of solids, as much as to their basic position in any rounded study of the physical sciences; we need mention only such objects as transistors, ferrites, lasers, electro-optic devices, etc., to make the point. The properties of solids are intimately bound up with their structures. If we are to be successful in modifying known structures and in designing new ones with various desirable properties we must first attempt to understand the factors which determine the structures and properties of solids. Good technology is based upon a sound knowledge of basic science, a point overlooked by politicians who snipe at 'ivory-tower' scientists and attempt to prune their budgets. The case for including some study of solids in chemistry courses today cannot therefore be challenged seriously.

If further reasons be needed to emphasize the value of studying the structural chemistry of the solid state, we need only remind ourselves of a few simple facts, such as the following.

(*a*) Many compounds exist only as solids, e.g. many oxides, halides and complex oxides and halides; many hydrated solids and all non-stoichiometric materials.

(*b*) The structure of a compound in the solid state may differ considerably from that which it adopts in other phases and we cannot claim to understand its bonding requirements unless we have covered all states. For example, aluminium trichloride exists in the vapour (at lowish temperatures) as Al_2Cl_6 dimers in which the metal atoms have approximately *tetrahedral* coordination. These dimers are also present in the melt but in non-aqueous solution in donor solvents, L, adducts of the types $AlCl_3L$ (tetrahedral) and $AlCl_3L_2$ (bipyramidal) may be formed where L may be pyridine, amines, etc. In the solid state the metal atoms are *octahedrally* six coordinated to chlorine atoms forming an array of infinite extent in which no molecules are present.

(*c*) Even if we know the structure of one compound of a given formula type, this does not necessarily tell us anything about the structures of other solids of this type; formula is not a reliable guide to structure. The point is well illustrated by considering the diversity of structures adopted by Group Ia and Ib metals in their iodides.

MI	Structure	Coordination of M
AuI	Chains (p. 142)	2, linear
AgI, CuI	Zinc blende (p. 60)	4, tetrahedral
LiI to RbI	Rock salt (p. 54)	6, octahedral
CsI	Caesium chloride (p. 50)	8, cubic

1.2 THEORIES OF CRYSTAL STRUCTURES

Since the discovery of X-ray diffraction by crystals, an enormous body of data has been accumulated on many thousands of crystalline solids. Quite naturally, scientists have sought to understand these results in terms of governing principles, and to use their knowledge in a predictive capacity.

Laue and his coworkers first demonstrated X-ray diffraction in 1912. By the mid-1920s there were already sufficient results for attempts to be made to understand them in terms of some generality. The most notable early survey was that of Goldschmidt whose lecture to the Faraday Society in London on 14th March, 1929, still makes fascinating reading. He sought to understand structures in terms of three factors: (*a*) the ratio of the components, that is the formula of the material; (*b*) the ratio of the radii of the components; (*c*) the effect of polarization. In this he was undoubtedly correct although the much greater number of structures now known allows us to refine and extend these principles considerably. We shall modify Goldschmidt's statement by rejecting the classical concept of polarizability in favour of equivalent, but more precise, description in terms of band theory. In describing dynamic processes, especially the interaction of radiation with matter, the concept of polarizability is invaluable, but we firmly reject its use in the explanation of an essentially *static* entity such as a crystal structure.

The theory of structures was further developed by Pauling (1927), whose best-known work on the subject is described in his book *The Nature of the Chemical Bond*, first published in 1940. Meanwhile, work on metals and alloys was revealing structures which were distinctly puzzling. Materials of composition KNa_2 or Cu_5Zn_8 could hardly be predicted, let alone understood. Great strides in understanding metallic systems were made over the next couple of decades from 1920, the work of Hume-Rothery and coworkers figuring prominently. A survey of this work is readily available in an idiosyncratic book, *Electrons, Atoms, Metals and Alloys* by W. Hume-Rothery, first published in 1947. Three factors were considered as fundamental in alloy structures: (*a*) size of the atoms; (*b*) their relative electronegativities; (*c*) the valence-electron concentration.

In an effort to bring the whole theory of solids, metallic and non-metallic, under one set of rules, Laves, in an important essay published in 1955 (*Crystal Structure and Atomic Size*) enunciated three principles of great generality. They are: (*a*) the space principle, requiring the most efficient use of space; (*b*) the symmetry principle, requiring the highest symmetry; and (*c*) the connection principle, aiming at the highest number of 'connections' between components. The term 'connection' is used here in a special sense which we shall discuss later. There are other factors to be considered in structures but we shall constantly be reminded of the essential validity of Laves's statement although sometimes we shall choose to express his principles in other terms. In the fifteen or so years since this work, arguably the most important advance has been Phillips's and van Vechten's quantitative demonstration of the effect of bond type on adoption of particular structures among AB compounds (1969). Their work was foreshadowed, qualitatively, by Mooser and Pearson (1959) whose ideas we shall also consider in Chapter 5.

1.3 CLASSIFICATION OF CRYSTAL STRUCTURES

As the number of known structures increased, the problem of their classification became important. In organizing the facts of inorganic chemistry, electronic structure, reflected in the Periodic Table, forms the natural basis. No such natural framework suggests itself for codification of structural results; consequently there is a variety of competing approaches, none of which is totally satisfactory. For our purpose we need not be unduly concerned with this matter, but the reader should be aware that it is an important issue, both for obvious practical reasons of codification and because many unsuspected structural interrelations are revealed by pursuit of any scheme of classification. In order to give some shape to our own investigation and to assist our memories, we shall adopt a simplified classification based upon closest-packed lattices; an elaborated version of this due to Lima-de-Faria and Figueiredo (1969) is also available. Other schemes of classification based, for example, upon nets or polyhedra are reviewed by W. B. Pearson (1972) whose own recent contributions to the subject deserve the closest attention.

1.4 IMPERFECTIONS AND NON-STOICHIOMETRY

We have spoken of crystals so far as if they were perfectly regular solids having the ideal composition implied by their formula. In fact, all crystals are imperfect in some degree.

Dislocations occur due to 'mistakes' during crystal growth. They are broadly of two types, screw and edge dislocations, although combinations of them are common. An edge dislocation can be considered as due to insertion of an extra plane of atoms some way into a crystal, much as in Figure 1-1. Distortion is severe where the discontinuity occurs and atoms close to it may not even have their correct coordination number. The lattice is said to be 'dilated' at the line where the new partial plane is inserted.

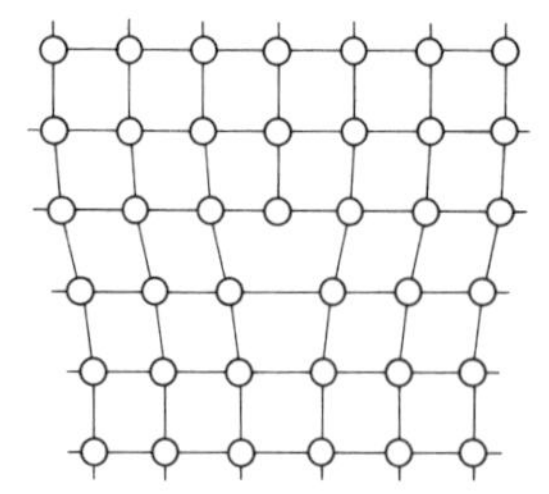

Figure 1-1 An edge dislocation in a crystal lattice

In contrast, a screw dislocation is associated with a sheared lattice, see Figure 1-2. Instead of atoms lying in planes they now lie on a single surface which spirals from one end of the crystal to the other, like a screw. Both the pitch and the chirality of the screw may vary.

The importance of dislocations is very great as many physical properties are affected by them. Furthermore, they figure prominently in reactions at surfaces of solids. Bulk properties of solids are affected by dislocations, as well as by the special properties of boundary regions between grains or crystallites in a compacted solid. Pursuit of these matters would take us straight into engineering aspects of materials science, an entrancing prospect but one which we shall resist.

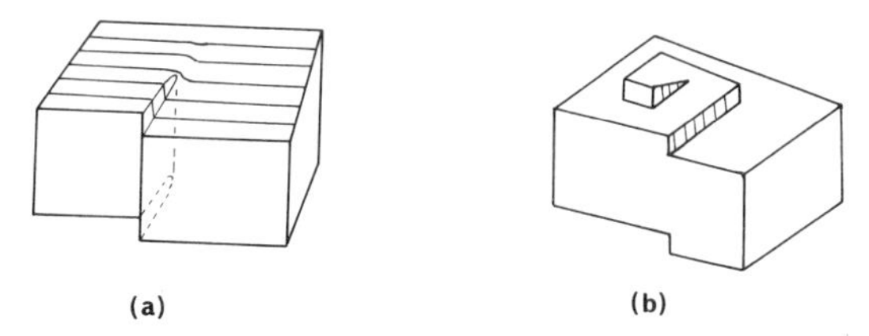

Figure 1-2 (a) A screw dislocation in a crystal, and (b) its further development. The step height is usually one lattice spacing

A further major class of imperfections is associated with *point defects*. Presence of impurity atoms, whether there by accident or design, clearly distorts the host lattice and is often revealed by changes in optical and other properties. Of rather more importance to our theme is the presence of lattice vacancies and interstitial species which are present as an inevitable concomitant of natural energy fluctuations. At all temperatures above the absolute zero all crystals have positive entropy, S, implying a corresponding degree of disorder given by

$$S = k \ln W$$

where k is Boltzmann's constant and W is a measure of the statistical order of the system. This disorder is achieved by the constant creation and removal of point defects of various types.

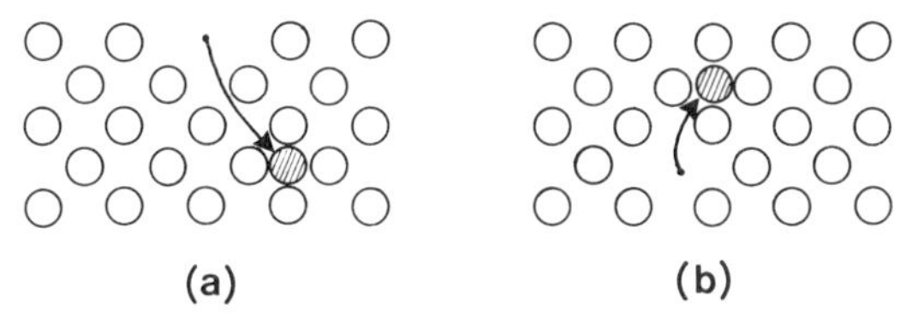

Figure 1-3 (a) Creation of an 'interstitial' defect by insertion of an atom from the surface; (b) creation of an 'interstitial' defect with simultaneous formation of a 'vacancy'

Vacancies are present in all crystals and are caused by removal of a lattice particle to the surface or to an *interstitial* position. Creation of a vacancy by removal to an interstitial position is known as a *Frenkel defect* (1926). These processes are depicted in Figure 1-3. Naturally, this process depends very much upon the relative sizes of the constituents in a non-elemental crystal and it is often found that the occurrence of Frenkel defects is restricted to the sublattice of the smaller component. The energy of formation of Frenkel defects, E_F, usually lies between 42 and 420 $kJmol^{-1}$.

In ionic crystals, vacancy formation occurs in such a way as to maintain electrical neutrality; thus anion and cation vacancies occur in pairs, viz., a *Schottky defect* (1930). The situation is represented in Figure 1–4. If one of the ions bears a multiple charge, it will be balanced by an appropriate number of counter-ion vacancies. The number of Schottky defects is extremely small; for example, one degree below its melting point sodium chloride has only 3×10^{-5} of its lattice sites vacant due to Schottky defects.

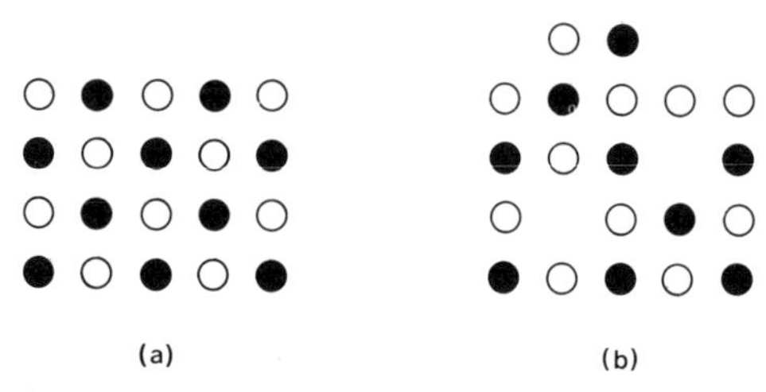

Figure 1-4 Creation of Schottky defects. At absolute zero the ideal crystal (a) is regular; at any other temperature, (b), equal numbers of vacancies are formed in anion and cation sublattices

In general E_F differs from the corresponding Schottky-defect creation energy, E_S, so that creation of defects by one of these processes usually predominates in any one crystal.

Although the above imperfections are all of enormous importance, we shall be more concerned in this book with various other departures (or apparent departures) from perfection. There are many materials that exhibit compositions characterized by formulae of the type AB_x where x takes some non-integral value. For example, the so-called 'ferrous oxide' exhibits a range of composition best represented by the formula $Fe_{1-x}O$ in which $1 - x$ varies from about 0·82 to 0·96. Such materials are termed *non-stoichiometric* to distinguish them from compounds having constituent atoms present in rational proportions A_nB_m where n and m are simple integers. Non-stoichiometry occurs widely among transition metal oxides, chalcogenides, etc.

Viewed from overall composition alone, AB_x, it appears that non-stoichiometric compounds exist in which departures from an apparently ideal formula are anywhere from less than 1% to as much as 15 or 20%. If we view these materials as formed from ideal stoichiometric crystals by removal of a proportion of one component or the insertion of atoms additional to the number of lattice points, the energy required would be enormous. Assuming, reasonably, that the energy of removal of an atom or ion must be comparable to that required for creation of a Frenkel defect it is clear that creation of a number of vacancies amounting to even half a percent would require prohibitive energy.

The whole subject of non-stoichiometry is currently undergoing rapid and exciting development, so much so that by the time this book is in print some of its contents on non-stoichiometry will almost certainly be out of date. But one main statement that is unlikely to change much with time can be made: that, for the most part, non-stoichiometry is not to be regarded as increasingly disorderly departure from structures of ideal composition and regularity but is to be seen in terms of the interplay of new forms and orders of regularity. Indeed, 'the perfect crystal structure has been replaced by an assembly of atoms in statistical mechanical equilibrium' (J. S. Anderson).

In order to understand these exciting new developments, we must first lay a basis by considering the structures of the parent lattices from which various non-stoichiometric systems may be considered as derived.

1.5 BOND TYPE IN CRYSTALS

It is self-evident that the crystal structure adopted by an element or compound is compatible with the bonding requirements of its constituents (atoms, ions or molecules). If we are to reach some understanding of any relations between bond type and crystal structure we should be asking constantly, 'what is the bonding like' in this, or that, crystal? This is not really a very helpful question because only rarely can a simple answer be given. What we would like to know is the electron-density distribution throughout the crystal, but this is not often known with precision. Instead we shall approach the whole problem in another way which has the added advantage of lending itself to a deductive treatment.

We shall ask 'what kind of structure would we expect to get *if* bonding were wholly ionic (or covalent or van der Waals)'? This has the advantage that we need only be concerned with the *dominant* contribution to the bonding, the part that determines the structure type. In this way a basis is laid upon which we can try out the effects of various modifications in bonding, such as the stereochemical effect of inert pairs or of non-bonding *d*-electrons. It sometimes happens that the bonding schemes used by atoms in crystals differ from those normally found in molecular situations. For example, carbon in metal carbides of the type MC is octahedrally coordinated to six metal atoms, using its three orthogonal 2*p*-orbitals as such, rather than hybridized with 2*s*.

Some structures are, in fact, associated with particular bond types. Structures in which only tetrahedral coordination is found (diamond and derivatives) are never adopted unless there is a large covalent contribution to the bonding. In contrast, the rock-salt (NaCl) structure is adopted by a large number of compounds, some of which are undoubtedly highly ionic; others are equally certainly covalently bonded (e.g. carbides, MC) or intermetallic. Since the rock-salt structure is compatible with this range of bond types we cannot take its adoption as indicative of any one type of bond. In such cases we need to seek other evidence.

1.6 ELECTRON CONFIGURATION

The number of valence electrons available for bonding, taken over all of the crystal constituents, broadly determines structure type as follows:

Number of bonding electrons	Structure type
Too few	Metals
Just right	Normal valence compounds; molecular crystals
Too many	Structures distorted by non-bonding electron pairs

In this context 'just right' means that there are enough electrons to form two-electron bonds between constituent atoms. For elements of Group VIIb (the halogens) this naturally means the formation of molecules, but for elements of other Groups in combination, typical ionic (e.g. Ia–VIIb) or directed-valence compounds (e.g. Groups IVb, IIIb–Vb) are formed. If there are electrons in excess of bonding requirements, structures are distorted by the stereochemical effect of these non-bonding pairs which have an influence in solids quite analogous to that well known in molecules (e.g. NH_3). Thus, the crystal structures of the Group Vb and VIb elements are determined in part by the need to accommodate non-bonding pairs. SnS, in which there is an 'inert pair' on tin, has a distorted rock-salt structure (p. 134).

'Too few' valence electrons in this context means fewer than are needed to form electron-pair bonds to all nearest neighbours. Thus in metals and alloys, electrons are shared over all atoms forming metallic bonds of a type which we discuss later.

Many subtler effects are associated with specific electron configurations; we shall meet some of them in due course and will discuss them as they arise.

1.7 THE PLAN OF THIS BOOK

We have already indicated our two-fold aim in this book: to study some of the results of X-ray (and other) diffraction methods in order to lay a foundation for our further work in chemistry or physics, and to try to understand the factors which influence adoption of particular structures. Before we can deal with the second part, which is best described as the 'natural philosophy' of solid-state chemistry, it is clear that we must deal with some of the basic facts upon which we shall later theorize.

There are many ways in which one might cover this ground and organize the material. Three quite different, but equally commendable, ways are represented by the works of Krebs (1968), Naray-Szabo (1969), and Wells (1962). The approach used here differs from all of these in being mainly deductive. I have chosen this way, partly because (so far as I am aware) it has not been tried before in this area, but mainly because it has the merit of causing one to think about the operative principles from the very outset and leads to a natural form of organization. I have done this in two steps: Chapter 3 gets us going with some basic facts, and is then used as a backdrop to Chapter 4 in which the subject is opened out more generally. By this stage some of the principles have emerged and the object of Chapter 5 is to extend this understanding. It is intended that the reader work through from Chapter 1 to Chapter 5 in the given order, but the remaining chapters can be taken in any order.

BIBLIOGRAPHY

Burke, J. G., *Origins of the Science of Crystals*, University of California Press, Berkeley, 1966.

Goldschmidt, V. M., *Trans. Faraday Soc.*, **25**, 253 (1929).

Pauling, L., *Nature of the Chemical Bond*, Cornell, University Press, Ithaca, 1940.

Hume-Rothery, W., *Electrons, Atoms, Metals and Alloys*, 3rd ed., Dover, New York, 1963.

Laves, F., pp 124–198 in *Theory of Alloy Phases*, American Society for Metals, Cleveland (Ohio), 1955.

Pearson, W. B., *The Crystal Chemistry and Physics of Metals and Alloys*, Wiley, London, 1972.

Lime-de-Faria, J., and M. O. Figueiredo, *Zeit. Krist.*, **130**, 41 (1969).

Krebs, H., *Fundamentals of Inorganic Crystal Chemistry*, McGraw-Hill, London, 1968.

Naray-Szabo, I., *Inorganic Crystal Chemistry*, Akademiai Kiado, Budapest, 1969.

Wells, A. F., *Structural Inorganic Chemistry*, 3rd ed., The Clarendon Press, Oxford, 1962.

CHAPTER 2

Some Basic Concepts

In discussing solids we shall frequently use a number of basic concepts and definitions. The object of this chapter is to gather these together.

2.1 LATTICE THEORY

2.1.1 Spatial Periodicity

A crystal is distinguished from a glass (short-range order only) or a fluid by its long-range order or spatial periodicity. In all crystals a small regular volume, the unit cell, can be identified which has all the distinctive features of the crystal structure. It is defined in terms of repeat distances along three axes ($\boldsymbol{a}$, $\boldsymbol{b}$, $\boldsymbol{c}$) and the angles between them (α, β, γ). If units cells were to be laid side by side along the three axial directions they would reproduce the crystal (assuming it is perfect, which it is not!). All points in a crystal can be related to a framework or *lattice*; note that a lattice is defined by the positions of points in space, and does not refer to the lines which we usually draw to assist visualization.

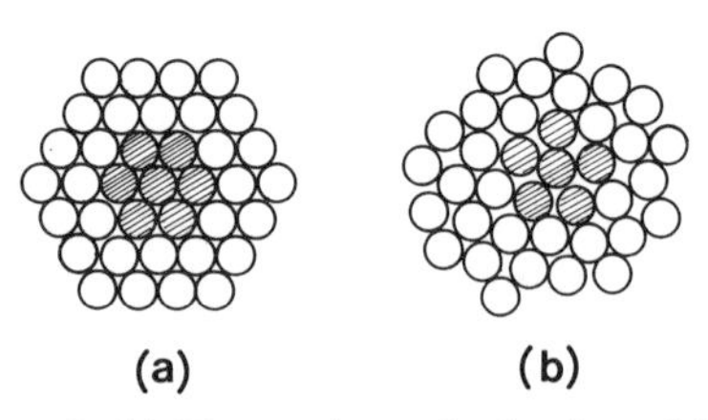

Figure 2-1 Illustration of: (a) Formation of a lattice with hexagonal symmetry. (b) The impossibility of forming a lattice with pentagonal symmetry. (Reproduced from B. L. Smith, *The Inert Gases*, Wykeham Press, 1971)

Spatial periodicity is compatible with a strictly limited number of ways of arranging points in space to give repeatable patterns. This automatically leads to restrictions in the ways in which atoms, ions or molecules, can be assembled to form a crystal. It underlies the fundamental 'symmetry principle' of Laves (p. 3). The restriction is ultimately a geometrical one. Before we formalize this statement (below), a glance at Figure 2-1 may help to make the point.

Development of even an outline of lattice theory would distort the balance of this book. We therefore introduce a few basic ideas which we need and refer the reader to more specialized works for further details, e.g. Phillips (1971).

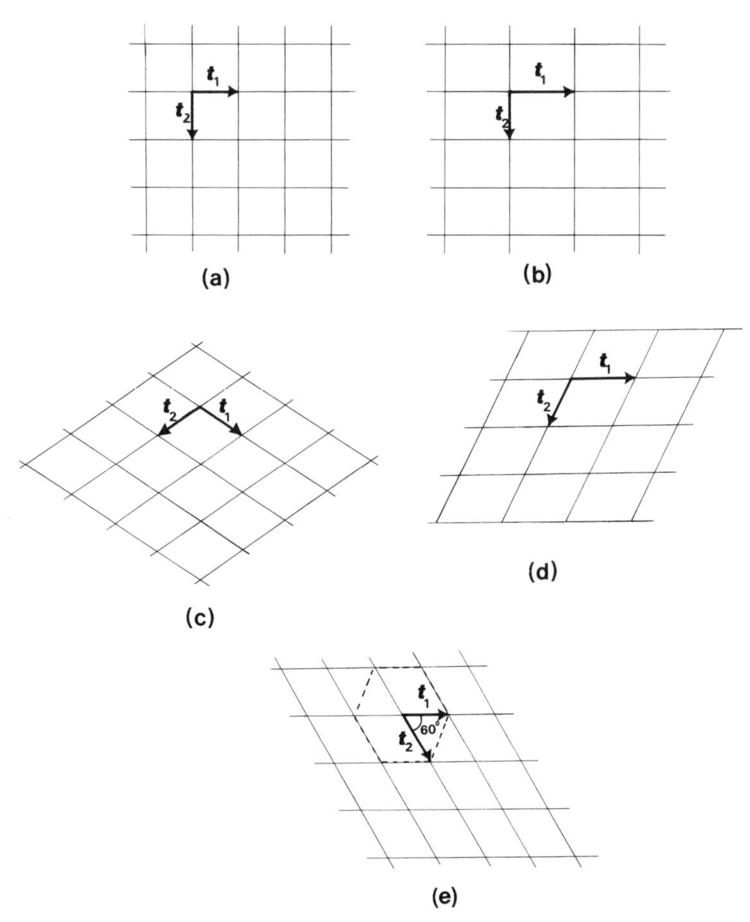

Figure 2-2 The five possible plane lattices: (a) Square, $t_1 = t_2$ and $\theta = 90°$. (b) Rectangular, $t_1 \neq t_2$ and $\theta = 90°$. (c) $t_1 = t_2$ and $\theta = \theta°$. (d) $t_1 \neq t_2$ and $\theta = \theta°$. (e) Hexagonal, $t_1 = t_2$ and $\theta = 60°$

In order to demonstrate just how restrictive the concept of lattice periodicity really is, consider the following argument. A row of points (or atoms) arranged in a straight line with distance $\boldsymbol{t}$ between their centres forms a one-dimensional array or lattice.

$$|\leftarrow \boldsymbol{t} \rightarrow|$$

$$\bullet \quad \bullet \quad \bullet \quad \bullet \quad \bullet \quad \bullet$$

Along the direction of the line there is spatial periodicity; every point in the array is said to be 'translationally equivalent' because a primitive translation of length $\boldsymbol{t}$ along the line brings us to a point identical with the one from which we started.

In a two-dimensional or plane lattice we need *two* repeat distances ($\boldsymbol{t}_1$ and $\boldsymbol{t}_2$) and an angle to define the array. Clearly there exist the possibilities (a) $\boldsymbol{t}_1 = \boldsymbol{t}_2$, ($b$) $\boldsymbol{t}_1 \neq \boldsymbol{t}_2$. Further, the angle θ between the two vectors may take any value, but we also recognize a special case, $\theta = 90°$. Combinations of these possibilities therefore lead to the four planar lattices shown in Figure 2-2a–d. These represent the filling of plane space by the square, rectangle, rhombus and parallelogram respectively. But space can also be filled by other figures, viz., the hexagon and the triangle. However, an equilateral triangle cannot be used as the basis of a lattice because a change of orientation is required as well as a translation if it is to fill plane space. An assembly of triangles is clearly equivalent to a hexagonal array and both can be made the basis of a plane lattice by choosing a 60° rhombus (i.e. $\boldsymbol{t}_1 = \boldsymbol{t}_2$, $\theta = 60°$) as the repeat unit, see Figure 2-3. There are therefore only *five plane lattices.*

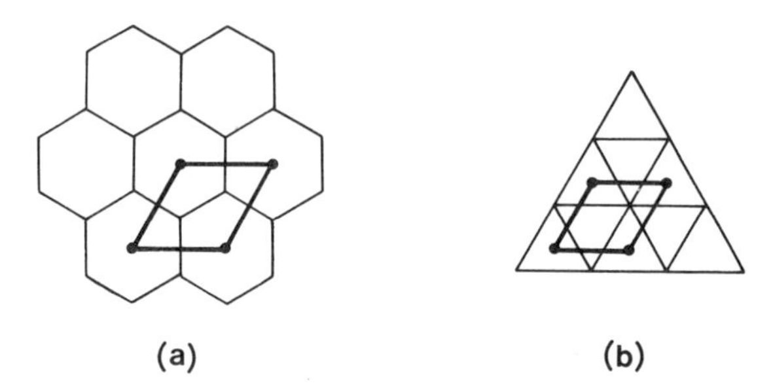

Figure 2-3 Demonstration of: (a) The equivalence of filling plane space with 60° rhombuses, and with hexagons. (b) The relation of this pattern to one based upon equilateral triangles in which, by joining centres of triangles of equivalent orientation, the same rhombohedral repeat is chosen. (After F. C. Phillips, *An Introduction to Crystallography*, Oliver and Boyd, Edinburgh, 1971)

There is no unique way of *choosing a unit cell*; alternative cells for a plane rhombohedral lattice are shown in Figure 2.4. A primitive cell (P) is one having lattice points at the corners only; it is the smallest possible unit cell. In general it is convenient to work with the cell of highest *symmetry* and this is not necessarily primitive. One such cell is shown in Figure 2-4: it is rectangular, has a lattice point at its centre, and is referred to as 'body-centred' (I, German *innenzentrierte*).

When we consider the extension of these ideas to three dimensions, life is a little more complicated but the same principles apply. We now need three vectors ($\boldsymbol{t}_1$, $\boldsymbol{t}_2$, $\boldsymbol{t}_3$) and three angles (α, β, γ) to describe solid figures and, as before, the vectors and the angles may or may not take identical values. Thus, for the special case of $\theta = 90° = \alpha = \beta = \gamma$, we have the following possibilities.

$\boldsymbol{t}_1 = \boldsymbol{t}_2 = \boldsymbol{t}_3 = \boldsymbol{a}$ (say) Simple cubic

$\boldsymbol{t}_1 = \boldsymbol{t}_2 = \boldsymbol{a}; \quad \boldsymbol{t}_3 = \boldsymbol{b}$ Tetragonal

$\boldsymbol{t}_1 = \boldsymbol{a}; \quad \boldsymbol{t}_2 = \boldsymbol{b}; \quad \boldsymbol{t}_3 = \boldsymbol{c}$ Orthorhombic

Now we need to consider all possible combinations of α, β, γ (i.e. $\alpha = \beta = \gamma = \theta$ or $60°$; $\alpha = \beta = 90°, \gamma = \theta$; $\alpha = \theta_1, \beta = \theta_2, \gamma = \theta_3$; etc.) with the three possible combinations of $\boldsymbol{t}_1$, $\boldsymbol{t}_2$ and $\boldsymbol{t}_3$. We shall not complete this exercise, partly because problems of visualization arise but mainly because there is an alternative and easier way of deriving these lattices. But it *is* important to see what happens if we pursue our simple approach.

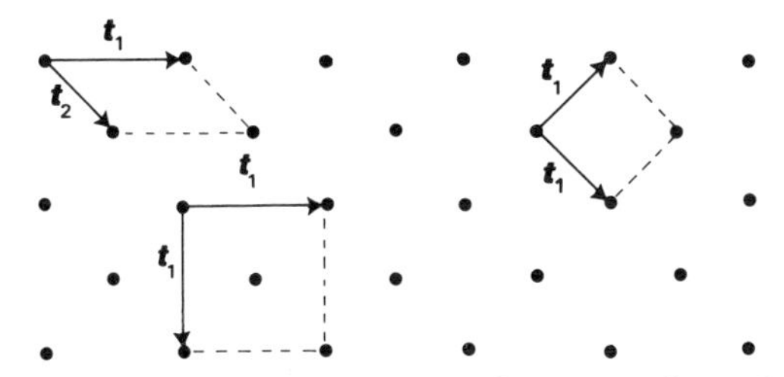

Figure 2-4 Three alternative unit cells for a two-dimensional lattice

The combination of $\alpha = \gamma = 90°$; $\beta = \theta$ with the set $\boldsymbol{t}_1 \neq \boldsymbol{t}_2 \neq \boldsymbol{t}_3$, is equivalent to giving the above orthorhombic lattice a push in one direction. The result is known as the primitive monoclinic lattice, Figure 2-5. This is simple enough but the result of such a procedure is not always obvious. Consider the deformation of the tetragonal lattice ($\boldsymbol{t}_1 = \boldsymbol{t}_2 = \boldsymbol{a}$; $\boldsymbol{t}_3 = \boldsymbol{b}$) such that $\alpha = 90°$, $\beta = \gamma = 60°$. This primitive unit cell is awkward to visualize and has low symmetry; but we *can* choose a body-centred version of it which has very high symmetry and this turns out to be none other than a body-centred cube. Their relationship is depicted in Figure 2-6. In a similar way we can relate the combination of the cubic lattice ($\boldsymbol{t}_1 = \boldsymbol{t}_2 = \boldsymbol{t}_3 = \boldsymbol{a}$) with the angle set ($\alpha = \beta = \gamma = 60°$) to a high-symmetry face-centred (F) cell, Figure 2-6.

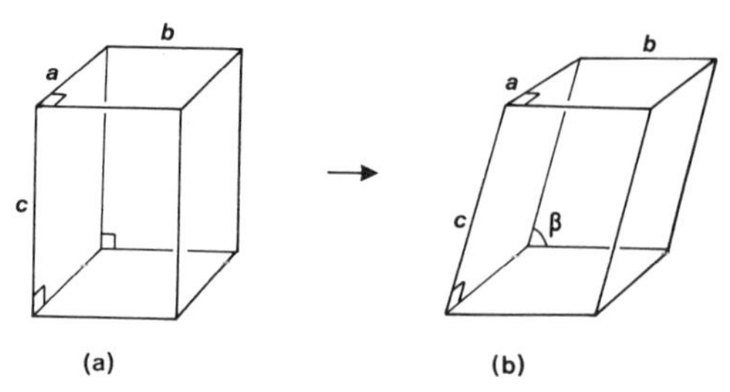

Figure 2-5 Relationship of the primitive orthorhombic lattice to the primitive monoclinic lattice

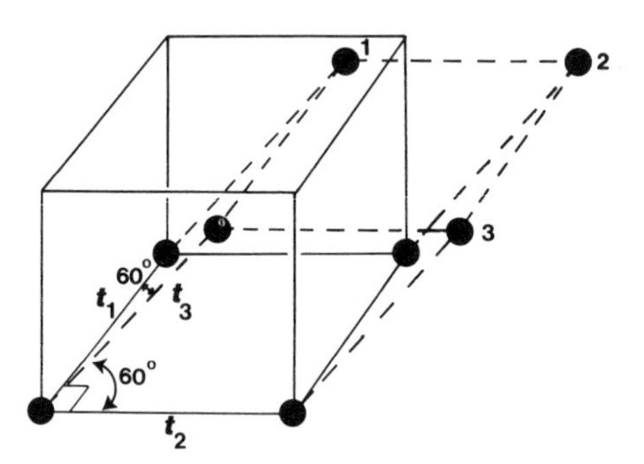

Figure 2-6 Primitive and body-centred unit cells for the lattice $t_1 = t_2 \neq t_3$, $\alpha = 90°$, $\beta = \gamma = 60°$. Atoms 1, 2 and 3 are at body-centre positions in neighbouring unit cells

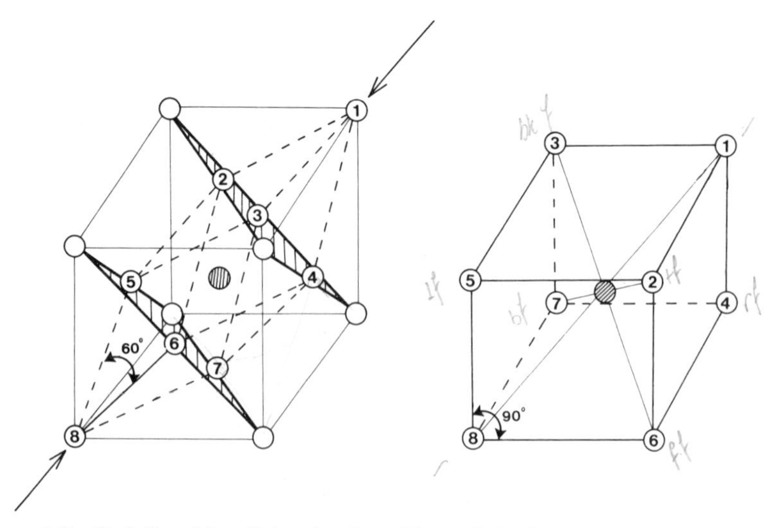

Figure 2-7 Relationship of the simple cubic and the face-centred cubic lattices. The close-packed planes are indicated for the f.c.c. lattice. The numbering shows the relationship explicitly

Note that there is a formal relationship between the simple cubic ($\boldsymbol{t}_1 = \boldsymbol{t}_2 = \boldsymbol{t}_3 = \boldsymbol{a}$; $\alpha = \beta = \gamma = 90°$) and face-centred cubic ($\boldsymbol{t}_1 = \boldsymbol{t}_2 = \boldsymbol{t}_3 = \boldsymbol{a}$; $\alpha = \beta = \gamma = 60°$) lattices in that they differ only in the angle of the rhombus (Figure 2-7). This is of especial physical significance as it suggests a *mechanism*

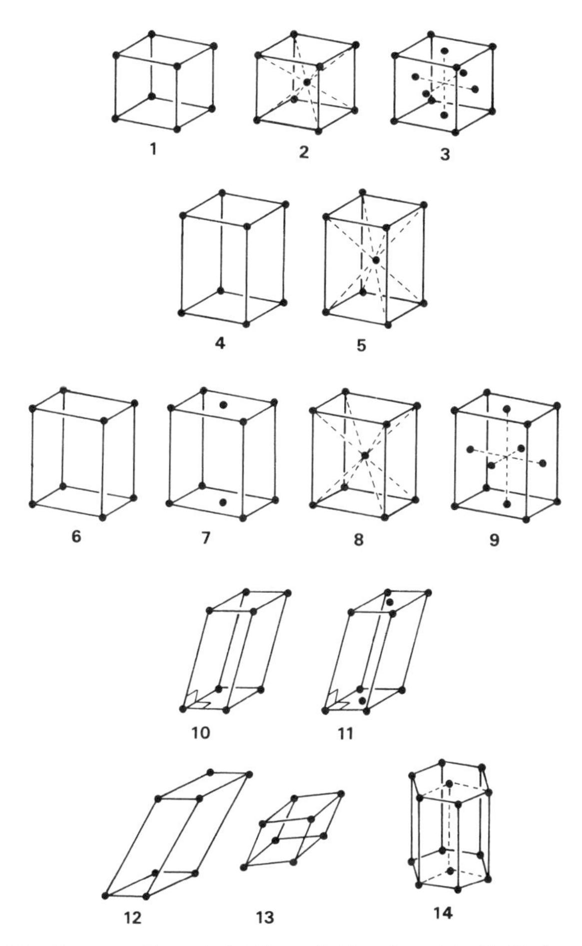

Figure 2-8 The fourteen Bravais lattices: 1. Simple cubic. 2. Body-centred cubic. 3. Face-centred cubic. 4. Tetragonal. 5. Body-centred tetragonal. 6. Orthorhombic. 7. End-centred orthorhombic. 8. Body-centred orthorhombic. 9. Face-centred orthorhombic. 10. Monoclinic. 11. End-centred monoclinic. 12. Triclinic. 13. Rhombohedral. 14. Hexagonal

whereby transformation of one lattice type into the other may be accomplished by the application of pressure.

We have now derived six space lattices. In all there are fourteen, Figure 2-8. They were originally derived by Frankenheim (1842) but Bravais published a more rigorous demonstration in 1848 (and showed that two of Frankenheim's lattices were equivalent) and his name has become attached to them. All points of a Bravais lattice are translationally equivalent.

2.1.2 Number of Particles in a Unit Cell

In determining occupancy of a cell the following self-evident rules must be obeyed.

(*a*) An atom at a corner counts $\frac{1}{8}$ since it is shared equally between eight adjoining cells.
(*b*) An atom in a face counts $\frac{1}{2}$.
(*c*) An atom on an edge counts $\frac{1}{4}$, whilst
(*d*) one in a body-centred position counts 1.

For example, a body-centred cubic cell contains $1 + (8 \times \frac{1}{8}) = 2$ atoms; a face-centred cubic cell has $(6 \times \frac{1}{2}) + (8 \times \frac{1}{8}) = 4$ atoms. Note that the *primitive* cell corresponding to each of these contains one *atom* only.

2.1.3 Symmetry

We have used terms such as high or low symmetry without explaining them fully. Symmetry can be handled quantitatively by means of a branch of mathematics called group theory. We shall not need to use symmetry theory explicitly in this book but it is necessary for our purpose to understand what is meant by 'space group' and to have some conception of how it is derived.

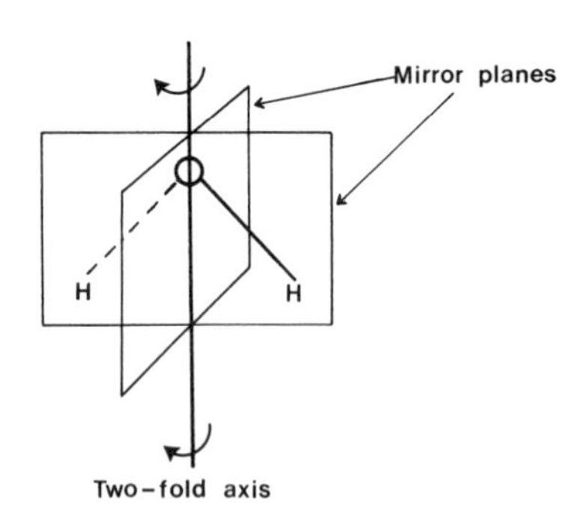

Figure 2-9 The symmetry elements of the water molecule

The symmetry of any object can be described by listing its associated *elements of symmetry*. Thus, the water molecule has a two-fold or diad axis of rotation (the symbol n or C_n is used for a rotation axis, where n is the order of the axis), and two mirror planes (a plane is given the symbol m), one bisecting the HOH angle, the other being in the molecular plane (Figure 2-9).

The result of applying any one of these operations to the water molecule is to transform it into a position indistinguishable from the original one. The diad axis exchanges H_1 and H_2 but there is no difference between the initial and final orientations. If we replace one hydrogen atom by deuterium the resulting molecule now only has the mirror plane which passes through the three atomic centres, thereby quantitatively describing the loss of symmetry.

Mathematically, these operations must occur in certain closed *groups* which are defined according to a set of four combinatorial laws, and each group has a label. (In fact there are two equivalent sets of group labels due to Hermann-Mauguin and to Schoenflies.) Water belongs to the point group *mm*2 (Hermann-Mauguin) ≡ C_{2v} (Schoenflies). The notation *mm*2 clearly indicates the presence of (*a*) the diad axis, 2; (*b*) two mirror planes, *m*. The term *point group* implies that all of the symmetry operations it contains pass through one point although, for groups of low symmetry, the 'point' may be replaced by a line. In the point group C_{2v} the two-fold axis represents the line of intersection of all the symmetry elements.

More complex structures require additional symmetry elements for their description. Consider a square–planar species, MX_4. It has a four-fold axis (4) normal to the plane. Rotation about this axis by $2\pi/4$ (given the symbol C_4^1) moves X_1 into X_2, etc.

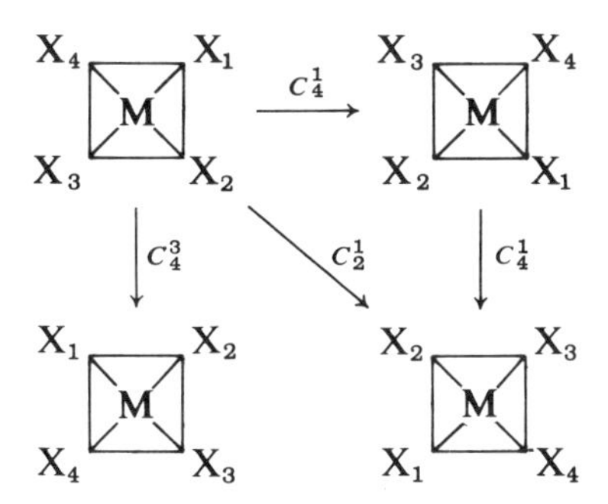

Repetition of this operation yields a result equivalent to a two-fold rotation used once, C_2^1. Further two-fold axes, C_2' and C_2'' exist as shown below.

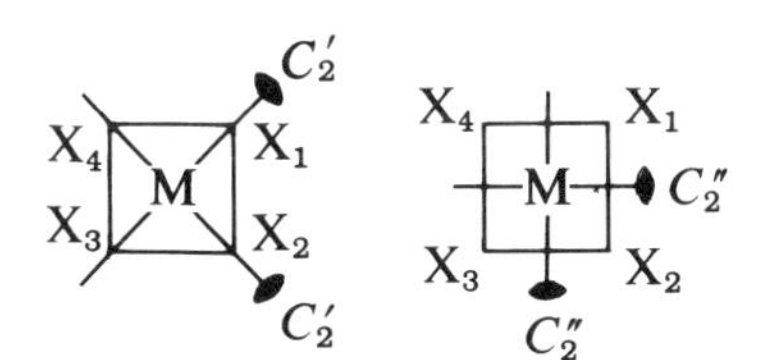

The operation C_2'', for example, interchanges X_1 with X_2, X_4 with X_3. There are also mirror planes, normal to the paper, associated with the two C_2' and the two C_2'' axes, and another in the plane of the paper and passing through all five atom centres. X_1 and X_3, X_2 and X_4 are interchanged also by inversion through the centre of the square, a so-called 'centre of inversion'. Square–planar MX_4 belongs to the *point* group $4/mmm$ ($\equiv D_{4h}$) where $4/m$ indicates the presence of one mirror plane normal to the four-fold axis (i.e. the plane of the paper).

It is commonly found that the result of one symmetry operation is indistinguishable from that of another. For example X_1 and X_2 are interchanged by C_2'' and by m, the mirror plane containing C_2''.

In order to describe the symmetry of crystalline solids, symmetry operations of a translational nature are also required. Consequently the symmetry elements no longer intersect at one point. The groups of symmetry elements are then referred to as *space groups*. Two symmetry operations of a special kind are necessary to complete our symmetry description of crystals: screw axes and glide reflection planes. Figure 2-10 shows a screw axis and compares

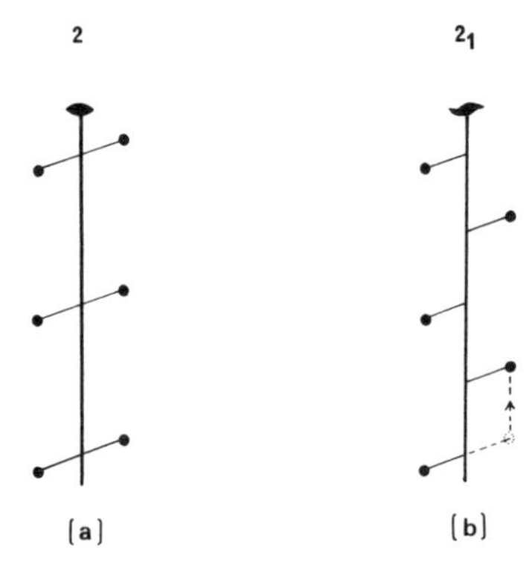

Figure 2-10 Comparison of: (a) A diad axis with (b) A two-fold screw axis. (From F. C. Phillips, *An Introduction to Crystallography*, Oliver and Boyd, Edinburgh, 1971)

it with a simple diad axis; a rotation of 180° is combined with a translation parallel to the axis. It is given the symbol 2_1. A glide reflection plane (g) is illustrated in Figure 2-11; in this operation, reflection in a mirror plane normal to the plane of the paper has been combined with translation in the plane of the paper.

Until now we have only considered lattices with dots or atoms at the points specified. We now recognize that in general each lattice point may represent a more complex grouping of atoms such as a complex ion or a molecule. It may also represent a combination of atoms or groups that are themselves related by symmetry operations. Not all symmetry operations are compatible with each Bravais lattice. For example, a four-fold rotation axis cannot be associated with a monoclinic Bravais lattice which itself has no axis of order greater than 2. When such restrictions are taken into account, it turns out that there are just 230 different combinations of Bravais lattices and symmetry operations. These are the 230 space groups. This remarkable result was arrived at independently by a Russian, E. S. Fedorov (1853–1919), a German, A. M. Schoenflies (1853–1928), and a London businessman, W. Barlow (1845–1934), all of whom published their findings in the period 1885–1894.

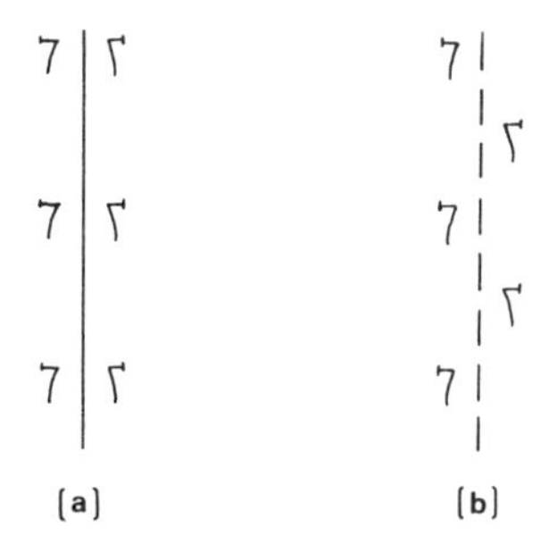

Figure 2-11 Comparison of: (a) A mirror reflection with (b) A glide reflection. (From F. C. Phillips, *An Introduction to Crystallography*, Oliver and Boyd, Edinburgh, 1971)

The demonstration of this result is quite straightforward but very lengthy and cannot be pursued here. It can be illustrated by reference to the restricted case of a plane group. Consider specifically combinations of the rectangular lattice of Figure 2-2b with symmetry elements.

Addition of mirror planes parallel to two opposite sides yields the *plane group* (i.e. the two-dimensional equivalent of a space group) *pm*, Figure 2-12a, but if, instead, glide lines are used, the plane group *pg* is formed, Figure 2-12b.

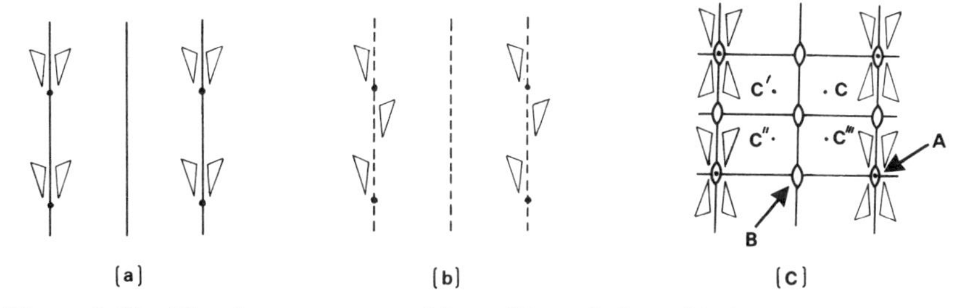

Figure 2-12 The plane groups *pm* (a), *pg* (b), and *p2mm* (c). (From F. C. Phillips, *An Introduction to Crystallography*, Oliver and Boyd, Edinburgh, 1971)

Use of *two* sets of mirror planes at right angles to each other creates the new pattern of Figure 2-12c, the plane group *p2mm* where the meaning of the symbols is: *p*—primitive lattice; 2—highest axis, of order two; *mm*—two mirror planes at right angles. In this pattern an automatic requirement arises for twofold axes (2) normal to the plane in the positions shown by the elliptical motifs. In all there are seventeen groups (Table 1).

Table 1. The plane groups

Lattice type	Plane group			
Parallelogram (p)[a]	$p1$	$p2$		
Rectangle (p) and body-	pm	pg	cm	
centred rectangle (c)	$p2mm$	$p2gg$	$p2mg$	$c2mm$
Square (p)	$p4$	$p4mm$	$p4gm$	
60° Rhombus (p)	$p3$	$p3m1$	$p31m$	
	$p6$	$p6mm$		

[a] p = primitive

2.1.4 Special and General Positions

There are 'special' and 'general' positions associated with all space and point groups. A 'special' position has one or more symmetry elements associated with it. For example, in Figure 2-12c the point A is a special position since a diad axis and two mirror planes pass through it. B is another special position. A point *not* associated with a symmetry element of the group is referred to as a 'general' position; C is one such point. Although general positions may be anywhere in the unit cell not on a symmetry element, they are required to be present in symmetry-related sets. Thus, C must be accompanied by C', C'' and C'''.

A crystal structure is accurately described by listing the sets of special and general positions which the atoms occupy. All possible sites in all space groups have been listed, along with their associated symmetry elements and coordinates, and given so-called Wyckoff labels (a, b, etc.); see *International Tables for X-ray Crystallography*. In general only a few of the available sets of Wyckoff sites are occupied in any one structure, and it is quite common for sets of general sites to be the only ones filled.

Postscript

The requirement of lattice periodicity imposes restrictions upon the ways in which points can be arranged in space. Sometimes this altogether precludes an apparently favoured coordination arrangement (p. 85). On other occasions it may suggest a different but equally acceptable geometrical solution to stereochemical requirement (p. 77). In a remarkable way, as we shall see in Chapter 6, the application of lattice theory assists our understanding of the structures of molecular crystals.

2.2 BAND THEORY

2.2.1 Bands and Conductivity

In describing bonding in terms of molecular-orbital theory, atomic orbitals are combined to give delocalized molecular orbitals extending over the entire molecule. Consider a homopolar diatomic molecule with one valence electron per atom; the formation of molecular orbitals can be represented by Figure 2-13a. Imagine now that we can add further atoms converting the molecule, successively, into triatomic, tetraatomic, etc. The energy level manifold is progressively modified as in Figure 2-13b. In the limit of an infinite array the atomic orbitals have become broadened into a 'band' of allowed energy levels, one for each valence orbital on the isolated atom. The core orbitals are not much affected by this process and remain localized on individual nuclei. In contrast, electrons within the band are delocalized over the entire assembly. The levels within a band are so close together that they may be regarded as forming an energy continuum.

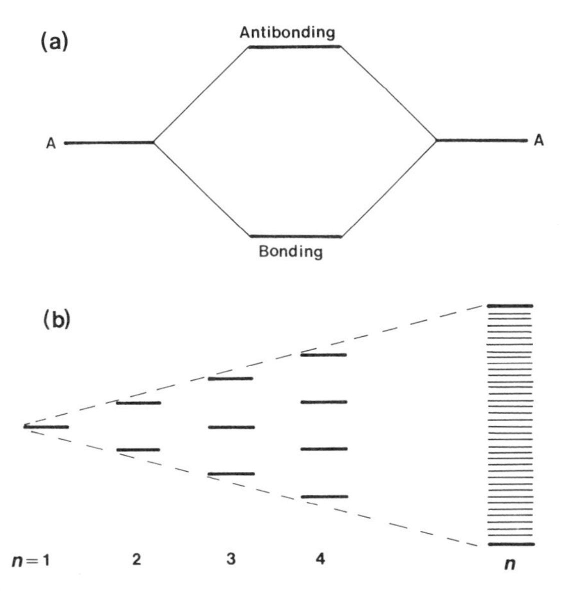

Figure 2-13 (a) Formation of a bonding and an antibonding molecular orbital from two atomic orbitals. (b) Formation of a 'band' by combination of n atomic orbitals on atoms in a periodic array

An alternative way to view this process is to imagine a set of atoms on lattice points so far separated from each other that their interaction energy is negligible; the atomic orbitals then have the same energy as in the isolated atoms. When the lattice is allowed to contract the interaction energy increases forming the orbitals into bands as before.

Just as the combination of two 1*s* functions yields a bonding and an anti-bonding level, so the higher-energy members of the set of levels in a band are of anti-bonding character.

A band formed from ground-state atomic orbitals is known as a 'valence' band. *s*- and *p*-valence bands are shown in Figures 2-14 and 2-15. Excited-state orbitals combine to form 'conduction' bands. Because of their width it is common for bands to overlap if the energy separation of the atomic orbitals of their constituent atoms is small enough. Figure 2-14 shows the band structure of metallic lithium, whilst Figure 2-15 illustrates the quite different case of diamond.

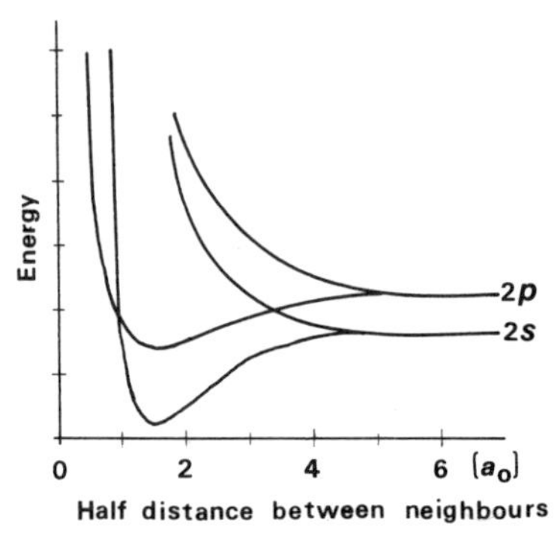

Figure 2-14 Energy bands in metallic lithium. (From C. A. Coulson, *Valence*, The Clarendon Press, Oxford, 1952)

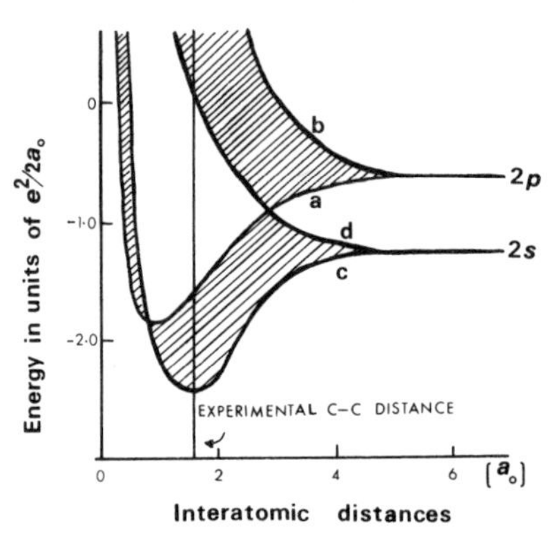

Figure 2-15 Energy bands in diamond. In addition to the shaded bands, there are bands of zero width following curves (a) and (b). (From C. A. Coulson, *Valence*, The Clarendon Press, Oxford, 1952)

Although energy levels within a band are so close that they effectively form a continuum they are nevertheless discrete and the Pauli principle is observed when feeding electrons into them. The number of electrons in a band is simply *NL* where *N* is the number of atoms and *L* is the number of electron states contributing. For example the 2*s*-valence band of lithium, formed from one 2*s*-orbital on each atom, has 2*N* levels, whilst the 2*p*-valence band has 6*N* levels. Since lithium has only a single 2*s*-electron it is evident that the 2*s*-valence band of lithium is only half full. The energy corresponding to the highest filled level in a partly filled band is known as the 'Fermi' level. It is usually defined at absolute zero since at higher temperatures Maxwell–Boltzmann distribution affects the energy distribution. Since Group IIa metals each have two *s*-electrons per atoms it might be thought that they should be insulators. However, this is not so because in these metals valence and conduction bands overlap, see Figure 2-16.

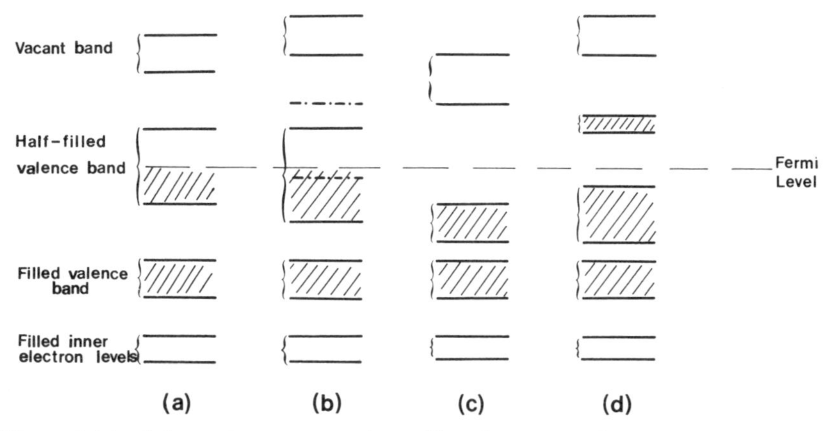

Figure 2-16 Schematic representation of band structure for: (a) A metal; (b) A metal with overlapping partially filled bands; (c) An insulator; (d) A semiconductor

The 'band gap' is the range of forbidden energies between non-overlapping bands; it is a quantity (in energy units) that has especial importance in many aspects of the theory of solids, and is of particular relevance to our enquiry. Together with the widths of the bands and the Fermi level, these quantities characterize the band structure of a solid. The size of the band gap depends upon the atoms involved. For elements, the gap shows a general decrease down any one group of the Periodic Table following the general trend towards metallic behaviour. For example: diamond (6·00), silicon (1·2), germanium (0·8), grey tin (0·1), where the figures in parentheses are values of the band gap in eV.

Band theory lends itself to a classification of solids. Figure 2-16 gives a schematic summary of various classes. Consider the case of a metal: the electron waves are delocalized and can be considered as occurring in pairs, each half travelling in opposite directions. Application of an electric field causes more waves to move in one direction than in the other. This is achieved by giving a small increment of energy to the waves flowing in the direction of the field, raising them into some of the previously vacant levels of the partly filled band. In contrast, an insulator is a solid in which this process cannot occur because the valence band is filled. Figure 2-16 also illustrates the case of a metal in which the partly filled valence band overlaps the conduction band.

Semiconductors are distinguished from insulators by the width of the energy gap, which is very roughly one-tenth that of insulators. Due to this small gap, electrons can be excited into the conduction band even at room temperature. Both the electrons in the conduction band and the 'holes' left in the valence band contribute to this 'intrinsic' conduction. In addition, unless extreme care is taken to ensure purity, semiconductors also exhibit conduction due to impurities, the so-called 'impurity' or 'extrinsic' conduction. By providing energy levels outside, but near, conduction or valence bands a mechanism is available for creating conduction within one or other band. Typical semiconductors are germanium, silicon, Cu_2O, selenium and a wide range of so-called 'III–V and IV–VI' materials in which elements from those groups of the Periodic Table are combined, e.g. PbS, PbTe, GaAs.

Before leaving this section we should look briefly at the rather special band structure of diamond, Figure 2-15. In addition to broad bands originating in $2s$ and $2p$ atomic orbitals, there are two other bands of zero width. The s and p bands can each accommodate two electrons per atom: they are therefore filled and the insulating properties of diamond are accounted for.

2.2.2 Brillouin Zones

We introduced the concept of bands in a very qualitative way, if only because we need only an outline understanding for our purposes. But we must consider some more detailed features which are of great importance in understanding metal structures. In a solid, the delocalized electron waves move in a periodic potential due to the positively charged atomic centres. In fact the wavelengths of electron waves at the Fermi energy are of the same order as atomic dimensions; the question of their diffraction therefore becomes important. As we shall see, this provides the mechanism whereby the highest allowed energy of a level in a band is set.

The close analogy with X-ray diffraction may help here. X-rays are not diffracted if their wavelengths are long in comparison with interplanar spacings in a crystal, but when they are of comparable dimensions interference occurs in accordance with the Bragg law,

$$n\lambda = 2d \sin \theta$$

where λ is the wavelength, d is the interplanar spacing, θ is the angle between the plane and the incident wave and n is an integer, the 'order' of the diffraction. In an exactly analogous manner, electrons of long wavelength (and hence low energy) pass readily through the lattice but, as higher energies and hence shorter wavelengths are reached, diffraction occurs and passage through the lattice is prohibited. The shortest interplanar spacing encountered sets the energy limit in any one direction but, since in general the interplanar spacings differ in various directions, these limits will enclose a polyhedral volume. For a simple cubic lattice this is a cube formed from the six equivalent planes (parallel to xy, yz, zx, and of unit intercept, ± 1, along the third axis). The first energy-restricted volume in a simple cubic lattice is therefore a cube. Such a volume is called a 'Brillouin zone'. It encloses a volume of space corresponding to the total band of the crystal. The shape of the Brillouin zone is different for each Bravais lattice, because they correspond to different arrangements of points in space. Since the order of diffraction n, can take values 1, 2, etc., there are also higher order zones. Figure 2-17 shows first and second Brillouin zones for some lattices.

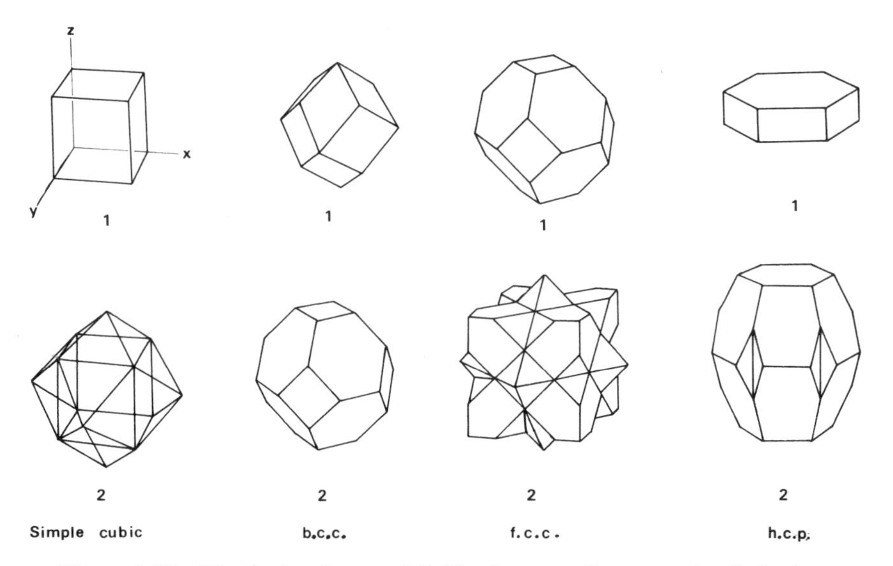

Figure 2-17 The first and second Brillouin zones for some simple lattices

2.2.3 Density of States

To complete this outline account we must describe the 'density-of-states' function, viz., the way in which electron density varies with band energy. Although at first sight this might be expected to be a simple linear relation it is in fact, far from it; the shape of density-of-states functions is of great importance in the theory of metal structures.

The kinetic energy of an electron in terms of its mass, m, and momentum, p, is given by

$$E = p^2/2m = \frac{h^2}{2m}\left(\frac{1}{\lambda}\right)^2$$

where h = Planck's constant and λ is the wavelength. For a free electron in a box this yields a simple parabolic dependence, Figure 2-18. However, the effect of introducing the periodic potential of the lattice is to break this curve up into a series of sections interspersed with forbidden gaps (the band

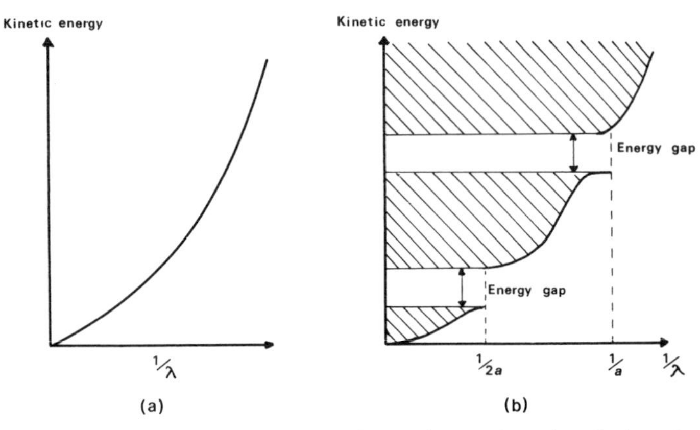

Figure 2-18 (a) The kinetic energy–reciprocal wavelength relation for a free electron in a potential energy box. (b) The same relationship as modified by the effect of a periodic potential. (From C. S. G. Phillips and R. J. P. Williams, *Inorganic Chemistry*, The Clarendon Press, Oxford, 1965)

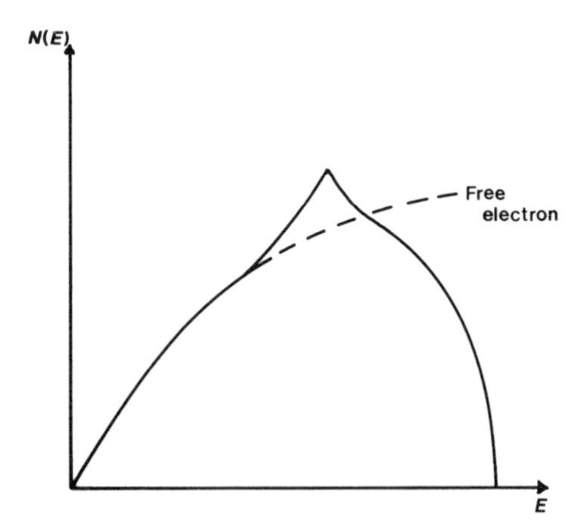

Figure 2-19 Density of states function for a typical band

gaps). As the wavelength approaches the diffraction limit the curve is seen to bend over, corresponding to a lowering of energy with increase in $1/\lambda$. The Fermi surface corresponding to free electrons is a sphere (i.e. the maximum allowed energy is the same in any direction). Since a decrease in energy is equivalent to an attractive force, we can describe the change in shape of the

E vs $1/\lambda$ curve as attraction of the free-electron Fermi sphere to the zone boundary, as the Fermi surface approaches that boundary. This causes a corresponding peak in the density-of-states function as shown in Figure 2-19. Since the shapes of the zones depend upon the Bravais lattice, the peak in the density-of-states function comes at different energies for different lattices.

2.3 ATOMIC SIZE

No precise physical significance can be attached to the concept of atomic or ionic radius since the electron density associated with atomic wave functions approaches zero asymptotically. Nevertheless, common sense suggests that observed internuclear distances may be divided according to some formula giving at least a relative scale of sizes. The idea goes back to the very earliest days of X-ray crystallography (W. L. Bragg, 1920; Wasastjerna, 1923; Goldschmidt, 1926; Pauling, 1927). On the other hand, the fact that new tables of radii of various types are *still* being compiled suggests that there are fundamental difficulties involved and remind us that we should treat *any* set of radii with caution. Unquestionably, atoms of high atomic number will be larger than those of small atomic number, although we could hardly be so simple-minded as to expect a monotonic change, and if we are to test any theory of crystal structures we shall almost inevitably find size considerations involved at some stage or other. It is desirable that any sets of radii obtained be transferable, thus yielding sets of self-consistent radii which can be summed to yield internuclear distances.

Before we delve into these murky waters, let us ask ourselves if we are really doing the right thing. We propose to compile tables of radii from observed internuclear separations and use them to see what relations (if any) exist between the relative sizes of the atoms in a crystal and the structure adopted. But these distances are themselves the result of a balance of all the forces operative in determining the structures. To what extent, then, is it meaningful to reverse the process and ask how size affects structure? If we were to begin with a set of atomic radii appropriate to the vapour we might then be able to relate their initial and crystal sizes, and possibly understand their effect upon structure adopted. To do this requires a set of agreed atomic radii and some means of dividing observed interatomic distances into radii associated with constituent atoms, and would be very difficult.

In order to make progress we recognize two distinctly different electronic situations. *Firstly*, the approach of atoms or ions with closed valence shells to form molecular and ionic crystals, respectively. Cationic sizes will inevitably be smaller than those of their neutral atoms due to orbital contraction resulting from Coulombic attraction, and anions will be correspondingly larger. *Secondly*, the formation of bonds by overlap of atomic orbitals on different atoms resulting in build-up of electron density between the atoms; this situation corresponds to bonding in covalent crystals and in metals. We are faced here with a much more difficult problem than in molecular or ionic situations because there is no obvious way of dividing interatomic distances (except in pure-element structures) between constituents. On the other hand this difficulty is to some extent compensated by the relative unimportance of size considerations in understanding metallic and covalent structures. We now consider each of these four cases in turn.

2.3.1 Van der Waals Radii

In crystals composed of molecules or noble gas atoms, non-bonded distances of closest approach may be divided to yield values of van der Waals radii, r_W, suitable for use in estimating various parameters of interest in crystal structure. Several investigators have sifted through large numbers of crystallographic papers to determine the most accurate values. A recent set due to Bondi is given in Table 2. A qualitative interpretation of van der Waals radii due to Morrison (1955) and Bondi (1964) shows that

$$r_W = C\lambda_B$$

where $\lambda_B = h\sqrt{M_e I_0}$, the de Broglie wavelength of the outermost valence electron (M_e = rest mass of the electron; I_0 = first ionization potential and C varies from 0·48 to 0·61 according to the group). Collecting constants we may write

$$r_W = k\sqrt{I_0}$$

a simple, logical and satisfying relation which can be used in a predictive capacity to obtain van der Waals radii of less commonly studied atoms in molecular crystals.

These radii are of value for computing volumes of molecules and groups as an aid to determining preferred orientations of bulky substituents, and for estimating modes of packing. Van der Waals radii are, naturally, larger than

Table 2. Van der Waals radii, Å. (A. Bondi, *J. Phys. Chem.*, **68**, 441 (1964.)

			H	He
			1·20	1·40
C	N	O	F	Ne
1·70	1·55	1·52	1·47	1·54
Si	P	S	Cl	Ar
2·10	1·80	1·80	1·75	1·88
	As	Se	Br	Kr
	1·85	1·90	1·85	2·02
		Te	I	Xe
		2·06	1·98	2·16

corresponding covalent radii. Although we list one value for each element in Table 2, small differences have been established for particular bond situations (e.g. double-bonded oxygen) and more detailed compilations should be consulted where necessary.

2.3.2 Ionic Radii

In an ionic crystal, the attractive Coulomb force is counterbalanced by a repulsive force due to interaction of the outer electron clouds of the ions, resulting in an equilibrium internuclear distance. Because of the short-range nature of the repulsive forces and their rapid increase as the ions are brought closer together it is possible to treat ions to a first approximation as elastic spheres with definite radii such that their sums reproduce the observed equilibrium internuclear separations. This does not rule out the possibility of slight mutual interpenetration; indeed it is occasionally necessary to invoke this to explain some radii. In ionic crystals, if the relative sizes of the anions and cations (which can be expressed as a 'radius ratio', r_+/r_-) are such that the anions are nearly in contact, mutual anion–anion repulsion must also be considered and will affect the equilibrium distance. *The basic problem* in estimating ionic radii from observed internuclear distances is to find a formula for apportioning contributions to each ion. Depending upon how this choice is made, various sets of radii result. Wasastjerna (1923) used data on molar refractivities and, by a method now discredited, obtained radii for eight cations and eight anions including F^- (1·33 Å) and O^{2-} (1·32 Å).

Pauling (1927) derived a more extensive set by a semi-empirical method starting with observed internuclear distances in five crystals: NaF, KCl, RbBr, CsI and Li_2O. Since the size of an ion is determined by the distribution of its outermost electrons this will be inversely proportional to the effective nuclear charge experienced by them (Z_{eff}). Thus, the radius of an ion

$$r = \frac{C_n}{Z - S} = \frac{C_n}{Z_{eff}}$$

where S is a screening constant and C_n a further constant determined by the quantum number. Pauling obtained a set of S values by theoretical methods combined with other experimental data from molar refraction and X-ray term values. C_n takes the same value for all members of an isoelectronic series. For neon-like ions, $S = 4{\cdot}52$. Thus, for NaF

$$r_{Na^+} = C_n/(11 - 4{\cdot}52)$$
$$r_{F^-} = C_n/(9 - 4{\cdot}52)$$

From experiment $r_{Na^+} + r_{F^-} = 2{\cdot}31$ Å. Solution of these equations yields $r_{Na^+} = 0{\cdot}95$ Å, $r_{F^-} = 1{\cdot}36$ Å. The radii obtained by this process are known as *univalent radii*. For multivalent ions having the rock-salt structure they correspond to radii the ions would have if they were univalent but otherwise retained their electronic distribution. Corrections of these radii for different charge and coordination arrangement are quite easily made; we shall not discuss them as our use of radii does not call for this kind of detail. Other physical properties have also been used in dividing internuclear distances to obtain ionic radii.

We begin to see the complexities involved. It is necessary to correct any set of ionic radii for both coordination number and geometry, and for radius ratio. Further, ions which can adopt more than one oxidation state will need to have a different radius determined for each such state and since the covalent contribution to bonding is bound to be greater in one oxidation state than in the other, one of the radii will be less related to physical reality than the other.

Even this is not the end of our troubles. Consider TiO (rock-salt structure) which has an electrical conductivity in the metallic range due to extensive overlap of *d*-orbitals throughout the crystal. Being at the beginning of the 3*d* series the *d*-orbitals on Ti are large and diffuse compared with those for elements later in the same series. Do we need *two* sets of radii for ions in this kind of situation, one for the *s*- and *p*-electrons, another for *d*-electrons? Or are we abusing the term 'ion' by using it in this connotation when, as we have admitted, there is extensive *d*-orbital delocalization in this crystal?

If our simple concept of ions as elastic spheres has any relation to reality it means that electron density should drop to zero (or very close to it) at some point along each cation–anion line; cannot we determine this experimentally? In fact it is only in recent years that X-ray techniques have improved to the point where the direct measurement of electron distribution in crystals is possible. It is, however, still fraught with experimental and other difficulties. By this means, ionic radii have been determined directly, although only a few solids have been treated in this way. We shall consider these results below when we have looked briefly at the traditional sets of figures.

Historically, several different sets of ionic radii have been compiled by investigators using different methods, although the most widely used continue to be those of Goldschmidt (1927), Pauling (1927) often with some later corrections, and Ahrens (1952). Although agreement between these scales is generally good there are some impressive discrepancies and they do not fare well on certain statistical tests. The most recent, and probably the most accurate, ionic radii are to be found in a massive compilation of data by Shannon and Prewitt (1969, 1970).

In an attempt to determine preferred values of ionic radii with good additivity (compared with experiment) Waddington (1966) used the method of the undetermined parameter. The basic idea is simple enough: there are more unknowns than there are equations (see below) but solutions can be obtained if we are prepared to accept answers in terms of an undetermined parameter, δ. For example, the radius of Li^+ is given as $2{\cdot}581 - \delta$ (in Å). The final stage is to decide on a value for δ by critical examination of the physical bases on which ionic distances have been divided previously (Wasastjerna, Pauling, etc.). The algebra runs as follows.

For a series of salts in which coordination numbers remain the same

$$r_i + r_j + E_{ij} = a_{ij}$$

where r_i, r_j = cation and anion radii, a_{ij} = measured internuclear distance and E_{ij} is the amount by which the sum of the assigned radii differs from a_{ij}. The best fit is obtained when the sum of the squares of E_{ij} are a minimum. Individual ionic radii are then given by

$$r_i = \bar{a}_i - \delta$$

where $\bar{a}_i = \frac{1}{4}(a_{i1} + a_{i2} + a_{i3} + a_{i4})$ and

$$r_j = \bar{a}_j - \bar{a} + \delta$$

where $\bar{a}_j = \frac{1}{4}(a_{1j} + a_{2j} + a_{3j} + a_{4j})$ and $\bar{a}$ is the arithmetic mean of all the a_{ij}.

Values of ionic radii obtained by this method are compared with others in Table 3, and a more extensive set due to Shannon and Prewitt given in Table 4. The meaning of any of the radii for multivalent ions is questionable, especially those of charge three and greater. However, they *may* be used to give approximate internuclear distances, although we must stress that even if they are used successfully in this respect the result does not validate the radii. Any radius for a multivalent ion (certainly for charges greater than $2e$) must be regarded as a sheer formalism (as are *all* radii, see beginning of this section) because the higher the *formal* charge on an ion the greater will be the covalent bonding between it and its ligands. With increasingly accurate evidence on

Table 3. Ionic radii (Å) from various sources

	Waddington	Shannon and Prewitt[a]	Pauling[b]	Goldschmidt	Ahrens
Li^+	0·739	0·74	0·60	0·78	0·68
Na^+	1·009	1·02	0·95	0·98	0·97
K^+	1·320	1·38	1·33	1·33	1·33
Rb^+	1·460	1·49	1·48	1·49	1·47
Cs^+	1·718	1·70	1·69	1·65	1·67
Tl^+	1·449	1·50	0·95	1·05	1·47
F^-	1·322	1·33	1·36	1·33	1·33
Cl^-	1·822		1·81	1·81	1·81
Br^-	1·983		1·95	1·96	1·96
I^-	2·241		2·16	2·20	2·20

[a] Based upon O^{2-} = 1·40 Å. See also Table 4.
[b] Crystal radii.

Table 4. Crystal radii for 'ions' in the formal valence states and coordination environments indicated based upon $r(F^-) = 1{\cdot}19$ Å (Shannon and Prewitt, 1969)

A: Non-transition Elements

	Li^+	Be^{2+}	B^{3+}						
C No. 4	0·73	0·41	0·26						
C No. 6	0·88								
	Na^+	Mg^{2+}	Al^{3+}			Si^{4+}			
C No. 4	1·14	0·63	0·53			0·40			C No. 4
C No. 6	1·16	0·86	0·67			0·54			C No. 6
C No. 8	1·30	1·03							
	K^+	Ca^{2+}	Sc^{3+}	Zn^{2+}	Ga^{3+}	Ge^{4+}	As^{3+}		
				0·74	0·61	0·54	0·475		C No. 4
C No. 6	1·52	1·14	0·87	0·885	0·76	0·68	0·64		C No. 6
C No. 8	1·65	1·26	1·01						
C No. 12	1·74	1·49							
	Rb^+	Sr^{2+}	Y^{3+}	Cd^{2+}	In^{3+}	Sn^{4+}	Sb^{5+}		
C No. 6	1·63	1·30	1·032	1·09	0·93	0·83	0·75		C No. 6
C No. 8	1·74	1·39	1·155						
C No. 12	1·87	1·58							
	Cs^+	Ba^{2+}	La^{3+}	Hg^{2+}	Tl^{3+}	Pb^{4+}	Bi^{3+}	Po^{4+}	
C No. 6	1·84	1·50	1·20	1·16	1·02	0·915	1·16	1·24	C No. 6
C No. 8		1·56	1·32		Tl^+	Pb^{2+}			
C No. 12	2·02	1·74	1·46		1·64	1·32			C No. 6

B: Transition Elements									
3d- series									
		Ti^{2+}	V^{2+}	Cr^{2+}	Mn^{2+}	Fe^{2+}	Co^{2+}	Ni^{2+}	Cu^{2+}
C No. 6	LS	1·00	0·93	0·87	0·81	0·75	0·79	0·84	0·87
	(HS)			(0·96)	(0·96)	(0·91)	(0·875)		
		Ti^{3+}	V^{3+}	Cr^{3+}	Mn^{3+}	Fe^{3+}	Co^{3+}	Ni^{3+}	
C No. 6	LS	0·81	0·78	0·755	0·72	0·69	0·665	0·70	
	(HS)				(0·79)	(0·785)	(0·75)	(0·74)	
		Ti^{4+}	V^{5+}	Cr^{6+}	Mn^{6+}				
C No. 4				0·44	0·41				
C No. 6		0·745	0·68						
4d- series									
		Zr^{4+}	Nb^{4+}	Mo^{3+}	Tc^{4+}	Ru^{3+}	Rh^{3+}	Pd^{2+}	Ag^{+}
C No. 6		0·86	0·83	0·81	0·78	0·82	0·805	1·00	1·29
			Nb^{5+}	Mo^{6+}		Ru^{4+}	Rh^{4+}	Pd^{2+}	
C No. 6			0·78	0·74		0·76	0·755	[0·78]	
5d-series									
		Hf^{4+}	Ta^{4+}	W^{4+}	Re^{4+}	Os^{4+}	Ir^{3+}	Pt^{2+}	Au^{3+}
C No. 6		0·85	0·80	0·79	0·77	0·77	0·87	[ ]	[0·84]
			Ta^{5+}	W^{6+}	Re^{7+}		Ir^{4+}	Pt^{4+}	
C No. 6			0·78	0·72	0·71		0·77	0·77	

[ ] = square planar

(continued)

Table 4 (cont'd)

C: Lanthanides and Actinides														
4f-series														
	Ce^{3+}	Pr^{3+}	Nd^{3+}	Pm^{3+}	Sm^{3+}	Eu^{3+}	Gd^{3+}	Tb^{3+}	Dy^{3+}	Ho^{3+}	Er^{3+}	Tm^{3+}	Yb^{3+}	Lu^{3+}
C No. 6	1·174	1·153	1·135	1·12	1·104	1·09	1·08	1·06	1·048	1·034	1·02	1·01	0·998	0·988
C No. 8	1·28	1·28	1·26		1·23	1·21	1·20	1·18	1·17	1·16	1·14	1·13	1·12	1·11
	Ce^{4+}													
C No. 6	0·94													
C No. 8	1·11													
5f-series														
	Th^{4+}	Pa^{4+}	U^{4+}	Np^{3+}	Pu^{3+}	Am^{3+}	Cm^{3+}	Bk^{3+}	Cf^{3+}					
C No. 6	1·14			1·18	1·14	1·15	1·12	1·10	1·09					
C No. 8	1·20	1·15	1·14											
			U^{6+}	Np^{4+}	Pu^{4+}	Am^{4+}	Cm^{4+}	Bk^{4+}						
C No. 6			0·89		0·94									
C No. 8				1·12	1·10	1·09	1·09	1·07						

electron-density distribution from many branches of both molecular and solid-state science, it is now becoming clear that electrical neutrality is much more closely approached in reality than has been generally recognized.

Others have modified the above story in detail, the most recent contribution being that of Shannon and Prewitt, some of whose effective ionic radii are given in Table 4. These were determined on the basis of a very large number of crystal data involving more than a thousand interatomic distances.

Empirical Ionic Radii

Figure 2-20 shows the experimentally determined electron density maps for NaCl and LiF. The electron density rises rapidly towards each nucleus but falls approximately to zero in between the ions. A plot of this function, Figure 2-21, shows this minimum very clearly for LiF. Only LiF, NaCl, KCl, CaF_2 and MgO, have been examined so far, Table 5. The radii of Table 5 do not

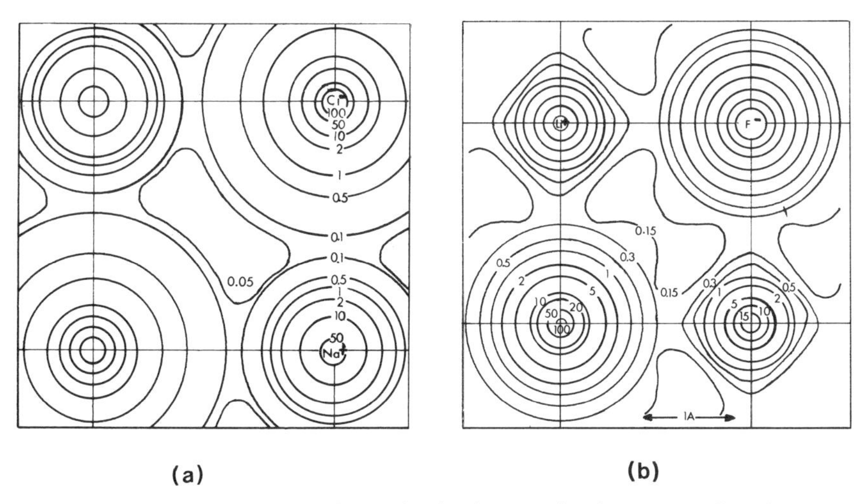

Figure 2-20 Experimentally determined electron-density maps (in the *x,y,o* plane) (a) Rock salt, NaCl, and (b) LiF. (From (a) H. Schoknecht, *Zeit. Phys. Chem.*, Frankfurt, **3**, 296, 1955, (b) J. Krug, H. Witte, and E. Welfel, *Ibid*, **4**, 36, 1955)

reproduce the internuclear separations for all of the alkali halides to the desired accuracy. It is believed that there is some mutual interpenetration of Li^+ and F^- ions as is suggested by the slightly different contour map (cf. NaCl); the radii necessary to reproduce crystal distances for the other alkali halides are larger, viz. Li^+ 0·94, F^- 1·16 Å. Taking these values, together with a marginally different one for Na^+, the 'corrected' radii of Table 6 were obtained by Gourary and Adrian (1960). They reproduce the internuclear separations for all alkali halides (other than LiF) within about one per cent.

Table 5. Ionic radii determined from electron density maps

	Cation, Å	*Anion*, Å
LiF	0·92	1·09
NaCl	1·18	1·64
KCl	1·45	1·70
CaF_2	1·26	1·10
MgO	1·02	1·09

Note that there is some deviation from spherical symmetry for Li^+ in the map. Further, there is the large difference between the value for F^- found in LiF and CaF_2. In the latter compound, the minimum electron density observed along the shortest Ca–F line is 0·23*e* $Å^{-3}$ compared with 0·19*e* $Å^{-3}$ for LiF. The difference in both these figures and the 'radii' is clearly associated with the different coordination arrangements about F^- in CaF_2 (4, tetrahedral) and LiF (6, octahedral), and with the higher formal charge on calcium. In short, we have here direct experimental indication that the

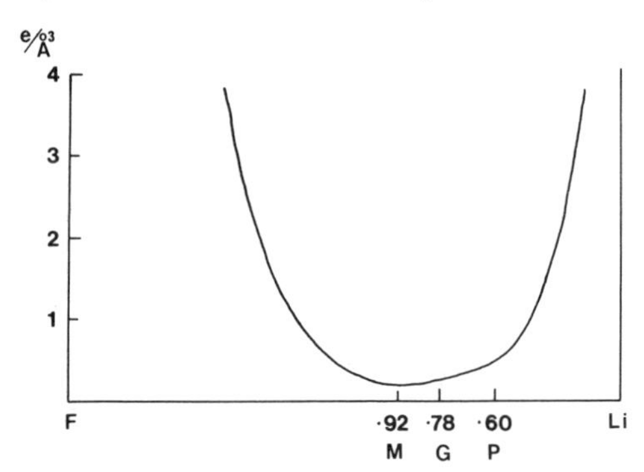

Figure 2-21 Variation of measured electron density with internuclear position for LiF. From *Fundamentals of Inorganic Crystal Chemistry*, by H. Krebs. © 1968. McGraw-Hill Book Company (UK) Ltd. Used with permission). M = minimum measured electron density. *G* and *P* represent ionic radius of Li^+ according to Goldschmidt and Pauling respectively

concept of 'ionic radius' is physically meaningless; it is a working approximation for univalent ions in the rock-salt structure. For even divalent ions, bonding is *seen* to depart from purely ionic, as may be expected from the electroneutrality principle.

The fascinating feature of these experimental radii is that they differ substantially from *all* of the currently accepted sets of radii which are themselves quite highly consistent, with a few understandable exceptions. And it is upon these older sets of radii that the extant discussions of 'ionic' crystals are based. The essential difference is that cation radii are *larger* than previously believed, whilst anion radii are correspondingly *smaller*.

Table 6. 'Corrected' ionic radii (in Å) from electron-density maps (Gourary and Adrian, 1960)

	'Corrected' radius	Waddington radius		'Corrected' radius	Waddington radius
Li^+	0·94	0·74	F^-	1·16	1·32
Na^+	1·17	1·01	Cl^-	1·64	1·82
K^+	1·49	1·32	Br^-	1·80	1·98
Rb^+	1·63	1·46	I^-	2·05	2·24
Cs^+	1·86	1·72			

Trends in Ionic Radii

Regardless of which set of ionic radii we work with, we note some more or less obvious trends in them.

(*a*) Within Groups Ia, IIa and IIIb–VIIb, radii increase with atomic number—simply because extra electron shells are added.

(*b*) Across each period of the Periodic Table, ionic radii decrease with increasing charge and atomic number, e.g. Na^+, Mg^{2+}, Al^{3+}, etc. This is due to the increasing effect of the nucleus on the outer electrons, coupled with some contraction in bond length due to the greater Coulomb attraction associated with multivalent ions.

(*c*) Increase in valency of a cation decreases its size (Å). For example,

$$Cr^{2+}\ 0{\cdot}84,\ Cr^{3+}\ 0{\cdot}69,\ Cr^{6+}\ 0{\cdot}52$$

(But note the qualifications expressed above regarding multivalent ions.)

(*d*) A series of ions of increasing atomic number but the same ionic charge shows a steady decrease in size. This is due to the steadily increasing effective nuclear charge. *d*- and especially *f*-orbitals do not shield the nucleus very effectively, resulting in the steady contraction observed in each of the three *d*-block transition metals series and in the lanthanides. The net result of this process can be seen by comparing the last member of a *d* series with the alkali-metal cation *preceding* it (Table 7).

The effect of *d*-electron configuration on ionic radii is quite dramatic; we defer its consideration to Section 5.4.4.

Filling the *f* shell in the lanthanide series causes such contraction that the early members of the third transition series [5*d*] have radii almost identical with those of their second-row congeners [4*d*]; this is reflected in the similarity

Table 7. Contraction in ionic sizes (Ahrens radii) due to filling $(n-1)$ *d* shells

Configuration	Å	Configuration	Å
Cu^+ [ ] $3s^23p^63d^{10}$	0·96	K^+ [ ] $3s^23p^6$	1·33
Ag^+ [ ] $4s^24p^64d^{10}$	1·26	Rb^+ [ ] $4s^24p^6$	1·47
Au^+ [ ] $5s^25p^65d^{10}$	1·37	Cs^+ [ ] $5s^25p^6$	1·67

of the chemistry of Zr and Hf, and of Nb and Ta. By the end of the series substantial differences have developed, e.g. Ag^+ 1·26, Au^+ 1·37 Å. It is interesting to note that some transition-metal ions have similar sizes to some non-transition-metal ions of the same charge. For example Ca^{2+} and Pt^{2+}; this if of value in investigating the relative contributions of covalent bonding in transition-metal compounds.

2.3.3 Covalent Radii

If ionic radii are applied to covalently bonded solids they fail to add to give the observed internuclear distances. In recognition of the different bonding involved, Pauling introduced a set of 'covalent' radii which do have correct additive behaviour. They were obtained from observed interatomic distances in crystals having diamond and related lattice types together with some others. However, as with ionic radii, covalent radii have their difficulties and other workers have produced different sets.

Do covalent radii have any *physical* meaning? Or are they simply numbers derived by a formalism? In seeking to answer this question we must consider only crystal situations, not molecular ones, as the bonding in the two cases is different because of 'end effects' present in molecules but not in crystals. This is illustrated by the differing values quoted for C—C single bond energies for diamond (336 kJ mol^{-1}) and for saturated hydrocarbons (356 kJ mol^{-1}). A particularly fruitful discussion of this problem by van Vechten and Phillips (1970) relates to crystals with tetrahedrally coordinated atoms (zinc blende or wurtzite structure). These form a favourable set since in them the bonding is formed basically of sp^3-hybridized σ-bonds, there are no π-bonding effects and lone pairs are absent. Further, in these crystals next-nearest neighbours are sufficiently well removed from nearest neighbours (ratio of the distances is $2\sqrt{2}/\sqrt{3} = 1{\cdot}63$) that no overlap between them takes place.

Covalent radii are considered as having two components arising from (*a*) the core (with radius r_c where r_c represents the size of the core as seen by the sp^3-hybridized electrons), and (*b*) the valence electrons. Note that in using this approach we are *not* retaining the elastic-sphere-with-definite-radius concept employed for ionic radii. What we *are* doing is recognizing that a contribution to the size comes from the closed shells of the core (and it can be estimated quite reliably); to this extent we retain some remnant of the elastic-sphere concept. But the valence-electron density and its distribution in many cases has a dominating influence on internuclear distances and must therefore be apportioned somehow. Crudely put (and so many models used in chemistry are crude!) it is as if we considered covalently bonded atoms in crystals as spheres bound together with large blobs of adhesive placed at tetrahedrally disposed sites on their surfaces. The problem is, how to divide the blob of adhesive between atoms A and B so that a correctly additive set of radii results.

The effective radius of the core as seen by the sp^3-hybridized bonding electrons is taken by van Vechten and Phillips as

$$r_c = \frac{R(n)}{Z_{\text{eff}}}$$

where Z_{eff} is the effective charge for the appropriate core state as given by Slater, n is the principal quantum number and $R(n)$ is a parameter which was fixed for one value of n and determined for all others by

$$R(n) = n^2 \left[\frac{Z_{\text{eff}}\,(\text{Carbon})}{Z_{\text{eff}}(\text{IV})}\right] 4a_0$$

where Z_{eff} (IV) is the value for the Group IV element in question. The constant $4a_0$ is an adjustable parameter, and Z_{eff} (Carbon) $= 5{\cdot}7e$. This approach is related to Waddington's use of an undetermined parameter for ionic radii in that all results are related to physically determined distances by means of a single parameter (a_0) which is then given a value.

Table 8 shows the results of this method. The radii are additive to about 1% and are of sufficient accuracy that they can be used to estimate the distorting effects on a lattice of replacing atoms by isoelectronic impurities.

Table 8. Covalent radii in Å (van Vechten and Phillips, *Phys. Rev. B.* **2**, 2160 (1970))

Tetrahedral							
Be	B	C	N	O	F		
0·975	0·853	0·774	0·719	0·678	0·672		
Mg	Al	Si	P	S	Cl		
1·301	1·230	1·173	1·128	1·127	1·127		
Ca	Ga	Ge	As	Se	Br	Cu	Zn
1·333	1·225	1·225	1·225	1·225	1·225	1·225	1·225
Sr	In	Sn	Sb	Te	I	Ag	Cd
1·689	1·405	1·405	1·405	1·405	1·405	1·405	1·405

Octahedral							
Be	B	C	N	O	F		
1·014	0·892	0·813	0·758	0·737 (0·811)[a]	0·737		
Mg	Al	Si	P	S	Cl		
1·367	1·296	1·239	1·236	1·236	1·236		
Ca	Ga	Ge	As	Se	Br	Cu	Zn
1·604	1·343	1·343	1·343	1·343	1·343	1·343	1·343
Sr	In	Sn	Sb	Te	I	Ag	Cd
1·771	1·541	1·541	1·541	1·541	1·541	1·541	1·541

[a] Value required when bonded to 3rd or 4th row element.

BIBLIOGRAPHY

Ahrens, L. H., *Geochim. Cosmochim. Acta*, **2**, 155 (1952)
Bondi, A., *J. Phys. Chem.*, **68**, 441 (1964).
Donaldson, J. D., and S. D. Ross, *Symmetry and Stereochemistry*, Intertext, London, 1972.
Goldschmidt, V. M., *Berichte*, **60**, 1263 (1927).
Gourary, B. S., and F. J. Adrian, *Solid State Physics*, **10**, 127 (1960).
Morrison. J. D., *Rev. Pure Appl. Chem.*, **5**, 46 (1955).
Pauling, L., *The Nature of the Chemical Bond*, Cornell University Press, Ithaca, 1940.
Phillips, F. C., *An Introduction to Crystallography*, 4th ed., Oliver and Boyd, Edinburgh, 1971.
Shannon, R. D., and C. T. Prewitt, *Acta Cryst.*, **B25**, 925 (1969); **B26**, 1046 (1970).
van Vechten, J. A., and J. C. Phillips, *Physical Review B*, **2**, 2160 (1970).
Waddington, T. C., *Trans. Faraday Soc.*, **62**, 1482 (1966).

CHAPTER 3

What Structures are Possible? I. Predictions on the Basis of Close-packing

3.1 THE SPACE-FILLING POSTULATE

The object of this chapter is to introduce some of the more important structure types. Later, we shall do our best to understand them, but first let us have a look at some of the raw material. Having said this, we shall not wander through the chemical catalogue, or even the Periodic Table, asking 'how about this one or that one?' We shall begin with a major postulate, one of Laves's principles, and follow its implications through.

Postulate: the most probable structures will be those in which the most economical use is made of space.

As a postulate, it needs no justification although commonsense suggests that it is not unreasonable. It does need some elucidation: the word 'probable' has been used in an evidently cowardly attempt to avoid the word 'stable' which no chemist worth his salt will let pass without some qualification. We shall come back to this problem later but just now we begin by taking the postulate at face value.

To make things easier for ourselves we impose the further restrictions that we shall compose our structures of spherical atoms or molecules of identical size, and that they have no directional bonding requirements. In other words, these restrictions ensure that the determining factor in structure adoption will be the postulated space-filling principle. Although this may seem a highly artificial exercise, we recall that molecular crystals composed of the noble gases and of some effectively spherical organic and inorganic molecules (e.g. CH_4) approximate to our requirements and, more to the point, so do most metals. How, then, do we expect identical spheres to pack, bearing in mind always that acceptable structures must show spatial periodicity?

3.2 CLOSEST-PACKED LAYERS

One way would be to line the spheres up as in Figure 3-1a to give layers which could then be stacked vertically in the same way. However, it is easily appreciated that the method of Figure 3-1b would be more economical in its use of space. Each sphere has *six* nearest neighbours (i.e. it has a coordination number of *six*) whereas in Figure 3-1a each sphere has a coordination number of *four* only. We can use Figure 3-1b also to suggest the nature of the vertical packing of the layers: each layer will fit into the one below, resting in the hollows as shown, each sphere in the second layer making tangential contact with three in the first layer.

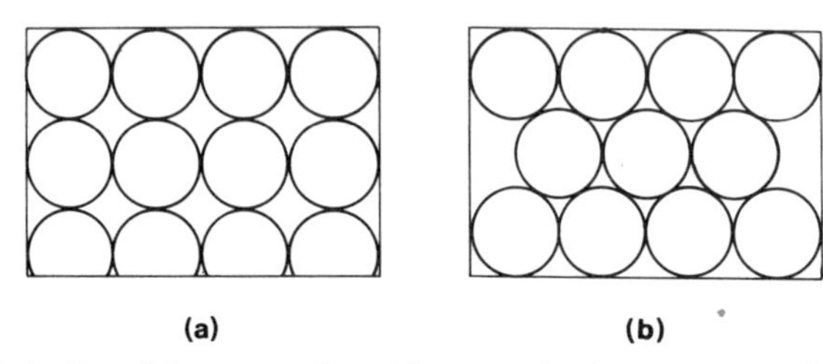

Figure 3-1 Possible ways of packing equal spheres in two dimensions

The third layer can be placed upon the second in two different ways such that:

(*a*) Each sphere in the third layer is directly above a corresponding sphere in the first layer; we may call this ABA packing.

(*b*) The third layer of spheres rests in the other set of hollows identified in Figure 3-2, so that it is not directly above either of the preceding layers; we call this ABC packing.

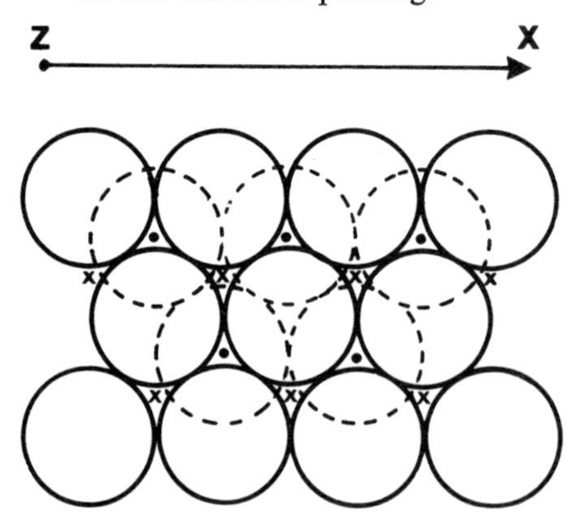

Figure 3-2 Two superimposed layers of close-packed spheres of the type shown in Figure 3-1b. Note the two sets of sites denoted by x and ●

Note that we are labelling the layers as A, B and C, in terms of *the relative positions of their projections* upon the plane of one of the layers. Thus, if the first closest-packed layer is in the *xy* plane (i.e. the plane of the paper) and the layers are piled up along the *z* direction, the planes which we use for our labelling purposes are a set of *xz* or *yz* planes. We have laboured this point because it is basic to a full understanding of close-packed systems.

Continuing these methods of packing to give the layer sequences

...$\overline{\text{AB}}$ABAB... and ...$\overline{\text{ABC}}$ABCABC..., we find that the structures have hexagonal and cubic symmetry respectively. They are usually referred to as hexagonal closest-packed (h.c.p.) and cubic closest-packed (c.c.p.). Note the following details:

(*a*) In *both* closest-packed arrangements every sphere has a coordination number of twelve, viz., six spheres arranged about it in the form of a planar hexagon, three more spheres above and three below, as in Figure 3-3.

(*b*) *Both* types of closest packing are equally economical of space, 74% being occupied.

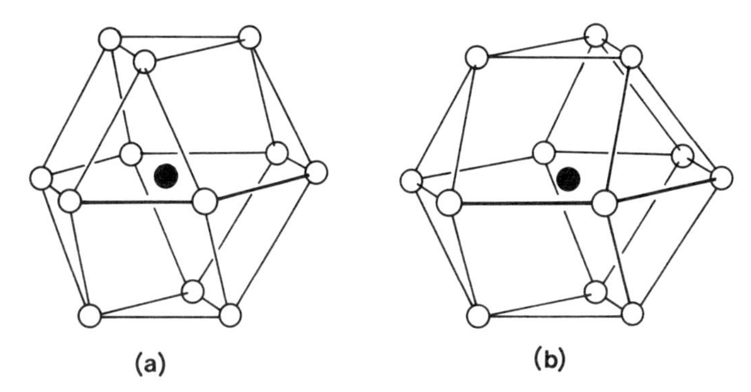

Figure 3-3 Arrangement of the 12 nearest neighbours about *one* packing atom for (a) the cubic closest packed, and (b) the hexagonal closest-packed arrangements

The symmetry of these structures is clearer if they are drawn as shown in Figure 3-4. Note that for the h.c.p. case the structure shown is three times the primitive unit cell. In the c.c.p. case the c.p. layers are stacked parallel to a cube diagonal. Further, arranging c.p. layers in the c.c.p. fashion automatically generates three other c.p. planes which, taken together with the original c.p. plane, intersect in tetrahedral fashion. These planes are shown in Figure 3-5.

Any sequence of layers other than those of h.c.p. (AB) and c.c.p. (ABC) will have lower symmetry. For example, we could imagine the sequence ...$\overline{\text{ABAC}}$ABAC... with a four layer repeat unit (neodymium adopts this structure). This is called double-hexagonal closest packing. It differs from c.c.p. and h.c.p. in one other important respect: there are now two sets of crystallographically distinct spheres. These can be identified by remembering that spheres in h.c.p. and c.c.p. structures each have different arrangements of nearest neighbours. There are a few very rare structures in which the repeat units are apparently many dozens, even a hundred or more layers. The nature of the long-range forces responsible for this ordering is not clear and the experimental facts themselves need extending.

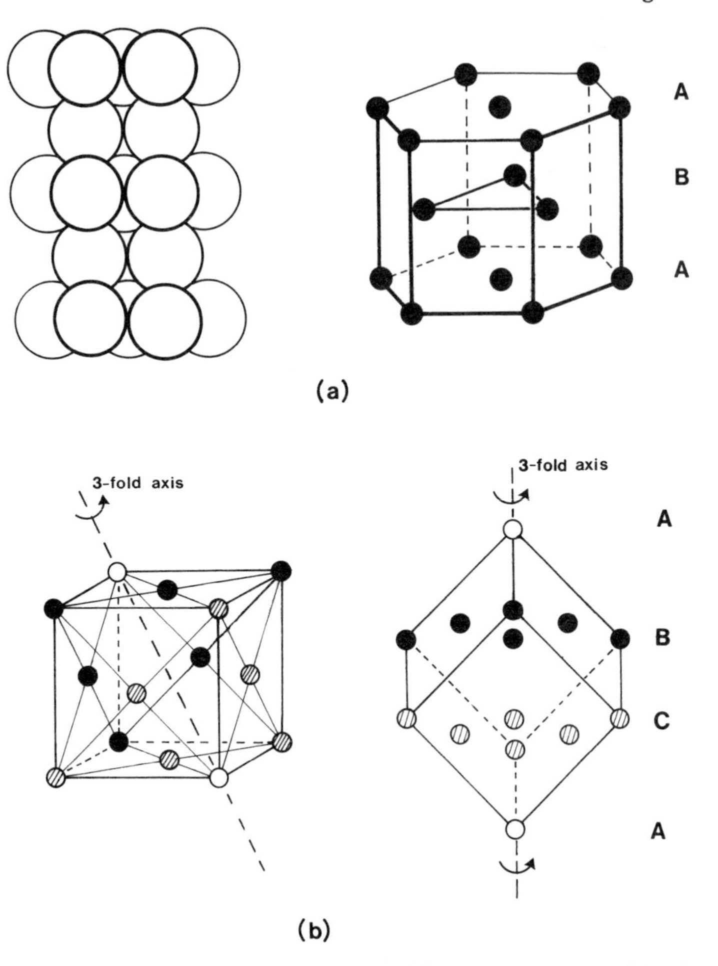

Figure 3-4 (a) The hexagonal closest packing arrangement. (b) The cubic closest-packing arrangement. Note that the closest-packed layers are stacked normal to the cube diagonal

3.2.1 Interstitial Sites

By their nature, spheres cannot fill all space. The interstitial volumes or sites are of great importance in understanding many crystal structures. Two types of site (or hole) are defined by *any two* closest-packed layers which are in contact.

Tetrahedral sites

Any sphere resting in the hollow formed by three spheres in an adjacent layer forms a 'tetrahedral' hole; the centres of the four spheres are at the vertices of a regular tetrahedron (Figure 3-6). The size of the site is considerably less than that of the surrounding spheres: $r/R = 0{\cdot}225$, where R = radius of the spheres forming the closest-packed layers, r = radius of the largest sphere which can fit into a tetrahedral site *without distorting* the structure. Two tetrahedral sites are associated with each atom in a closest-packed layer, one above and one below the packing atom.

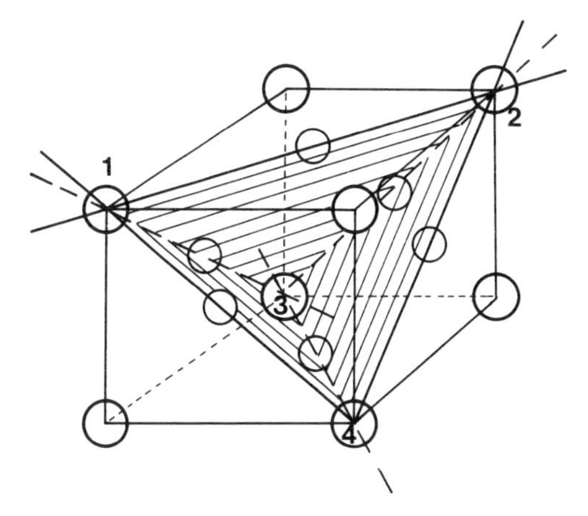

Figure 3-5 Four planes in a face-centred cubic lattice intersecting to form a regular tetrahedron. The planes are defined by atoms (1,2,3), (1,2,4), (2,3,4), and (1,3,4)

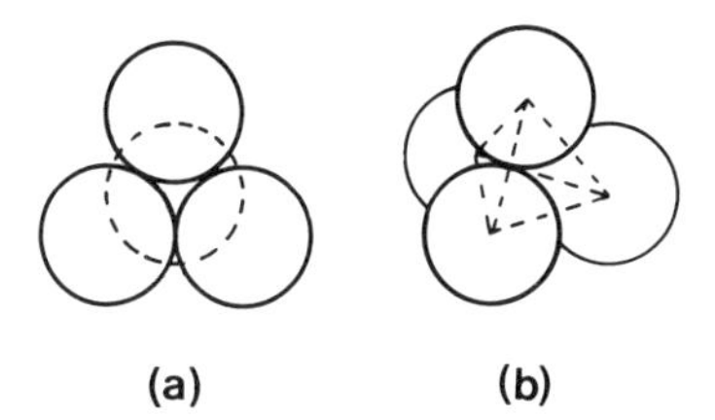

Figure 3-6 A tetrahedral site formed by closest-packed spheres. (a) Top view. (b) Perspective

Octahedral sites

These sites are found at the centres of regular octahedra, the vertices of which are defined by two sets of three spheres in adjacent layers. They are identified and illustrated in Figure 3-7. An octahedral site is considerably larger than a tetrahedral site, and has $r/R = 0{\cdot}414$. One octahedral site is associated with each sphere (each sphere is surrounded by six octahedral holes, each of which is surrounded by six spheres).

N.B. Both octahedral and tetrahedral sites are present in *any pair* of adjacent closest-packed layers and are therefore to be found in closest-packed structures *with any layer sequence.* They are identified for c.c.p. and h.c.p. structures in Figure 3-8.

Each sphere in either an h.c.p. or a c.c.p. array has as nearest neighbours eight tetrahedral sites and as *next*-nearest neighbours six octahedral sites. However, the arrangement of these sites is entirely different in the two cases, see Figure 3-9. **Note especially** that in the h.c.p. case interstitial sites lie

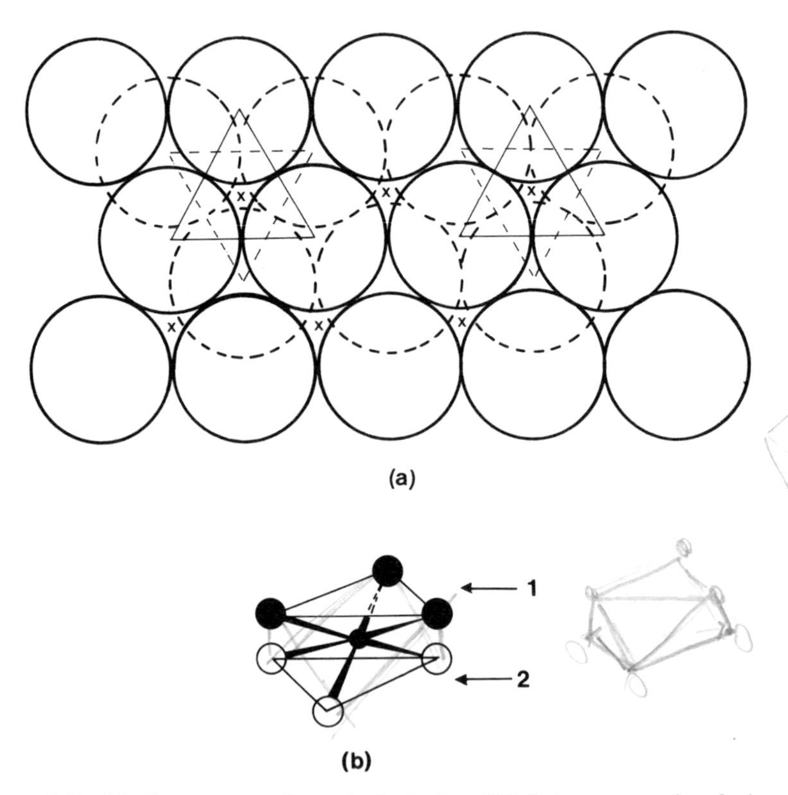

Figure 3-7 (a) Occurrence of octahedral sites (X) between a pair of closest-packed layers. (b) Detail of one such site. Note that three of the six atoms defining each site come from each layer

directly above each other forming rows parallel to the c-axis (i.e. the stacking direction). This has most important consequences. One of these is that a third kind of interstitial site is present in h.c.p. structures, at the centre of a trigonal prism formed from three packing atoms in one layer and one each from the layers above and below. In other words, it lies at the centre of the face common to two joined tetrahedral sites, see Figure 3-10.

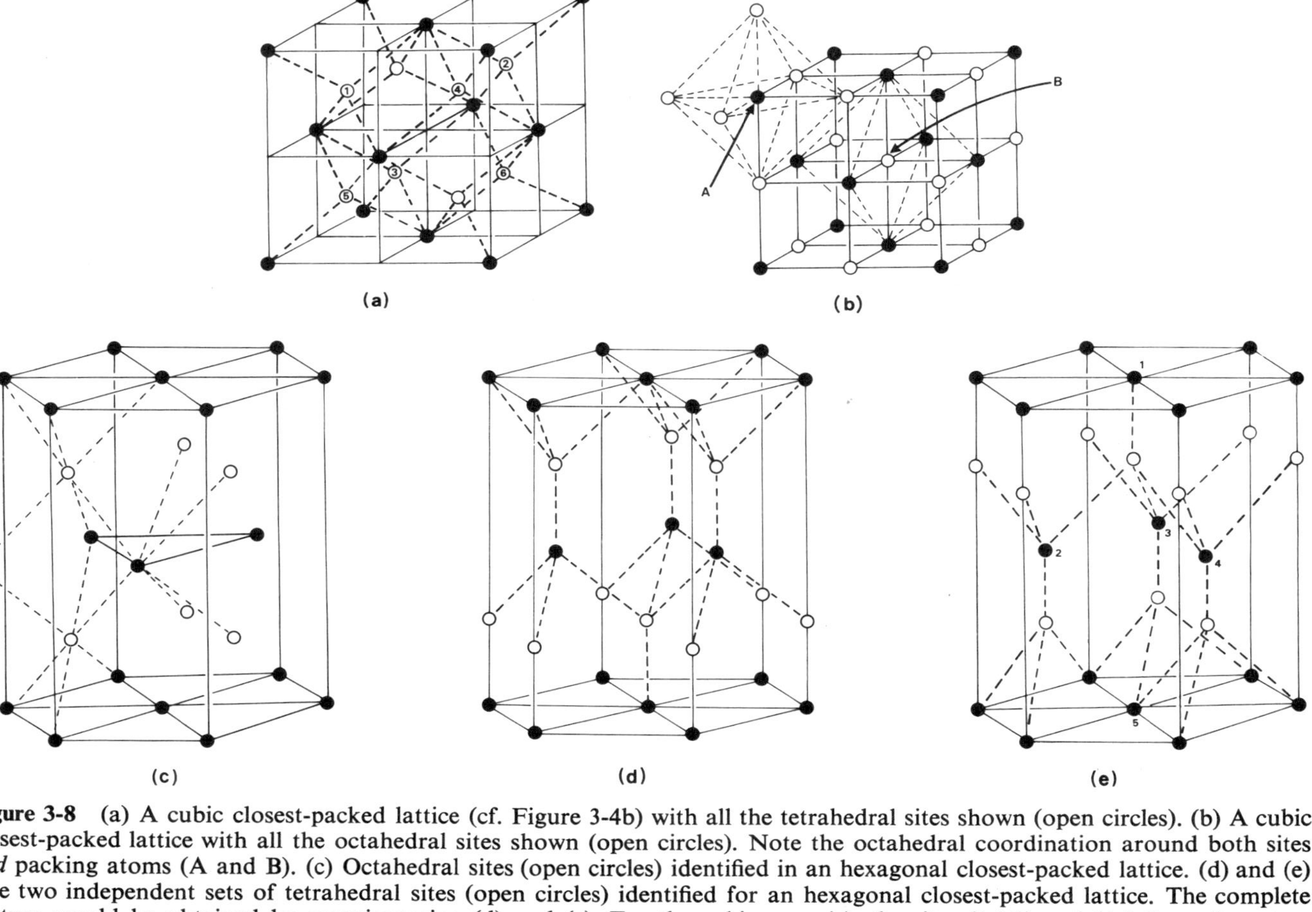

Figure 3-8 (a) A cubic closest-packed lattice (cf. Figure 3-4b) with all the tetrahedral sites shown (open circles). (b) A cubic closest-packed lattice with all the octahedral sites shown (open circles). Note the octahedral coordination around both sites *and* packing atoms (A and B). (c) Octahedral sites (open circles) identified in an hexagonal closest-packed lattice. (d) and (e) The two independent sets of tetrahedral sites (open circles) identified for an hexagonal closest-packed lattice. The complete picture would be obtained by superimposing (d) and (e). For the cubic case, (a), the sites (1,2,3) and (4,5,6) are from two similar independent sets

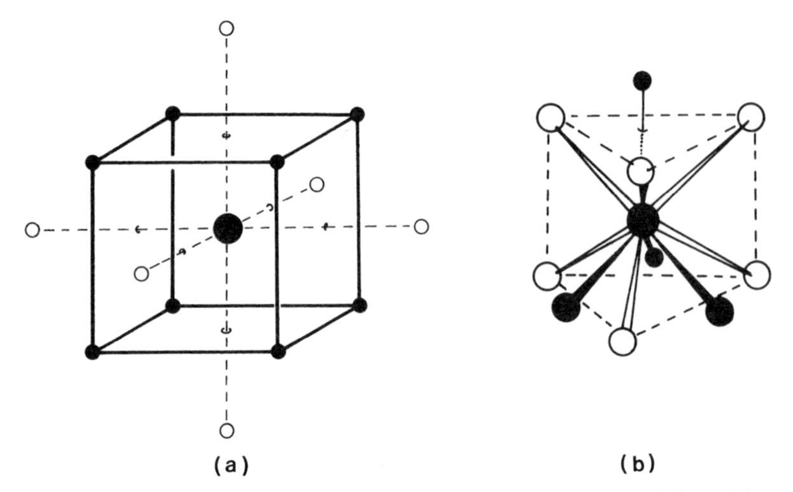

Figure 3-9 The arrangement of the *nearest* octahedral (open circles) and tetrahedral (small black spheres) sites around *one* packing atom in: (a) a cubic closest-packed lattice, (b) an hexagonal closest-packed lattice

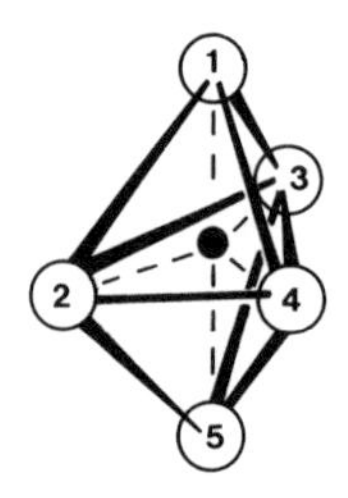

Figure 3-10 A trigonal bipyramidal site in an hexagonal closest-packed lattice. The numbering of the packing atoms is the same as in Figure 3-8e to facilitate comparison

3.2.2 An Overall View of Closest-packed Structures

In order to be sure that we are clear about the spatial relationships between the close-packed layers and the interstitial sites, let us summarize what we have found by stating it in another way.

Figure 3-11 shows an edge-on view of two layers of a close-packed array; the layers are shown separated for the sake of clarity. The planes containing

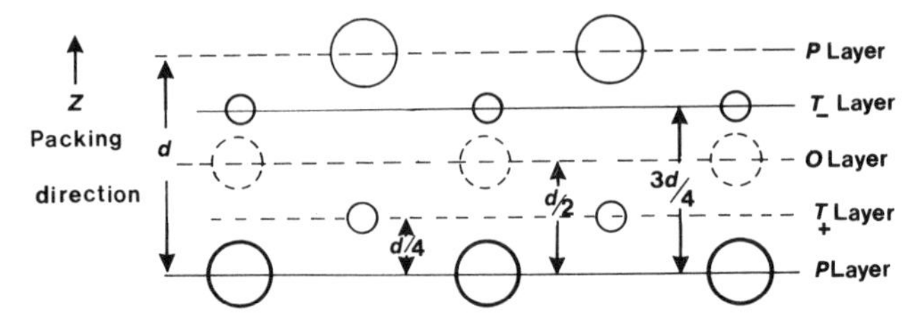

Figure 3-11 An edge-on view of two closest-packed layers. (From S. M. Ho and B. E. Douglas, *J. Chem. Educ.*, **45**, 475, 1968)

the packing spheres in layers A and B are a distance d apart. Between them lie the octahedral and tetrahedral sites. But in fact these sites themselves form arrays identical with those of the close-packed layers: we can treat them as layers. In other words we may consider close-packed structures to be built up from a basic block of four layers PT^+OT^-, where P = close-packed layer of spheres (which we shall in future refer to as the 'packing' layer), O = the layer of octahedral sites, T^+ and T^- = layers of tetrahedral sites with apices pointing in the $+z$ or $-z$ directions respectively. This nomenclature is due to Ho and Douglas (1968).

As we shall see later, more often than not the atoms occupying interstitial sites are larger than r/R ratios allow, with the consequence that the packing spheres are pushed apart. Indeed, in some examples, the so-called interstitial atoms may be larger than the spheres of the host layers. It therefore makes sense sometimes to lift the interstitial sites into prominence by considering them in terms of layers.

3.3 COMPARISON OF CLOSEST-PACKED STRUCTURES WITH THOSE FOUND EXPERIMENTALLY

We have derived closest-packed structures by considering the packing of spheres held together by non-directional forces. The noble gases Ne, A, Kr and Xe, do in fact crystallize in the c.c.p. form, whilst He is h.c.p. We might reasonably expect to find similar structures for other elements which are not held together by directional forces in the solid state, that is for many metals. Figure 3-12 gives a form of the Periodic Table which shows clearly that the majority of metals crystallize with one of three structures: h.c.p., c.c.p. or body-centred cubic (b.c.c., see below). We shall return to the structure of metals in Chapter 8. It is not entirely clear why He is h.c.p. whereas the other noble gases are c.c.p. In both lattices there are twelve nearest neighbours at distance d (the sphere diameter), and six next-nearest neighbours at $d\sqrt{2}$. However there *is* a difference for third-nearest neighbours: the h.c.p. lattice has only two at $1{\cdot}64d$, whereas c.c.p. has twenty-four at $1{\cdot}73d$. This leads to a free energy for the h.c.p. lattice which is $0{\cdot}01\%$ lower than for c.c.p., implying that the h.c.p. structure should be preferred. It is possible that the different symmetry of the arrangement of next-nearest neighbours allows multiple interactions to take place which just tip the balance the other way.

Closest-packed structures are also adopted by molecules which are either close to spherical in shape (e.g. methane), or which can become spherical by rotation, for example H_2 (h.c.p.), HCl, H_2S (c.c.p.). There is often also a low-temperature phase of lower symmetry; it is most probable that there the molecules have ceased to rotate.

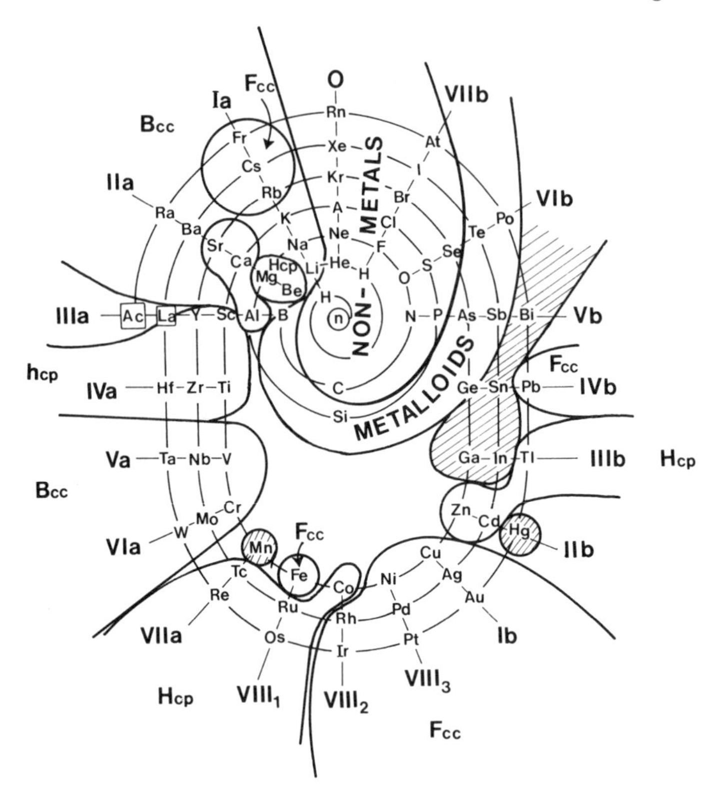

Figure 3-12 A version of Periodic Table (after Tomkieff) showing the distribution of structures adopted by the metallic elements

3.4 THE BODY-CENTRED CUBIC AND CAESIUM CHLORIDE STRUCTURES

We did *not* predict the existence of b.c.c. structures on the basis of the space-filling postulate: it comes up here because we have now realized that a good many metals adopt it. Accounting for its existence obviously requires another principle that outweighs the importance of packing alone.

The b.c.c. structure is shown in Figure 3-13. It is not strictly a closest-packed structure, although it can be related to them (see below). It can be considered as two interpenetrating simple cubic lattices. Each sphere has eight nearest neighbours at the corners of a cube with the atom centres a distance d apart, and six *next*-nearest neighbours at $(2/\sqrt{3})d = 1{\cdot}15d$. The coordination number (c.n.) is therefore effectively 14. It occupies 68% of the available space and is thus only slightly less economical of volume than closest packing. As you have probably realized, it is just what we would have predicted had we pursued the idea of structures based upon the layer of Figure 3-1a.

Tetrahedral sites are present with $r/R = 0{\cdot}291$, see Figure 3-13b and c, there being six per packing atom (cf. c.p. lattices). *Distorted* octahedral sites are found at the centres of the cube faces and at the midpoints of the cube edges. The distances are such that large spheres cannot be accommodated. There are three 'octahedral' sites per packing atom. This difference from c.c.p. is well illustrated by the rarity of interstitial structures based upon b.c.c., and

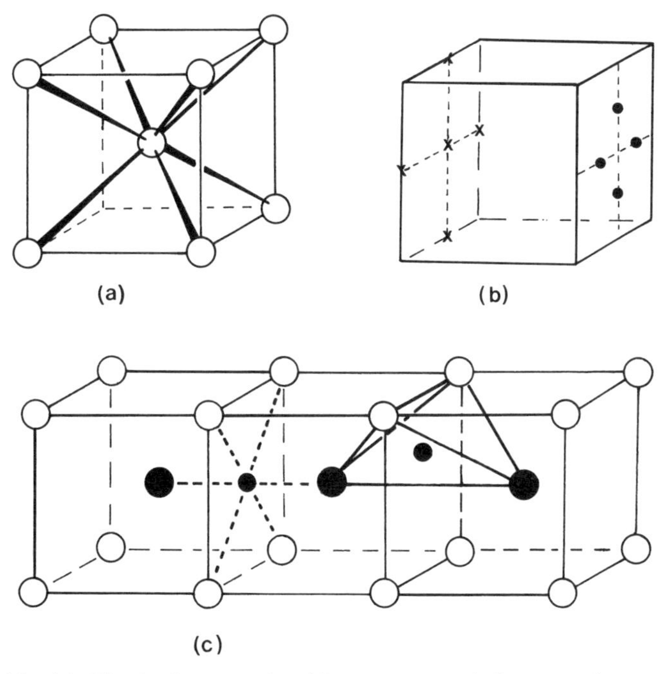

Figure 3-13 (a) The body-centred cubic structure. If the central atom is of a different type from the others, the lattice is known as the CsCl type. (b) Positions of 'tetrahedral' sites (●) and 'octahedral' sites (X) are shown for one cube face each. (c) Detail of one 'octahedral' and one tetrahedral interstitial site. Here one set of packing atoms is distinguished from the other: this picture therefore represents the CsCl structure

by the fact that carbon is much less soluble in α-iron (0·1 atom % of C in b.c.c.) than in γ-iron (3·6 atom % of C in face-centred cubic), a point of great relevance in steel making.

The b.c.c. structure can be related to the c.c.p. structure by considering it as c.c.p. with all the octahedral and tetrahedral sites filled with the same type of atom as the close-packed layers. However, this is a formalism and it is more to the point to recall (p. 15) that both c.c.p. and b.c.c. structures are based upon rhombohedral unit cells which differ only in the angle (90° or 60°); see Figure 2-7.

If one of the two interpenetrating cubic lattices which make the b.c.c. structure is formed from atoms of type A and the other from type B the '*caesium chloride*' structure results. The c.n. of each ion is 8, cubically. Although this lattice type is apparently stuck with its label, it is in fact of extremely rare occurrence except among intermetallic compounds in which quite different factors favour its adoption. The structure is shown in Figure 3-13c. Apart from intermetallic compounds it is adopted by:

(*a*) CsCl, CsBr, CsI, TlCl, TlBr.
(*b*) NH_4X (X = Cl, Br, I).
(*c*) CsX (X = CN, SH, SeH, NH_2) and TlCN.

The situation is much more complex than has been indicated here. All of these compounds undergo changes to other structures (i.e. phase changes) either with thermal (up or down) or pressure change. Forces responsible for adoption of the caesium chloride structure by non-metallic compounds are evidently very finely balanced.

3.5 DEVELOPMENT AND CLASSIFICATION OF STRUCTURES BASED UPON CLOSEST-PACKING

A very large number of structures can now be developed from closest-packed lattices by consideration of the *geometrical* possibilities. We stress that in developing this line of argument we are saying *nothing* about bond type: it is an exercise in geometry. Later, we shall see how bond type and crystal structure are related.

Three ways of development immediately suggest themselves:

(*a*) We can form c.p. layers of more than one type and size of sphere.
(*b*) We can fit smaller spheres into octahedral and/or tetrahedral sites in close-packed layers composed of one type of sphere only.
(*c*) Both variations can be used together.
(*d*) These schemes may be further developed by omitting a proportion of the spheres in the c.p. layers and by filling varying proportions of the interstitial sites.
(*e*) Finally, we can develop related structures which are distorted from closest-packed by the presence of non-spherical groups.

This is, of course, a very simple scheme of classification and much more elaborate varieties are available. However, this one serves to introduce us to quite a varied selection of structures and forms a good background for more advanced treatments.

3.5.1 Closest-packed Structures with more than one Type and Size of Atom

Some possible closest-packed arrangements of (approximately) equal spheres are shown in Figure 3-14a–c. The question also arises whether in any of these structures each sphere can be completely surrounded by spheres of the opposite type. For a 1:1 lattice AB, it is not possible to find such an arrangement, as can be readily seen in Figure 3-14d. If A is surrounded by six spheres B in a close-packed manner, then each B already has two other B

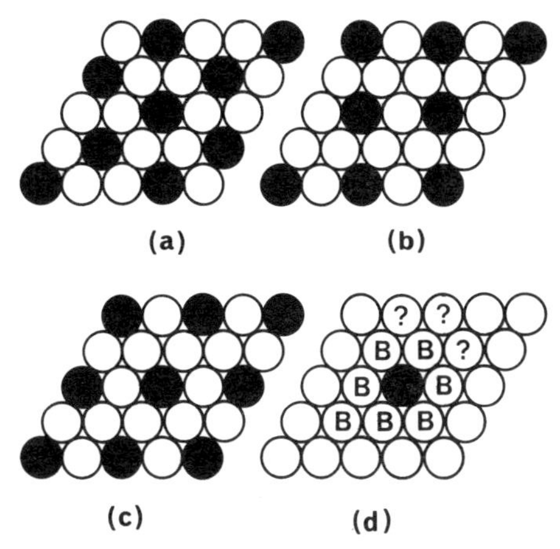

Figure 3-14 (a) One possible close-packed layer of type AB_2. (b) and (c) show two possible AB_3 close-packed layers. (d) See text. (After A. F. Wells, *Structural Inorganic Chemistry*, The Clarendon Press, Oxford, 1962)

as neighbours. For AB_2, the arrangement of Figure 3-14a has each A completely surrounded by B, but this layer could not be stacked and still keep A surrounded by B. WAl_5 has layers WAl_3 of the type shown in Figure 3-14b alternating with layers of aluminium. Layers of composition AB_3, are found in some alloys, e.g. $TiAl_3$ (layer b), $ZrAl_3$ (layers of types b and c), which are superstructures of the aluminium structure.

3.5.2 Closest-packed Structures with Filled Interstitial Sites

Filling interstitial sites in c.p. structures leads to a variety of possibilities, depending upon the stoichiometry AB_n, and upon whether tetrahedral, octahedral, or both types of site are used. Since one octahedral site is associated with each atom, filling all the octahedral sites leads to a compound of stoichiometry AB. Two tetrahedral sites are associated with each sphere A in a c.p. structure; if all tetrahedral sites are filled the compound has stoichiometry AB_2. Some simple possibilities are shown in Table 1 and related to the examples by which they are generally known.

Table 1. Some common structures formed by filling octahedral or tetrahedral holes in close-packed lattices

Type and fraction of sites occupied	C.c.p.	H.c.p.
All tetrahedral	Fluorite, CaF_2	(Not known)
All octahedral	Rock salt, NaCl	Nickel arsenide, NiAs
$\frac{1}{2}$ tetrahedral	Zinc blende, ZnS (sphaelerite)	Wurtzite, ZnS
$\frac{1}{2}$ octahedral	$CdCl_2$	CdI_2
$\frac{1}{3}$ octahedral	$CrCl_3$	BiI_3; β-$ZrCl_3$

The structures below the dotted line (other than β-$ZrCl_3$) are layer structures in which alternate sets of octahedral sites are occupied. We shall consider the structures of Table 1 in detail as each is adopted by many other compounds. It is usually found that the interstitial atoms are larger than can be accommodated by the available sites without distortion of the c.p. layers. It is therefore not strictly correct to talk of closest-packed layers but rather of close-packed layers. In particular, the filled c.c.p. structure is often referred to as face-centred cubic (f.c.c.).

Rock salt, NaCl

Geometrically this is a very simple structure, Figure 3-8b, which we all heard about at mother's knee. There is, in fact, much more to it than appears from a superficial glance. It is generally thought of as an f.c.c. lattice of Cl^- (the host lattice) with Na^+ (*not* the guest!) in all of the octahedral sites, although it can also be correctly described as two interpenetrating cubic close-packed lattices. Note, however, that if *all* of the atoms in the NaCl structure are identical we have the primitive cubic lattice. The coordination number (c.n.) of Na^+ = c.n. of Cl^- = 6 octahedrally. $r_{Na^+}/r_{Cl^-} = 0{\cdot}71$, compared with the maximum value of 0·414 if the host lattice is to remain undistorted.

This structure is extremely common. It is assumed by:

(*a*) Alkali metal hydrides.
(*b*) All alkali halides other than CsCl, CsBr and CsI (even CsCl has this structure above 445°).
(*c*) The oxides, sulphides, selenides and tellurides of Mg, Ca, Sr and Ba (except for MgTe).
(*d*) First-row transition metal oxides, MO (M = Ti, V, Mn, Fe, Co, Ni).

(*e*) Interstitial carbides and nitrides, MC and MN (M = Ti, Zr, V, etc.), and some phosphides, such as GeP, LaP.
(*f*) Alloy phases ME where M is a lanthanide or actinide and E = As, Sb, Bi.

The bond types covered by this list range from ionic to covalent to metallic, and the properties (electrical, magnetic, mechanical) are also very varied.

Nickel Arsenide, NiAs.

Nickel is in octahedral sites in an h.c.p. host lattice of arsenic and hence has a c.n. = 6. Each arsenic atom is also six-coordinate but the nickel atoms are at the vertices of a trigonal prism (Figure 3-15). Note, however, that by

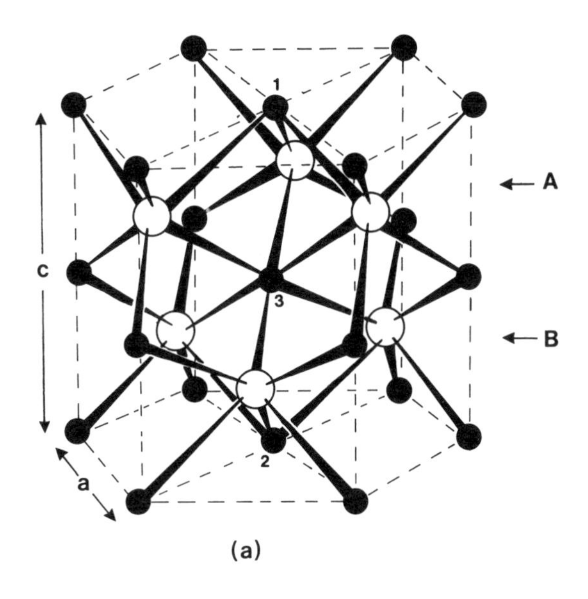

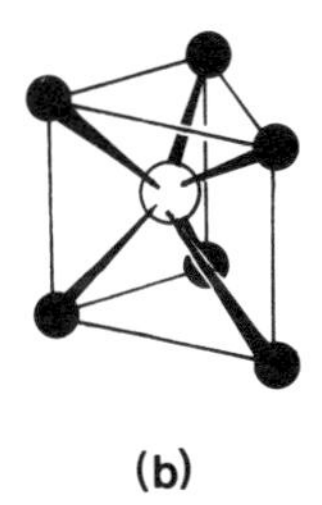

Figure 3-15 (a) The nickel arsenide structure. (b) The coordination arrangement about one As atom. (Ni atoms are shown in black in both pictures.) Compare the octahedral coordination around Ni atom 3 with that of Figure 3-7b

reducing the axial ratio, c/a, two more metal atoms (1 and 2 in Figure 3-15) could be brought quite close to the metal atom 3 giving it a c.n. approaching 8 (i.e. 6As + 2M). Chains of metal atoms would therefore be formed, the extra M—M bond strength compensating for the compression. Evidence for this effect is provided by the axial ratios of some compounds which adopt this structure; the ratios (Table 2) suggest that with decrease in ionic character, metal atoms are less likely to repel each other and more likely to form metal–metal bonds.

Table 2. Axial ratios for some solids with the nickel arsenide structure

Compound	FeS	FeSe	FeTe	FeSb	CoTe	NiTe
Axial ratio, c/a^a	1·68	1·64	1·49	1·25	ca. 1·40	ca. 1·36

[a] For normal h.c.p. of rigid spheres, $c/a = 1{\cdot}633$.

The flexibility of the NiAs structure† makes it particularly suitable for compounds with bond types intermediate between ionic and intermetallic. It is only adopted by transition-metal compounds. Besides many sulphides, selenides and tellurides, ME, it is also adopted by intermetallic compounds such as MSn (M = Mn, Fe, Ni, Cu, Pt) and, in a related distorted form, by phosphides, arsenides and germanides. It is further discussed in Section 9.2.8.

β-$ZrCl_3$ Structure

A structure closely related to that of NiAs is typified by β-$ZrCl_3$. It is adopted by trihalides of transition metals having d^1, d^3 or low-spin d^5 configurations (see Chapter 5 for discussion of *d*-electron configuration). Metal atoms are in one-third of the octahedral holes in an h.c.p. lattice of halogen atoms, but these are so occupied that chains of metal atoms run through the lattice parallel to the *c*-axis, Figure 3-16. The consequent strong bonding between metal atoms distorts the h.c.p. arrangement making the occupied holes smaller than the unoccupied holes.

Table 3. Some compounds adopting the β-$ZrCl_3$ structure

β-$TiCl_3$	β-$ZrCl_3$	β-$RuCl_3$	β-$TiBr_3$	β-$ZrBr_3$
$MoBr_3$	TiI_3	ZrI_3	MoI_3	HfI_3

An alternative and equivalent description is to consider the structure as composed of chains of MX_6 octahedra sharing opposite faces (see Section 7.2.1 for development of this theme).

† By this form of words we refer to the flexibility of the structure *type*, not to any supposed variability in lattice constants of the particular compound NiAs.

Layer Structures in which Octahedral Sites are Used

In CdI_2 octahedral sites between *every other* pair of h.c.p. layers of iodine atoms are occupied by cadmium atoms. Each Cd is therefore six-coordinate (octahedrally), but each iodine has only three nearest neighbours—at the vertices of a trigonal pyramid, Figure 3-17; compare with Figure 3-15a.

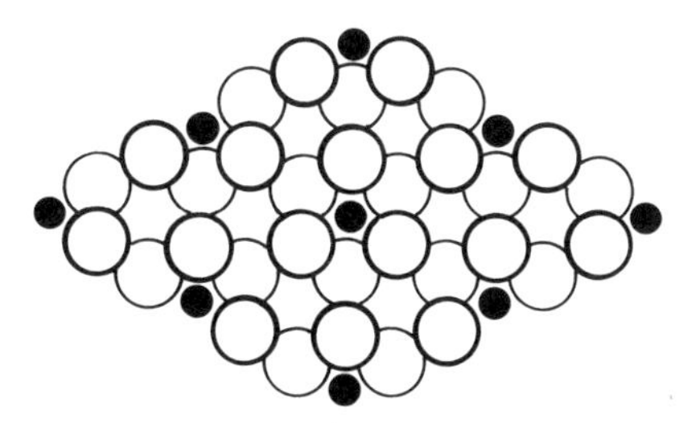

Figure 3-16 The structure of β-$ZrCl_3$. Two superimposed close-packed layers are shown. Zr atoms (in black) occupy chains of octahedral sites running parallel to the *c*-axis or layer-stacking direction. (After L. F. Dahl, T. Chiang, P. W. Seabaugh and E. W. Larsen, *Inorganic Chemistry*, **3**, 1239 (1964))

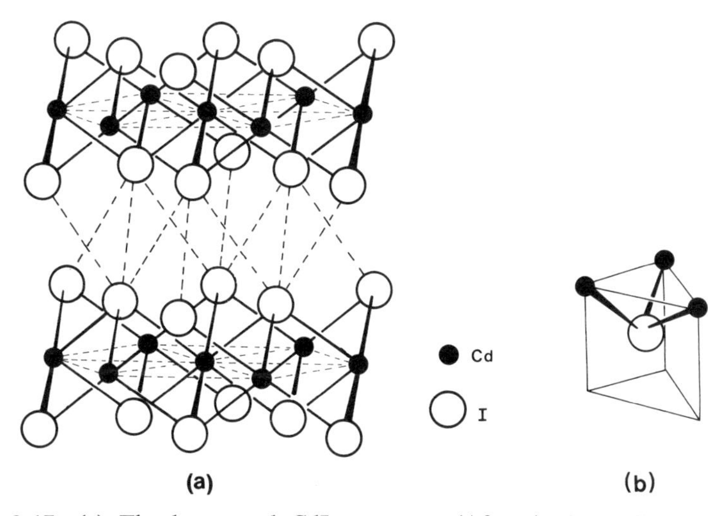

Figure 3-17 (a) The hexagonal CdI_2 structure (After A. F. Wells, *Structural Inorganic Chemistry*, The Clarendon Press, Oxford, 1962). (b) The coordination around one iodine atom

CdI_2 is termed a 'layer structure' because it consists of repeat units (or sandwiches) I—Cd—I···I—Cd—I···I—Cd—I (where the symbols Cd and I represent layers of those atoms) with contact between adjacent iodine layers. This is clearly not an ionic structure; bonds to the metal will be semi-ionic,

Table 4. Some compounds adopting the $CdCl_2$ and CdI_2 structures

$CdCl_2$	MCl_2 (M = Mg, Mn, Fe, Co, Ni, Zn, Cd)
	$NiBr_2$ NiI_2; $ZnBr_2$, ZnI_2
CdI_2	MCl_2 (M = Ti, V)
	MBr_2 (M = Mg, Fe, Co, Cd)
	MI_2 (M = Mg, Ca, Ti, V, Mn, Fe, Co, Cd, Ge, Pb, Th)
	$M(OH)_2$ (M = Mg, Ca, Mn, Fe, Co, Ni, Cd)
	MS_2 (M = Ti, Zr, Sn, Ta, Pt)
	MSe_2 (M = Ti, Zr, Sn, V, Pt)
	MTe_2 (M = Ti, Co, Ni, Rh, Pd, Pt)

but the I---I attraction is principally of the van der Waals type. The structure is easily cleaved between the I---I planes, which accounts for the flaky nature of the crystals.

The analogous c.c.p. layer structure is typified by $CdCl_2$. Some compounds adopting these structures are shown in Table 4, from which it is very clear that there is something especially favourable about the hexagonal structure so far

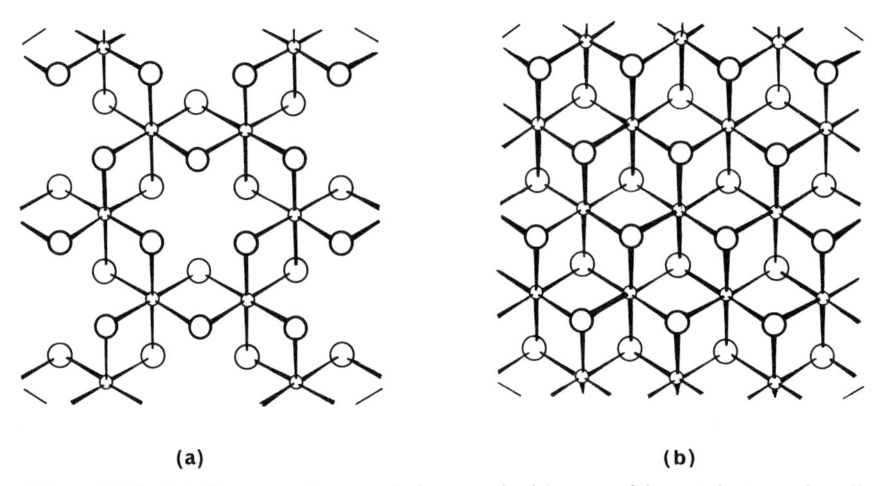

(a) (b)

Figure 3-18 (a) Two superimposed close-packed layers with metal atoms (small circles) in two-thirds of the octahedral holes. Layers of this type occur in the $CrCl_3$ (c.c.p.) and BiI_3 (h.c.p.) structures. (b) Two superimposed close-packed layers with all octahedral sites filled. Layers of this type are found in the $CdCl_2$ (c.c.p.) and CdI_2 (h.c.p.) structures; compare with Figure 3-17a. (After A. F. Wells, *Structural Inorganic Chemistry*, The Clarendon Press, Oxford, 1962)

as compounds of metals with the heavier non-metals are concerned. These structures are further discussed in Section 7.1.4.

With one-third of the octahedral holes occupied the $CrCl_3$ (c.c.p.) and BiI_3 (h.c.p.) layer structures are formed, in which two-thirds of the octahedral sites between *every other* c.p. layer of host lattice are filled (Figure 3-18a).

Note that in BiI_3, the iodine atoms have only two neighbouring bismuth atoms—at two corners of a trigonal prism. The α-forms of the trihalides described above (β-$ZrCl_3$ type) mostly have the $CrCl_3$ or BiI_3 structure.

Lead Chlorofluoride, PbFCl

This structure is about as common as that of CdI_2, being adopted by:

(*a*) Oxide halides MOCl where M is a large trivalent cation, Bi^{3+}, a rare earth or actinide; and some corresponding bromides and iodides.
(*b*) Mixed halides or hydride halides of divalent metals, MHX, where M = Ca, Sr, Ba and X = Cl, Br, I; and PbFCl, PbFBr.
(*c*) Oxide chalcogenides of large tetravalent ions, e.g. MOS (M = Th, Pa, U, Np, Pu) and UOSe, ThOSe, ThOTe.

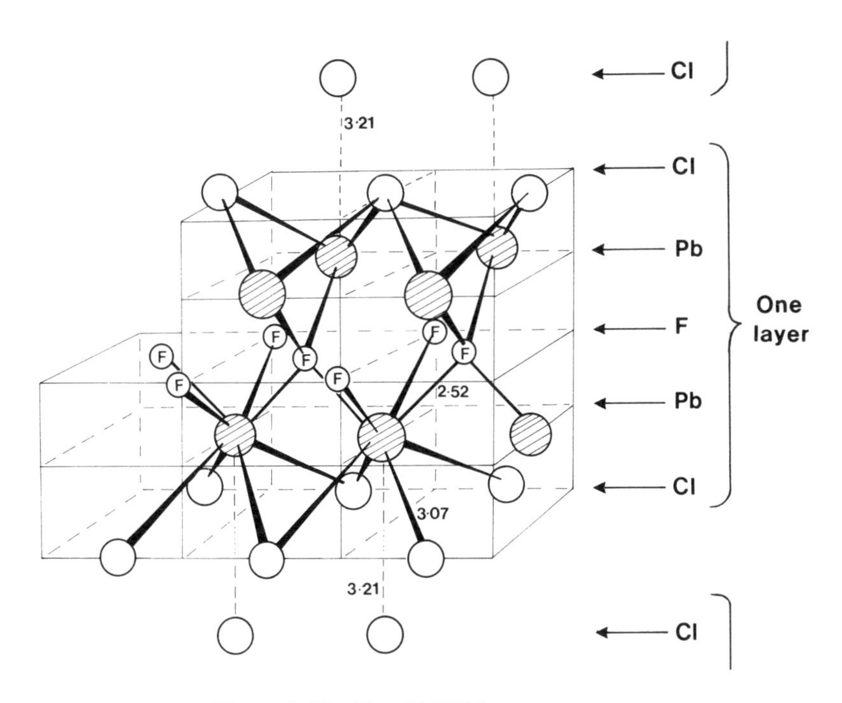

Figure 3-19 The PbFCl layer structure

Note how the valence of the cation is balanced by the correct number of charges of the anions. This is a rather complex double-decker sandwich structure in which the layers repeat in the order ClPbFPbCl, ClPbFPbCl, etc. The central layer of the sandwich is an accurately co-planar layer of fluorines, with layers of chlorine either side. The cations are coordinated to four fluorines and four chlorines in the shape of a somewhat distorted square antiprism. However, the distances between the layers are shorter than would be expected on the basis of ionic radii, suggesting appreciable bonding between metal ions in one layer and chlorines in the neighbouring layers. The structure is shown in Figure 3-19. See also p. 85.

Li_2O and Fluorite, CaF_2

In fluorite, all the tetrahedral sites in an f.c.c. lattice of Ca^{2+} are filled with F^- (Figure 3-8a). Since $r_{F^-}/r_{Ca^{2+}} = 0{\cdot}945$ (using the radii of Table 4, Chapter 2), widely different from the maximum r/R of 0·225, it is little more than a formalism to speak of an f.c.c. array of Ca^{2+}. The coordination number of Ca^{2+} equals 8, i.e. the fluoride ions are disposed at the corners of a cube, whilst the c.n. of F^- equals 4, tetrahedrally. The fluorite structure is adopted by fluorides and oxides of very large divalent or tetravalent ions, Table 5.

Table 5. Some compounds adopting the fluorite (CaF_2) and Li_2O (or anti-fluorite) structures

Fluorite	MF_2 (M = Ca, Sr, Ba, Ra, Pb, Cd, Hg, Eu)
	MO_2 (M = Ce, Pr, Tb—i.e. lanthanides, and Th, Pa, U, Np, Pu, Am, Cm)
	MMg_2 (M = Ge, Sn, Pb); AuM_2 (M = Al, Ga, In)
Anti-fluorite	M_2O (M = Li, Na, K, Rb)
	M_2S, M_2Se, M_2Te (M = Li, Na, K)

When the more electronegative constituent occupies the f.c.c. positions, a structure anti-isomorphous with fluorite is obtained, known as 'anti-fluorite', typified by Li_2O. It is widely adopted by many alloy-like phases and by crystals containing complex ions (see p. 62).

ZnS: Zinc Blende and Wurtzite

In zinc blende (Figure 3-20) one-half of the tetrahedral sites in an f.c.c. array of sulphur atoms are filled with zinc; the arrangement of the filled sites is such that the structure may also be described as two interpenetrating f.c.c. lattices, one of zinc, one of sulphur. The c.n. of S = c.n. of Zn = 4, tetrahedrally. If carbon replaces all the zinc and sulphur, the diamond structure is obtained.

Wurtzite is the h.c.p. analogue of zinc blende and has the structure shown in Figure 3-8d. Coordination spheres of both zinc and sulphur remain tetrahedral; the only structural change is in the sequence of the layers. The subtle differences determining adoption of the wurtzite, as opposed to the zinc blende, structure will be dealt with in Chapter 5. Nevertheless, the data of Table 6 indicate a slight tendency for the more ionic materials to adopt the wurtzite structure. Note (*a*) the adoption of the zinc-blende structure by

Table 6. Some compounds crystallizing with the zinc blende or wurtzite structures (* indicates diamorphism)

Zinc blende (f.c.c.)	CuX (X = F, Cl, Br, I) ZnO* MS, MSe (M = Be, Mn*, Zn, Cd*, Hg) MTe (M = Be, Zn*, Cd, Hg) MP, MAs, MSb (M = Al, Ga, In) SiC
Wurtzite (h.c.p.)	MO (M = Be, Zn*) MnS*; MSe (M = Mn, Cd*) MTe (M = Mg, Zn*) MN (M = Al, Ga, In) NH_4F

compounds such as gallium phosphide and indium antimonide which are of enormous importance in the electronics industry; (*b*) only *non-transition metal* compounds adopt these structures, with the exception of the spherically-symmetric Mn^{II}, d^5.

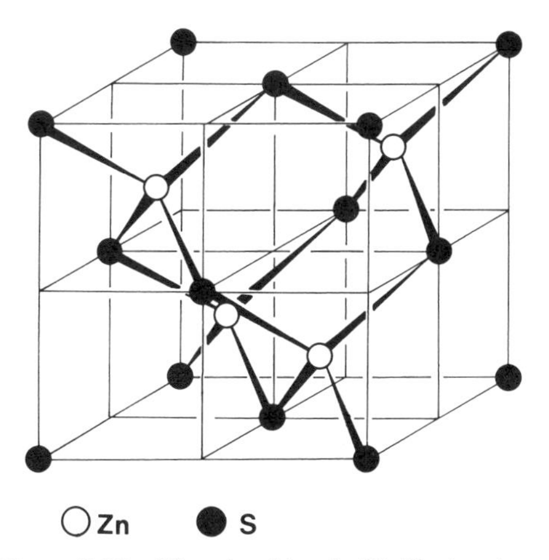

Figure 3-20 The zinc blende (ZnS) structure

3.5.3 Structures with Mixed Close-packed Layers and Filled Interstitial Sites

We have already seen that many structures based upon c.p. layers can be derived by (*a*) using c.p. layers of two or more components, (*b*) filling a proportion of the octahedral or tetrahedral sites in c.p. layers composed of one type of host atom. A great many more compounds may be simply included in our classification by combining these two variables. The geometrical possibilities are too numerous to list here; we can, however, illustrate the procedure with reference to the specific examples of complex halides and oxides of the general type ABX_3 and ABO_3.

There are many complex halides of formula A^IBX_3, where X = halogen, in which the A and X atoms in the ratio 1:3 pack together to form an effectively c.p. array. Ionic radii (Table 6, Chapter 2) are almost identical for Rb^+ and Cl^- (1·63, 1·64 Å) and for Cs^+ and Br^- (1·86, 1·80 Å), whilst those for K^+, F^- and Cs^+, Cl^-, are not very different; these pairs can therefore form approximately c.p. layers with relatively little distortion and many complexes KMF_3, $RbMCl_3$, $CsMCl_3$, etc., are known.

We can derive geometrically possible structures as follows:

(*a*) Draw out all possible c.p. layers of the AX_3 type.
(*b*) Stack these in different sequences.
(*c*) Identify the interstices and show which, if any, are completely surrounded by A only or X only. In practice the only AX_3 layer commonly found is that of Figure 3-14c. If we further restrict ourselves to structures in which every B is surrounded only by X (e.g. the layer of Figure 3-14c) then the pattern of occupation of the holes is shown in Figure 3-21 for the compounds of Table 7.

Table 7. Structures based upon AX_3 close-packed layers

Layer sequence	Fraction of octahedral X sites occupied by B atoms		
	All	$\frac{2}{3}$	$\frac{1}{2}$
AB (hexagonal)	$CsNiCl_3$	$Cs_3Tl_2Cl_9$	K_2GeF_6
ABC (cubic)	$RbCaF_3$		K_2PtCl_6

From a geometrical viewpoint, the only difference between these structures is in the occupation of the octahedral sites. All their formulae may be rewritten in the form $A^IM_nX_3$ to emphasize this point. In K_2GeF_6 and K_2PtCl_6, discrete $[GeF_6]^{2-}$ and $[PtCl_6]^{2-}$ ions are present; in $Cs_3Tl_2Cl_9$ the sites are occupied in pairs, giving binuclear $[Tl_2Cl_9]^{2-}$ ions which may also be described as two $(TlCl_6)$ octahedra sharing a common face; in $CsNiCl_3$ and $RbCaF_3$ all the X_6 sites are filled with B, so that no discrete ions are possible and the compounds must be viewed as three-dimensional complexes of infinite extent.

The description which we have given of these structures is perfectly valid but we should note alternative and equivalent statements. In K_2PtCl_6, platinum is in the +4 oxidation state and will therefore form bonds to chlorine which are quite highly covalent; the ion $[PtCl_6]^{2-}$ has an autonomous existence in solution. It is therefore not unreasonable to think of the structure of K_2PtCl_6 in terms of the packing of very large and approximately spherical

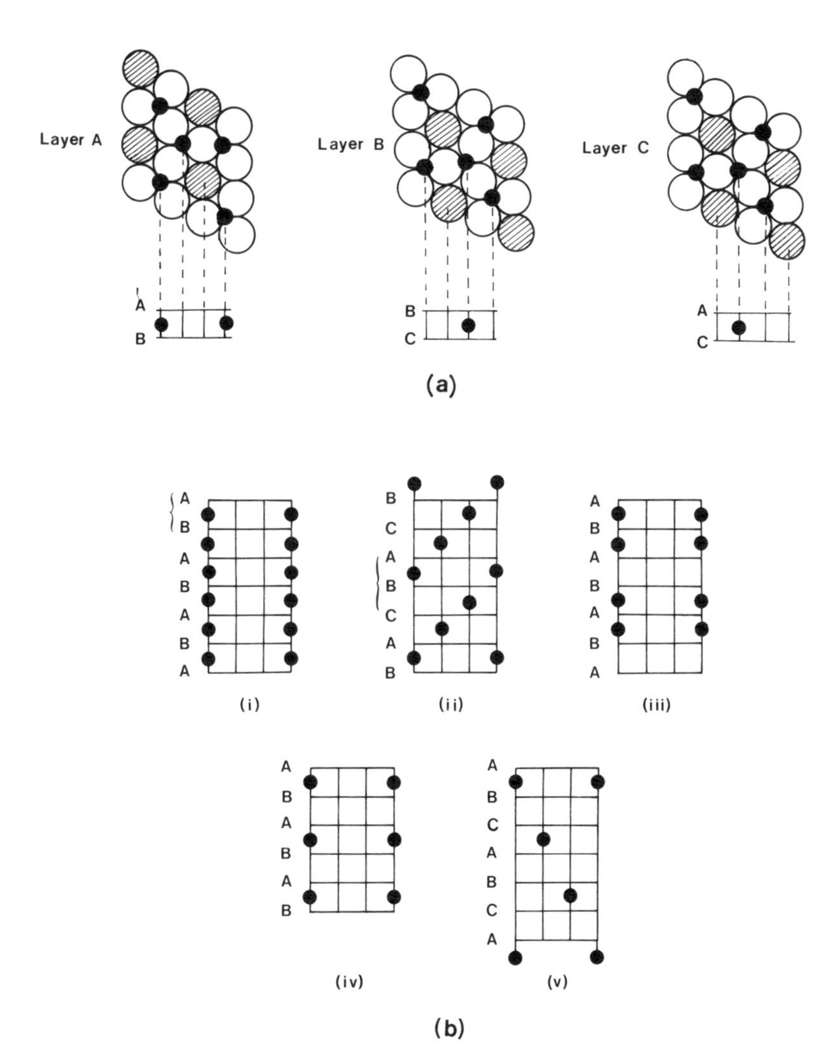

Figure 3-21 (a) Close-packed layers AX_3 of the type shown in Figure 3-14c showing the positions for metal atoms (black circles) on either side of each layer. Beneath each is shown the possible positions in which metal atoms may be accommodated between *pairs* of layers stacked in the sequence AB, BC, or CA. (b) Elevations of the structures of Table 3-7, showing the occupation of octahedral holes. (i) $CsNiCl_3$ (ii) $RbCaF_3$ (iii) $Cs_3Tl_2Cl_9$ (iv) K_2GeF_6 (v) K_2PtCl_6. (After A. F. Wells, *Structural Inorganic Chemistry*, The Clarendon Press, Oxford, 1962)

$[PtCl_6]^{2-}$ ions and potassium ions; the anti-fluorite structure (p. 60) is ideal for this purpose. Draw out the anti-fluorite structure and convince yourself of the equivalence of the two descriptions. A version of this is shown in Figure 3-22.

In $CsNiCl_3$ the octahedral sites are so occupied that chains of Ni atoms run through the crystal along the *c*-axis (cf. β—$ZrCl_3$, p. 56). It is therefore accurately described as chains of $(NiCl_6)$ octahedral sharing opposite faces (see Figure 5-39). Although such chains cannot be obtained free in solution, crystals such as $[(CH_3)_4N]^+MnCl_3^-$ can be grown which have analogous structure, and in these there is no doubt that the best description is in terms of $(MCl_3^-)_n$ chains.

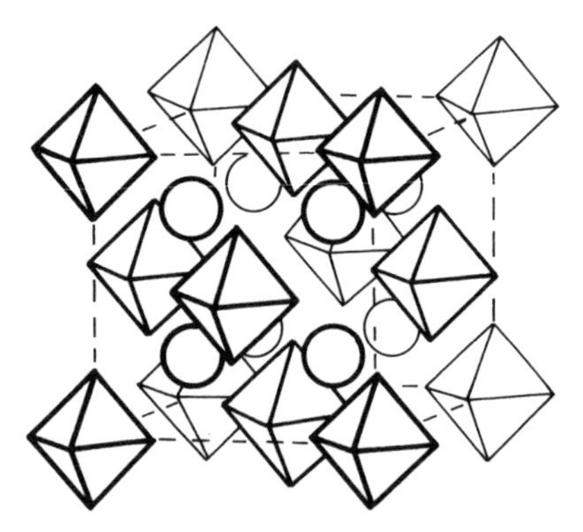

Figure 3-22 The K_2PtCl_6 structure considered as an array of K^+ (circles) and $[PtCl_6]^{2-}$ ions packed in the anti-fluorite manner. (From A. F. Wells, *Structual Inorganic Chemistry*, The Clarendon Press, Oxford, 1962)

An interesting variant of the $RbCaF_3$ (perovskite) structure is shown by $CsAuCl_3$ (Figure 3-23) which is, in fact, composed of square-planar $[AuCl_4]^-$ and linear $[Cl—Au—Cl]^-$ ions, i.e. it is really $Cs_2[Au^{I}Cl_2]\,[Au^{III}Cl_4]$.

Complex oxides ABO_3 are of two basic types. (*a*) Those in which A and B are small compared with O^{2-}; both A and B fit into octahedral sites in c.p. arrays of O^{2-}. (*b*) Those in which A and O^{-2} are of comparable size and form mixed c.p. arrays analogous to those of the complex halides above. Examples of (*b*) are $BiNiO_3$ (cf. $CsNiCl_3$) and perovskite, $BaTiO_3$ (cf. $RbCaF_3$). The perovskite structure is capable of various distortions.

3.5.4 Structures Based upon More-complex Variants

As suggested above, still more compounds may be drawn into our classification by allowing:

(*a*) gaps in the c.p. host arrays,
(*b*) filling of some proportion of both octahedral and tetrahedral sites in one and the same host lattice.

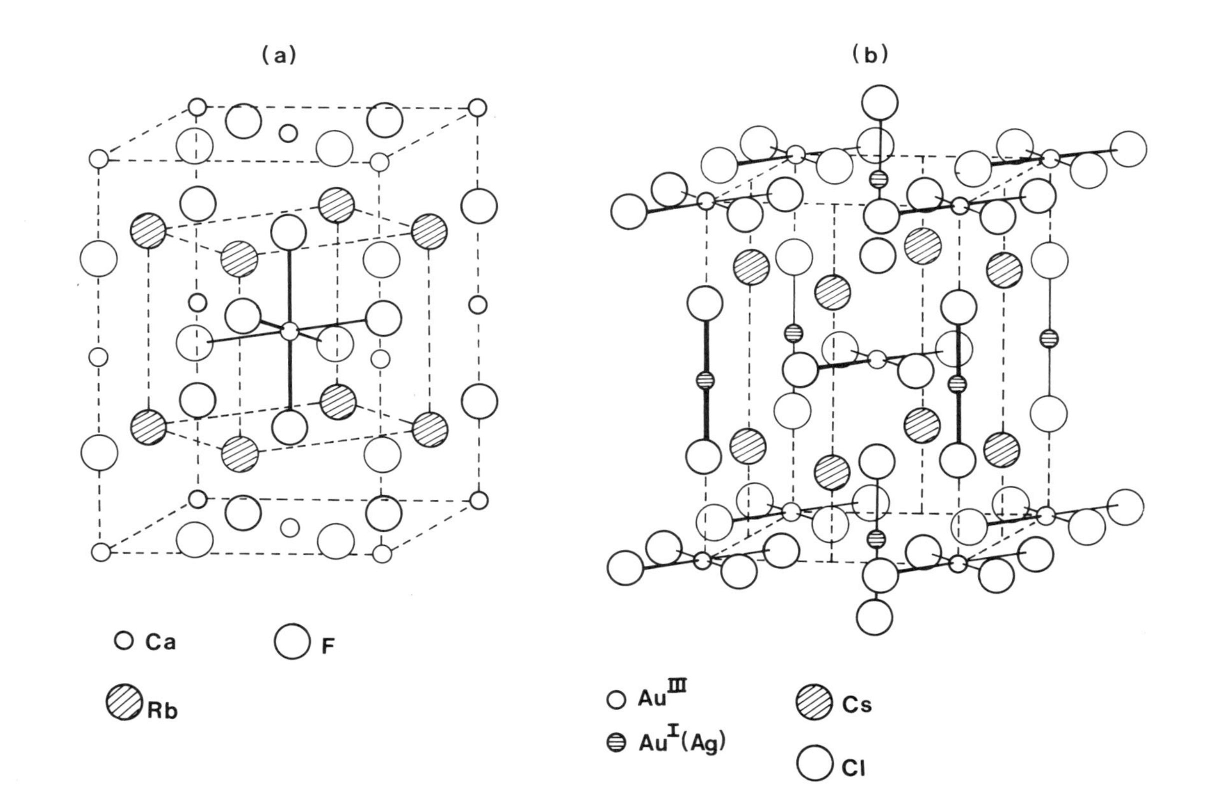

Figure 3-23 Comparison of (a) the $RbCaF_3$ (perovskite), and (b) the $Cs_2[Au^{I}Cl_2]\,[Au^{III}Cl_4]$ structures. (After A. F. Wells, *Structural Inorganic Chemistry*, The Clarendon Press, Oxford, 1962)

We can best illustrate these procedures by example.

(*a*) *Rhenium trioxide*, ReO_3, consists of an incomplete f.c.c. host lattice of O^{2-} with rhenium in one quarter of the octahedral sites (Figure 3-24). It is closely related to perovskite, $CaTiO_3$, Figure 3-23a. Just as many complex oxides and complex fluorides of similar formula type have related structures, we find that some trifluorides adopt the ReO_3 structure: viz. ScF_3, NbF_3, TaF_3, MoF_3. The metal is octahedrally coordinated to six fluorines; each fluorine is linearly coordinated to two metal atoms. The structure may alternatively be described as consisting of MF_6 octahedra sharing all six vertices (see also p. 235). Note in contrast that the $CrCl_3$ and BiI_3 structures adopted by trihalides *other than* fluorides may be said to consist of MX_6 octahedra sharing edges. This property is general; when present in bridging situations fluorine is only ever found in linear or angular *two*-coordination. The other halides frequently occur in double or triple bridges.

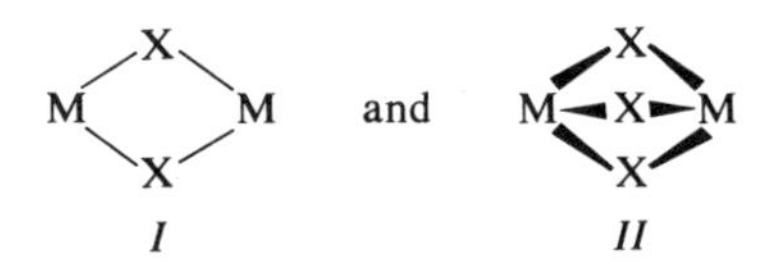

Part of the reason for this is that fluorine is too small to bridge in forms I and II, because in so doing it would increase metal–metal repulsion to an extent which would destabilize the structure.

Many other trifluorides crystallize in related structures which are not cubic but have become distorted to accommodate *bent* linear F—M—F bridges with angles of 140° (M = V, Fe, Co, Ru) or 132° (M = Rh, Pd, Ir). The reason for this particular kind of distortion is not understood and neither is the similar observation for the tetrameric molecular pentafluorides $(MF_5)_4$; bridges are linear (180°) for M = Nb, Ta, Mo, W, but bent (ca 135°) for M = Ru, Os, Rh, Ir; see p. 186.

(*b*) Two structures of particular interest are formed by simultaneous occupation of two types of interstitial site. Li_3Bi, Figure 3-25, is formed by filling all the octahedral *and* all the tetrahedral sites in an f.c.c. bismuth array with lithium. It can therefore be considered as related to fluorite, since the Li_3Bi lattice is formed from that of fluorite by filling all the octahedral sites. Each lithium atom in an octahedral site, Li_O, and in a tetrahedral site, Li_T, has the following neighbours:

$$Li_O\begin{cases}6\ \text{Bi (octahedrally) at } a/2\\ 8\ \text{Li}_T\ \text{(cubic) at } \sqrt{3}a/4\end{cases}$$

$$Li_T\begin{cases}4\ \text{Bi (tetrahedrally) at } \sqrt{3}a/4\\ 4\ \text{Li}_O\ \text{(tetrahedrally) at } \sqrt{3}a/4\end{cases}$$

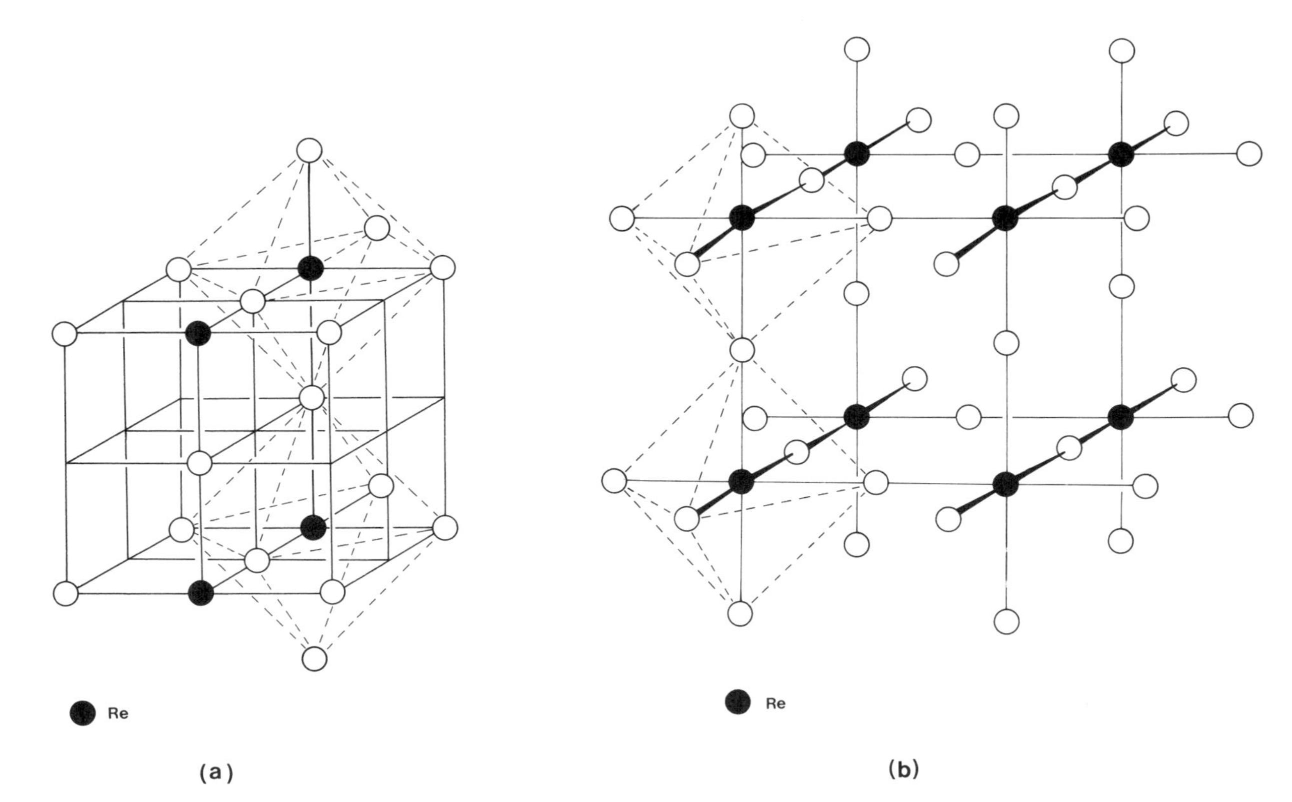

Figure 3-24 The ReO_3 structure. (a) Showing the relation to an oxygen-deficient c.c.p. array. (b) Represented as ReO_6 octahedra sharing all vertices

That is, each lithium has eight nearest neighbours in cubic coordination. Bismuth has coordination 8 + 6. The Li_3Bi structure is also adopted by Li_3Sb, Li_3Pb, and the ordered form of Fe_3Al, *and* by BiF_3. The basic reason for Li_3Bi and F_3Bi being isostructural is that both Li and F have a strong tendency to adopt tetrahedral coordination where possible, with octahedral coordination as a second best (see Section 5.4.2).

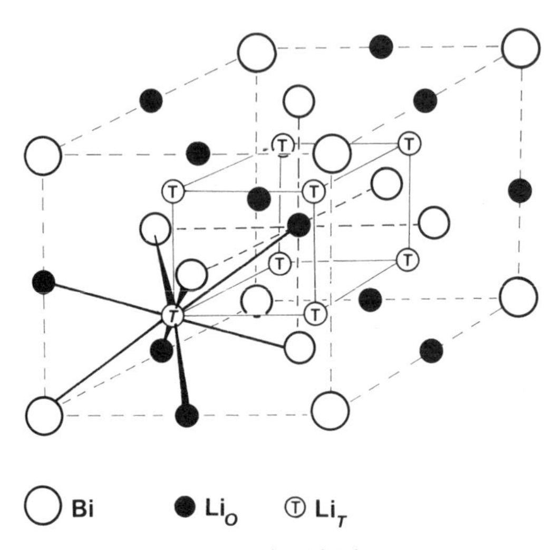

Figure 3-25 The Li_3Bi structure

In the extensive complex oxide series known as the *spinels* a proportion of *both* types of interstitial site are filled. In addition to oxides AB_2O_4, this structure is also adopted by some selenides, sulphides, fluorides and cyanides of the same formula type, some 135 compounds in all. Spinel itself is $MgAl_2O_4$, Figure 3-26, the structure being based upon a c.p. O^{2-} lattice. There are thirty-two O^{2-} ions in the unit cell. Electrical neutrality of the structure AB_2O_4 can be satisfied in the following ways:

(*a*) A^{2+}, B^{3+} (*b*) A^{4+}, B^{2+} (*c*) A^{6+}, B^{+}

In the 'normal' type (*a*), B^{3+} ions occupy one-half of the octahedral sites and A^{2+} ions one-eighth of the tetrahedral sites in the c.p. array. An alternative arrangement has one-half of the B^{3+} ions in the tetrahedral sites, with the other half of B^{3+} together with A^{2+} distributed at random amongst the octahedral sites; this is the so-called *inverse spinel* arrangement, $B(AB)O_4$.

The structure of a spinel is conveniently described by a parameter γ which equals the fraction of A^{2+} in octahedral sites. Thus, for a 'normal' spinel, $\gamma = 0$, for an 'inverse' spinel $\gamma = 1$. A random distribution of A^{2+} amongst the two groups of sites gives $\gamma = \frac{2}{3}$. There are many points of interest in regard to the spinels, an obvious one being why some have the normal and some the inverse arrangement. We shall return to this question in Chapter 5. All of the type (*b*) spinels so-far studied have the inverse structure, for example Co(SnCo)O_4. Type (*c*) spinels include Na_2MoO_4 and Na_2WO_4. Magnetite, Fe_3O_4 (and Co_3O_4) is an inverse spinel $Fe^{III}(Fe^{II}Fe^{III})O_4$. Compare Mn_3O_4 which is a *tetragonally distorted* (i.e. the cubic spinel structure has been elongated in one axial direction) normal spinel $Mn^{II}(Mn^{III})_2O_4$.

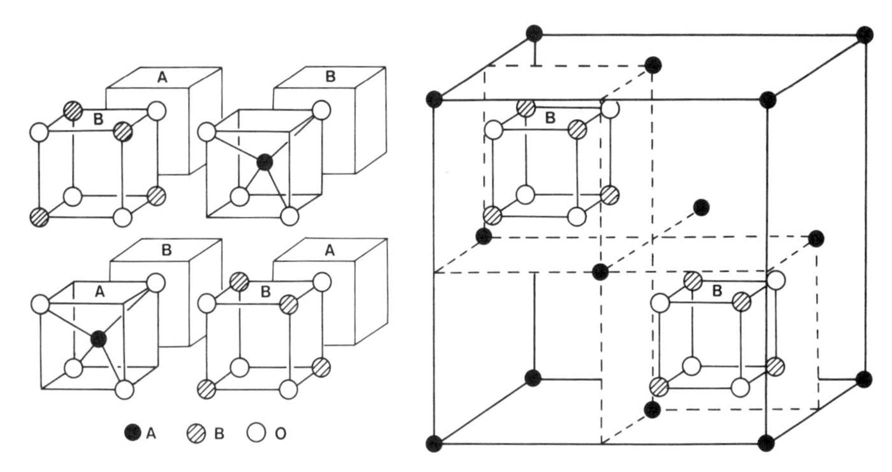

Figure 3-26 'The spinel structure, AB_2O_4. The structure can be thought of as eight octants of alternating AO_4 tetrahedra and B_4O_4 cubes as shown in the left-hand diagram; the O have the same orientation in all eight octants and so build up into a face-centred cubic lattice of 32 ions which coordinate A tetrahedrally and B octahedrally. The four A octants contain four A ions and the four B octants sixteen B ions. The unit cell is completed by an encompassing face-centred cube of A ions (●) as shown in the right-hand diagram; this is shared with adjacent unit cells and comprises the remaining four A ions in the complete unit cell $A_8B_{16}O_{32}$. The location of two of the B_4O_4 cubes is shown for orientation'. (From N. N. Greenwood, *Ionic Crystals, Lattice Defects, and Non-Stoichiometry*, Butterworths, London, 1968)

3.5.5 Structures Containing Complex Ions

So far we have only considered crystals composed of monatomic components which either form part of a c.p. array or fit into interstices. The crystal structures of ionic substances containing polynuclear components are often found to be very simply related to some of the standard structures which we have already considered. Indeed, we have already encountered examples in K_2PtCl_6 and related types. We can distinguish three main possibilities.

(*a*) The 'standard' structure is expanded by the presence of complex ions but its symmetry is not lowered.

(*i*) One component of a c.p. structure is replaced by a complex ion of approximately spherical shape.† A very large number of complexes of the type $M^{I}_2[M'X_6]$ crystallize with the *anti-fluorite* structure, e.g. $K_2[PtCl_6]$ and $[NMe_4]_2[CeCl_6]$. The *fluorite* structure is adopted by $[ML_6]X_2$ types, typically $[Co(NH_3)_6]I_2$. In compounds such as $(NH_4)_2[FeCl_5(H_2O)]$ and the analogous indium salt, the anion is also sufficiently close to spherical to allow a tetragonally distorted variant of the K_2PtCl_6 structure to be used.

(*ii*) Both components of a structure are replaced by approximately spherical ions, e.g. $[Co(NH_3)_6][TlCl_6]$, NaCl structure; $[Ni(H_2O)_6][SiF_6]$, CsCl structure.

(*iii*) One of the components of the crystal can attain or approach spherical symmetry by rotation. Thus, CsCN and NH_4NO_3 (above 125°C) have the CsCl structure, both of the complex ions rotating in the latter case.

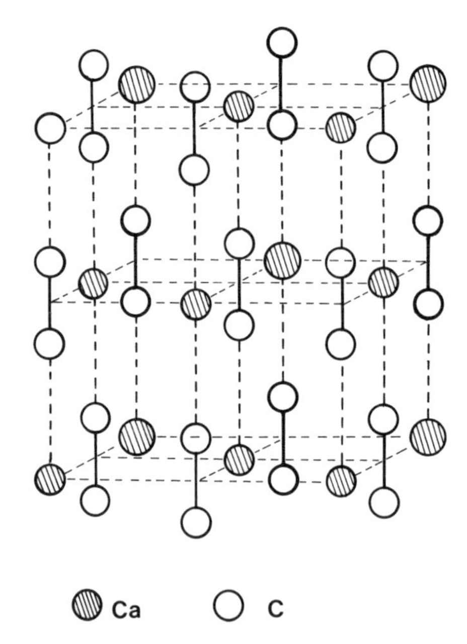

Figure 3-27 The structure of CaC_2. (From A. F. Wells, *Structural Inorganic Chemistry*, The Clarendon Press, Oxford, 1962)

† The shapes of some complex ions are as follows:
Linear: O^{2-}, C_2^{2-}, CN^-, N_3^-, CNS^-, ICl_2^-
Angular: NO_2^-, ClO_2^-, ICl_2^+
Planar: CO_3^-, NO_3^-
Trigonal pyramidal: SO_3^-, ClO_3^-, BrO_3^-
Tetrahedral: ClO_4^-, SO_4^{2-}, PO_4^{3-}, $[MnO_4]^-$, $[FeCl_4]^-$
Square-planar: $[ICl_4]^-$, $[PtCl_4]^{2-}$, $[Ni(CN)_4]^{2-}$
Octahedral: $[MX_6]^{n-}$, (X = F, Cl, Br, I), $[M(H_2O)_6]^{2+}$, $[M(NH_3)_6]^{n+}$.

(*b*) The 'standard' structure is distorted by an unsymmetrical ion which does not rotate. Thus, in CaC_2 the carbide ions are aligned parallel to each other so that the NaCl structure is tetragonally distorted (i.e. along one axis). The structure is shown in Figure 3-27.

Calcite, $CaCO_3$, has a structure related to NaCl but distorted by the planar anions, Figure 3-28. *Aragonite*, another form of $CaCO_3$, is similarly related to NiAs.

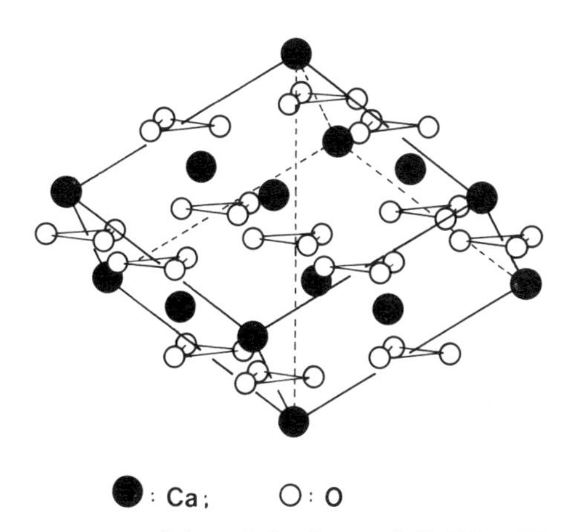

Figure 3-28 The structure of the calcite form of $CaCO_3$. (From Evans, *Crystal Chemistry*, 2nd ed., Cambridge University Press, 1964)

(*c*) In cases where the complex ion is large and far from spherical the structure of the crystal will be principally determined by the way in which the complex ions can pack. Interstices in the resulting structure will be filled by counter-ions, water molecules, etc. Examples of this behaviour are found amongst silicates, heteropolyacids, and other complex systems, some of which are treated in succeeding chapters.

3.6 RUTILE AND α-Al_2O_3

There are many structures which can be regarded as distorted variants of the basic ones which we have considered in this chapter, and of a good many more which we have not had space to deal with. Rather than pile on the facts at this stage we shall introduce some of these, along with explanations for them, in later pages. *Rutile*, TiO_2, is an important structure for many binary oxides and fluorides. Thus:

MO_2 (M = Ge, Sn, Pb, Ti, Cr, Mn, Ta, Tc, Re, Ru, Os, Ir, Te)
MF_2 (M = Mg, Mn, Fe, Co, Ni, Zn, Pd)

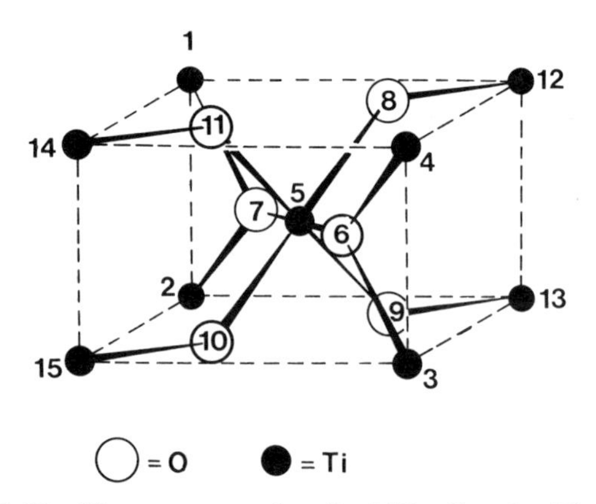

Figure 3-29 The structure of rutile, TiO_2. See also Figure 3-30

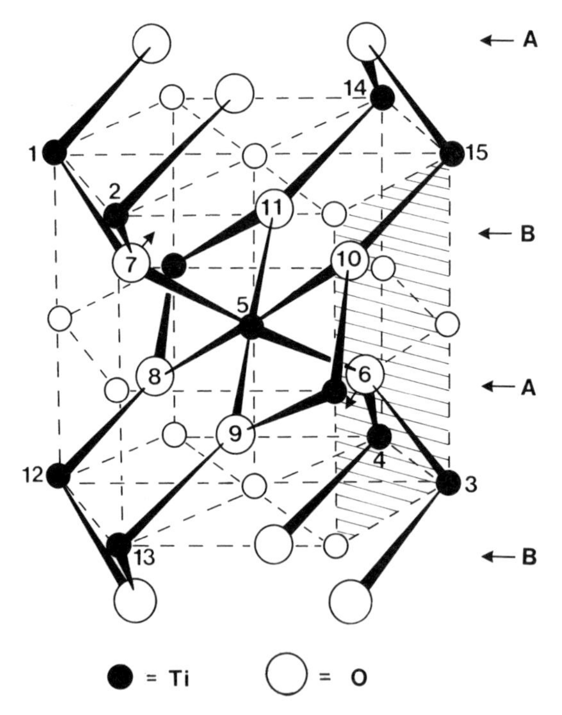

Figure 3-30 Structure of the hypothetical precursor to rutile; this is basically the NiAs lattice (Figure 3-15) with half of the octahedral sites vacant. Note the three-coordination of oxygen (e.g. atoms 7, 10) which is *pyramidal* as the close-packed oxygen layers are still planar. Following distortion of the close-packed layers (in directions indicated by arrows), coordination about oxygen becomes *planar*, and atoms 1 to 7 are all co-planar. The numbering corresponds to that of Figure 3-29

(a)

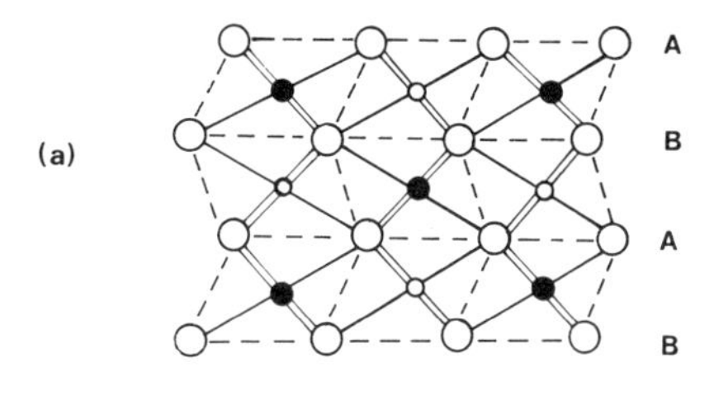

(b)

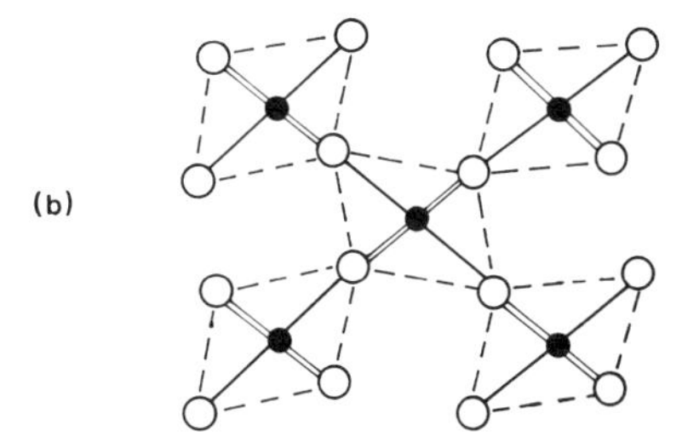

Figure 3-31 If the structure of Figure 3-30 is viewed along an axis normal to the shaded plane, it is seen to consist of chains parallel to that axis. These are indicated in (b) of this Figure and compared with (a), the equivalent diagram for NiAs. Note the buckling of the close-packed layers in rutile due to making three oxygens co-planar with Ti. (From A. F. Wells, *Structural Inorganic Chemistry*, The Clarendon Press, Oxford, 1962)

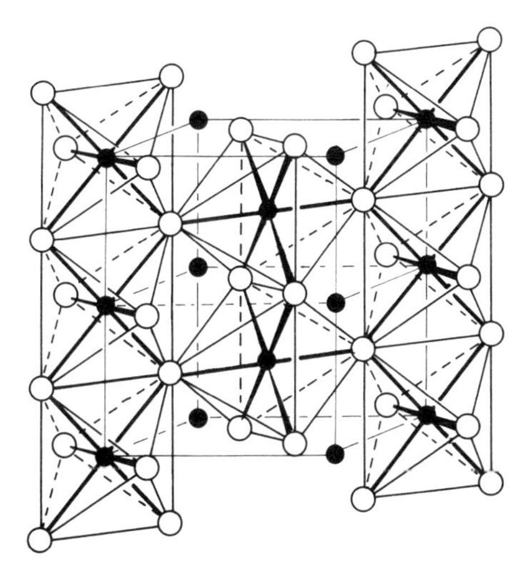

Figure 3-32 An overall view of rutile emphasizing the arrangement of the octahedral chains, and showing the relation to Figure 3-29

TiO_2 has three polymorphs (anatase, brookite, rutile) of which rutile is the best known. It is used in enormous tonnages in paints and floor polishes. The structure is shown in Figure 3-29 from which it can be seen that the metal is octahedrally (slightly distorted) coordinated to six oxygens and that each oxygen is in planar three-coordination with titanium atoms at the corners of a nearly equilateral triangle.

Rutile can be related to a regular h.c.p. lattice and its distinctive structure rationalized as follows.

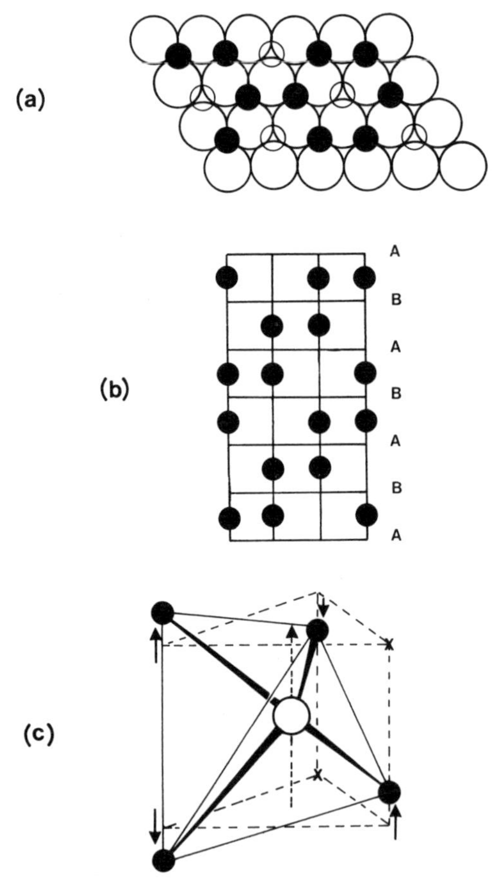

Figure 3-33 (a) A close-packed oxygen layer showing the pattern of octahedral sites occupied in α-Al_2O_3. Small open circles represent vacant sites. (b) Elevation showing pattern of octahedral sites filled. (From A. F. Wells, *Structural Inorganic Chemistry*, The Clarendon Press, Oxford, 1962.) (c) The actual coordination about O^{2-} in α-Al_2O_3, related to that in the 'ideal' precursor structure. (From *Fundamentals of Inorganic Chemistry* by H. Krebs. © 1968 McGraw-Hill Book Company (UK) Ltd. Used with Permission)

Consider the positions of the octahedral sites in an h.c.p. lattice: note that they run in lines parallel to the *c*-axis. If they were to be occupied by highly charged ions (such as Ti^{4+}) these ions would repel each other and destabilize the structure. Rutile is related to a lattice in which half of the octahedral sites are occupied in a symmetric manner, which causes just such a distortion, see Figure 3-30. Note, in passing, that the different layer sequence in c.c.p. and the resulting difference in octahedral site arrangement precludes distortion of the type which leads to rutile.

A second distortion now pushes the oxygen atoms into the plane defined by the three nearest titanium atoms, Figure 3-30. Why should this happen; why not stop after the first part of the distortion? In the observed rutile structure each oxygen is sp^2 bonded and this is energetically more favourable than the half-way stage. See also Figures 3-31 and 3-32.

A similar and slightly more subtle kind of distortion is encountered in *corundum*, α-Al_2O_3. This is basically an h.c.p. oxygen lattice with Al^{3+} in two-thirds of the octahedral sites (i.e. $\equiv Al_{2/3}O$). Hence, c.n. Al = 6 octahedrally. The pattern of octahedral sites occupied results in each O^{2-} having four nearest neighbours, see Figure 3-33. The coordination about oxygen would be strictly analogous to the coordination arrangements about arsenic in NiAs but with two vacant prism corners. The Al^{3+}—Al^{3+} distance is reduced by displacements along the *c*-axis as shown in Figure 3-33c. This is equivalent to puckering the Al^{3+} layers (recall, p. 49, that interstitial sites may be considered as layers in their own right). The oxygen ends up in distorted tetrahedral coordination.

This structure is also adopted by oxides M_2O_3 where M = Ti, V, Cr, Fe, Rh and Ga.

BIBLIOGRAPHY

Ho, S. M., and B. E. Douglas, *J. Chem. Educ.*, **45**, 474 (1968); **46**, 207 (1969).

CHAPTER 4

What Structures are Possible? II. Predictions on the Basis of Bond Type

In Chapter 3 we examined the implications of the space-filling postulate and derived closest-packed lattices, specifically restricting ourselves to the case of spherical atoms without directional bonding requirements which we found to be a tolerable basis for understanding simple molecular crystals. Following *purely geometrical* reasoning we related quite a selection of structures to the two basic closest-packed types and in doing so met examples of materials covering the whole range of bond types, including those which are evidently directional, diamond for example. Evidently what has happened is that many of these geometrically simple structures are also compatible with directional bonding requirements. Or have we got the problem the wrong way round? Should we not be trying to predict structures compatible with the bonding requirements of the crystal constituents and with lattice periodicity? If we do this then it may turn out that efficient use of space is automatically taken care of.

In this chapter we examine the problem of crystal structure from the viewpoint of *extreme bond type*. We ask, what kind of structure would be compatible with, say, purely ionic or purely covalent bonding? In the real world most bond types are of some intermediate character; in Chapter 5 we see how this may be quantified and related to structure type. But let us see just how far we can get on the basis of a few more simple postulates.

4.1 NON-DIRECTIONAL BONDING: METALS AND MOLECULAR CRYSTALS

We have covered this topic in an introductory manner in Chapter 3 in deriving the closest-packed structures. Further consideration of metals is deferred until Chapter 8 where more complex factors will be dealt with. In order to deduce the structures of molecular crystals (other than those formed from noble gases or very simple nearly spherical molecules) we need to relax the shape restriction and enquire how best to pack arbitrarily shaped molecules.

This theme is developed in Chapter 6, but we can summarize the conclusions of that chapter by saying that packing considerations in combination with the requirements of lattice theory determine the structures of molecular crystals. We should add that the term 'packing considerations' really covers two aspects: (*a*) the efficient use of space; (*b*) the detail of the relative orientations of molecules in a unit cell. All this applies to non-polar molecules. If they are polar then directional interactions must be taken into account as well. The most important of these is the hydrogen bond which sometimes exerts a determinative effect upon the structures of the solids in which it occurs. Due to its importance both to inorganic and biological systems we treat it in more detail in Chapter 6.

4.2 DIRECTIONAL BONDING: COVALENT CRYSTALS

When considering possible structures for covalently bonded solids the problem must be resolved into two quite distinct parts.

(*a*) The valence requirements of the individual atoms: this is what determines the number and disposition of the *nearest* neighbours.

(*b*) The arrangement of the more-distant neighbours in ways that are compatible with the bonding requirements (*a*). These are matters of a geometrical and topological nature. Such problems are not restricted to covalent lattices but they arise naturally in this context, which is why we consider them here.

We can illustrate the two aspects of these problems by reference to diamond. Valence theory requires that carbon be joined by sp^3 hybrids to four other carbon atoms disposed tetrahedrally, but this tells us nothing about the arrangement of the more-distant neighbours. The structure of diamond, Figure 4-1a, shows that each carbon atom has the environment required by valence theory, but is this the only way in which this requirement could have been satisfied? It turns out (see below) that there are *two* different lattices which have tetrahedral coordination about each point, Figure 4-1a and c, which then raises a whole new problem—why is one found in nature but not the other? Diamond should either undergo a phase transition to a hexagonal form under appropriate (and probably very extreme!) conditions, or, more probably, should sometimes be found in hexagonal form. Why this doesn't happen is not known.

4.2.1 Valence Requirements

It is assumed that the reader is familiar with the basic quantum mechanical treatments of bonding, and with hybridization. From this it is clear how electronic configuration in the valence shell is related to the number and the disposition of the nearest neighbours of an atom. For the Groups IV to VII of the Periodic Table the number of covalent bonds formed is given by the so-called 8–N rule, which is simply a version of the octet rule familiar in molecular situations. N is the group number. Thus, for the halogens, Group VII, only a single neighbour is expected; molecules are therefore formed and give, in the solid state, crystals of the molecular type. For Group VI the two bonds from each atom may be used to form either rings or chains, as in the S_8 ring or the chain form of selenium (p. 190). When two or more different types of atom are involved, the number of structures possible is considerably increased.

Transition metals on the other hand often require other, different, stereochemical arrangements. These depend in detail upon: the oxidation state of the metal and hence upon the *d*-electron configuration; the nature of the other atom(s) with which it is combined; the atomic number, especially whether or not it is a 3*d*-transition series metal. These matters are dealt with in standard texts on inorganic chemistry.

In Summary, valence theory specifies the number and disposition of the *nearest* neighbours in a covalently bonded crystal. A small number of coordination arrangements are of wide occurrence: octahedral, tetrahedral, square-planar, trigonal-planar or pyramidal, and linear. The disposition of non-bonding electron pairs is also of importance in crystal structures of some elements and compounds.

4.2.2 Geometrical and Topological Requirements

We must now see how valence requirements may be accommodated in repeatable three-dimensional structures. In terms of topology, valence theory specifies the 'connectedness' of each atom. Topology is concerned with the connectedness and environment of each point without reference to the numerical values of the bond lengths and angles. The detailed *geometry* of a structure will be determined partly by bonding requirements and partly by packing considerations—which are intimately associated with the energetics of crystal stability.

The topological requirement of an atom may, in general, be satisfied in more than one way. Take phosphorus as an example. In accord with the 8–N rule it will form three covalent bonds which will be directed approximately at right angles to each other as the orbitals used are predominantly p_x, p_y, and p_z. The simplest solution is seen in the structure of white phosphorus which consists of tetrahedral P_4 molecules. This is unstable with respect to other modifications as the bond angle of 60° is considerably less than the preferred 90°, although this is almost certainly compensated by the increased opportunity for interaction of electrons with all four atoms.

Black phosphorus, formed by heating the white allotrope under very high pressure, has a polymeric layer structure, Figure 7-8, in which the bond angles are close to 90°. The structures of the other forms of phosphorus (in all some eleven modifications are known) are all lacking in some degree of detail, but the available evidence suggests that the bond angle is usually near 100° in several complex structures, and all are three-connected networks.

The general aim of topological enquiries is to find out what types of structure are *possible* in principle. It is an aspect of crystal chemistry which has received far less attention than it warrants. Nearly all current discussion of crystal structures is based upon known structure types. It is important also to ask whether there are any other structures which are topologically and geometrically possible for a *given coordination type.* Knowledge of their possible existence would in all probability point to some factors not clearly understood before in the context of their relative stabilities. For example, there is an allowed dihalide structure MX_2 in which M and X have the same immediate environment as in the $CdCl_2$ structure (p. 58), but no dihalide adopts this structure; the reason is not known.

To pursue these topological enquiries we need a suitable language. A structure such as diamond may be regarded as a four-connected net, i.e. a net which has four connections (bonds) from each point (atom) to others. There is a branch of mathematics which deals with nets (and polyhedra—which can be regarded as nets); its application to crystal chemistry has been pursued in recent years, notably by Wells. Families of nets in two (planar) and three dimensions have been identified, in each of which each point is three-connected, four-connected, etc. Many are exemplified in nature but others, so far, have not been identified in crystal structures. This is because in some cases n-connectedness can be achieved in other ways (e.g. white phosphorus, P_4), whilst in other cases one particular net from an n-connected family may represent the best in terms of energetics under many circumstances (e.g. hexagonal nets, see below).

A set of three-dimensional four-connected nets is shown in Figure 4-1. (a) and (c) are tetrahedrally connected, (a) being the familiar diamond structure. As noted above diamond does not adopt structure (c) but it *is* found for some other compounds. Placing zinc and sulphur on alternate points of nets (a) and (c) yields the zinc blende and wurtzite structures, respectively.

Two of the polymorphs of silica, SiO_2, have structures which are simply related to the two tetrahedrally connected nets: silicon atoms are on all the four-connected points, oxygen atoms midway between silicon atoms. Cristobalite has the structure of net (a), trydimite that of net (c). The density of cristobalite is slightly greater than that of trydimite, a property that is turned to good advantage in the pottery industry. The transition between the two polymorphs occurs at 1470°C.

If we require half of the atoms in a four-connected net to be tetrahedrally connected, half to be square planar, there turns out to be only one simple net which can accommodate these requirements, Figure 4-1d. PdO, PtO and PtS have this structure which may be regarded as mainly covalent. The hybrid orbitals used are dsp^2 (M) and sp^3 (O, S).

There are many structures known which consist of covalently bonded sheets which may be either planar or non-planar. The bonding between the sheets is weak compared with that within the sheets, resulting in ready cleavage of such crystals along planes parallel to the sheets. It is profitable to ask

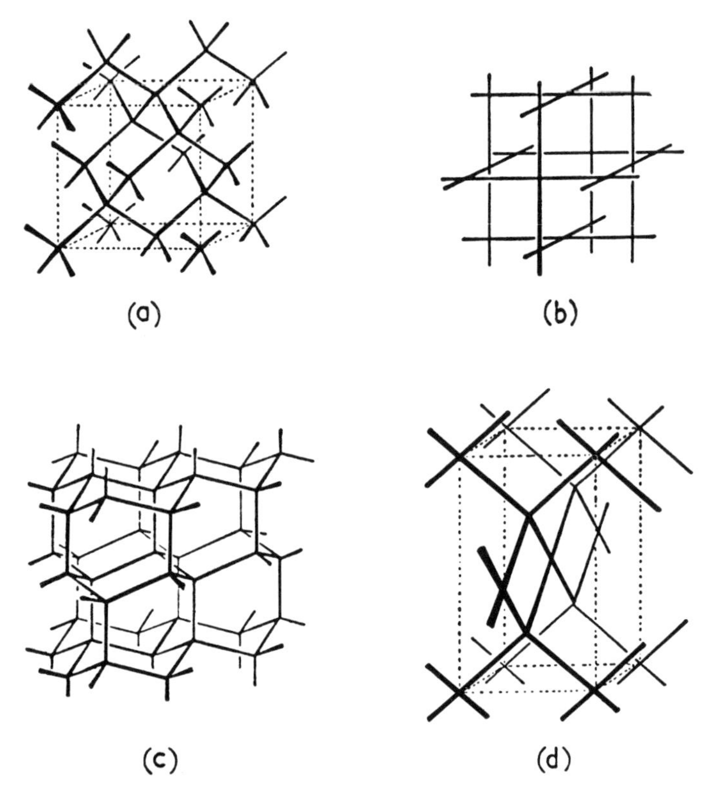

Figure 4-1 Three-dimensional four-connected nets. (After A. F. Wells, *Structural Inorganic Chemistry*, The Clarendon Press, Oxford, 1962)

what the topologically *possible* sheet structures are before looking at some known ones.

Wells has worked out the possible nets for several of the important co-ordination arrangements. Taking the three-connected net system as an example, three are shown in Figure 4-2. Net (a) is commonly found both in its planar (graphite, BN) and its puckered (black phosphorus, 'metallic' arsenic, As_2S_3) forms. Net (b) occurs in some silicate minerals (apophyllite and gillespite), but no structure corresponding to (c) is known.

Finally, we note that important negative results can follow from these topological considerations. Thus, a simple layer structure A_2B_3 in which A is six-coordinated, B four-coordinated, is topologically impossible.

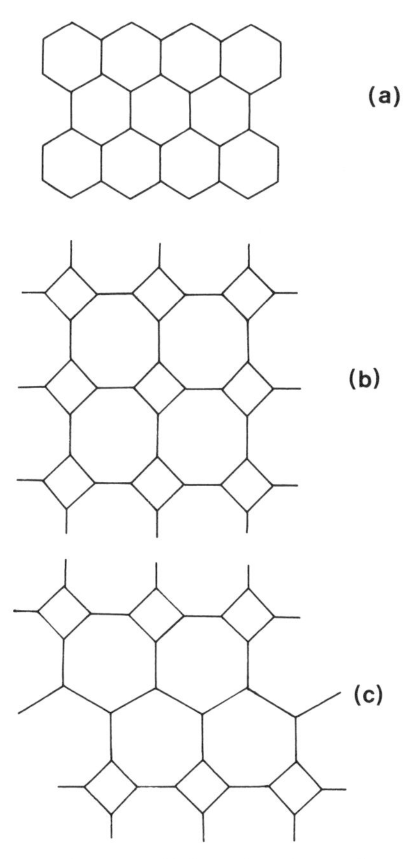

Figure 4-2 Three-connected nets. (After A. F. Wells, *Structural Inorganic Chemistry*, The Clarendon Press, Oxford, 1962)

4.3 NON-DIRECTIONAL BONDING BETWEEN CHARGED SPECIES: IONIC CRYSTALS

For purely ionic bonding the electrostatic (Coulombic) forces holding the crystal together are non-directional, but because we are dealing with ions and not neutral atoms we must clearly introduce an additional requirement that cations be surrounded by anions and *vice versa* thus preserving local electrical neutrality. For such an arrangement to be stable, any given ion must be in contact with its oppositely charged neighbours and they must be so disposed that the electrical repulsion energy is minimized. Further, there will be a limit on the number of ions which can be packed about one of opposite charge, especially for cations which are often smaller than the anions, with which they are associated in crystals. By working out the implications of these very simple ideas we shall be able to rationalize much that is known about solids in which the bonding is mainly ionic. These ideas go back to the earliest days of X-ray crystallography but a particularly complete treatment was given by Pauling (1938) who enunciated several 'rules', and an updated account is found in the well-known 'Once upon a time' review of Dunitz and Orgel (1960). We summarize the basic postulates of this simple theory of ionic structures and then discuss them.

4.3.1 A Simple Theory of Ionic Structures

Postulates

(*a*) Ions can be treated as charged, incompressible, non-polarizable spheres.

(*b*) A coordinated polyhedron of anions will be formed around each cation; the structure will be stable only if the cation is in contact with each of its neighbours.

(*c*) Subject to (*b*) the c.n. will be as large as possible and will be determined by the ratio of the ionic radii.

(*d*) In a stable ionic structure constructed according to these postulates, the valence of each anion will be equalled (with a change of sign) by the sum of the electrostatic bonds to it from adjacent cations.

(*e*) The disposition of the coordinating ions will be such as to minimize the electrostatic repulsion energy between them.

Postulate (*a*) uses the term 'polarizable' which we have agreed to reject (p. 3) as an explanation for the existence of a structure. In the context of this statement it means simply that the bonding is regarded as wholly ionic with no covalent contribution.

Postulate (*b*) suggests that we focus attention on the immediate cationic environment; this is especially important in dealing with highly complex solids such as silicates, in which it becomes difficult to appreciate what is going on without reference to coordination polyhedra. There are some instances in which an ion is apparently not in contact with all of its immediate neighbours. A good example is provided by V_2O_5. The metal 'ion' is in so distorted an octahedron of oxygen that the coordination is generally referred to as trigonal bipyramidal (i.e. c.n. = 5). A view of the structure in Figure 4-3 shows one very short V—O distance (1·54 Å) corresponding effectively to formation of a vanadyl (V=O) group by $p\pi$ donation from oxygen to vacant

$3d\pi$-orbitals on vanadium (a favoured process early in a d-block transition series). Clearly the bonding to this oxygen is far from ionic so one of the postulates (*a*) has broken down. Orgel has suggested that distorted octahedral coordination in their oxides is exhibited by transition-metal ions of a size too small to form regular octahedral structures but not small enough to be restricted to tetrahedral ones. The metal 'ions' have been described as 'rattling' inside their octahedral cages, but the description is not to be taken too literally!

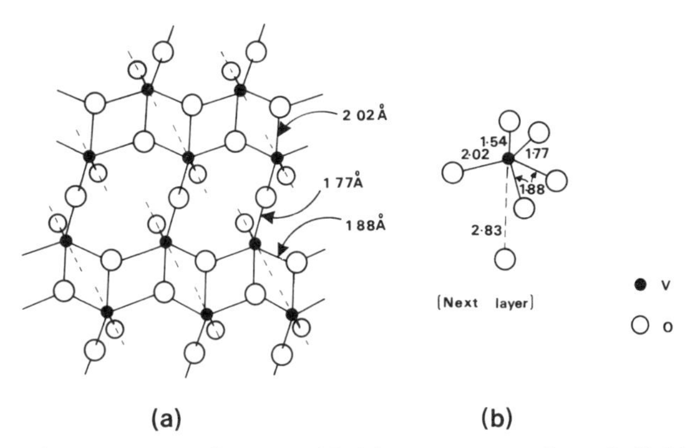

Figure 4-3 The structure of V_2O_5. (a) The structure of an individual 'layer' (After A. F. Wells, *Structural Inorganic Chemistry*). (b) Detail of the coordination at vanadium. N.B. Although it is convenient to describe this solid in terms of layers, it is not a layer structure in the sense used in Chapter 7

Postulate (*c*) leads to the so-called *radius-ratio rules*, hallowed by long (but uncritical) usage. Simple geometry allows us to compute the maximum size a cation M can have if the anion array (X) is not to be distorted. If M were smaller than this critical value postulate (*b*) would not be obeyed. In Figure 4-4 is shown a tetrahedral MX_4 arrangement inscribed in a cube of side d with M having the critical size for occupation of the site without distortion. Thus:

the cube face diagonal XX $= 2R_X = d\sqrt{2}$
the cube body diagonal XZ $= 2(R_M + R_X) = d\sqrt{3}$

Hence:

$$R_M/R_X = \sqrt{\tfrac{3}{2}} - 1 = 0{\cdot}225$$

Similarly, for an octahedral site,

$$2(R_M + R_X) = \sqrt{2}(2R_X)$$

Therefore,

$$R_M/R_X = \sqrt{2} - 1 = 0{\cdot}414$$

Table 1. Minimum radius-ratio values for several types of ionic coordination

Coordination arrangement	c.n.	Minimum ratio
Linear	2	No geometrical limit
Trigonal planar	3	0·155
Tetrahedral	4	0·225
Octahedral	6	0·414
Cubic	8	0·732

Table 1 summarizes these predictions for several coordination arrangements. We emphasize that these are *minimum* radius ratios; we have *not* predicted maximum allowed ratios, although it would not be unreasonable to expect that when R_M/R_X for (say) a tetrahedrally coordinated structure reaches 0·414 there might be a transition to a structue based upon octahedral six-coordination.

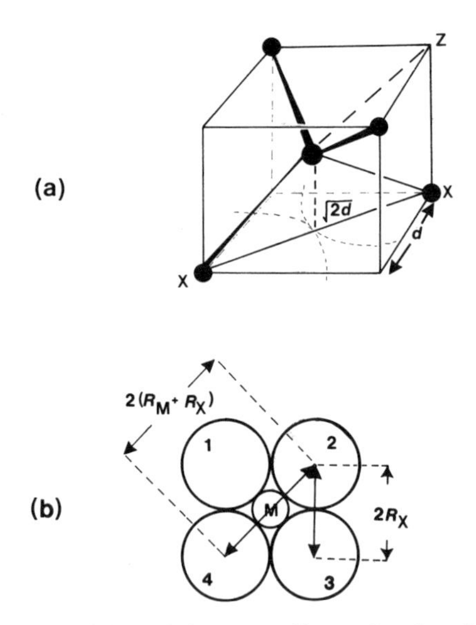

Figure 4-4 Calculation of the minimum radius ratios for: (a) a tetrahedral site, (b) an octahedral site. In (b), which shows a cross-section of an octahedral site, atoms, 1,2 come from one c.p. layer, atoms 3 and 4 from the next one

Postulate (*d*) is one version of the Pauling 'electro-neutrality' principle. It requires local electrical neutrality. This is obviously met in simple structures where, for example, Ca^{2+} in fluorite is accompanied by two F^- ions but each Na^+ in rock salt has only one halide anion. More to the point, it helps us to understand details of more complex structures, such as the various ways of balancing the charge of the oxygen lattice in spinels (p. 68). It is of particular value in dealing with silicates where isomorphous replacement is very common. In the context of this postulate electrostatic bond strength, E, is given by Ze/n where Ze = charge on cation, n = c.n. Thus, for Ca^{2+} in fluorite, c.n. = 8 and $E = \frac{1}{4}$. Each fluorine bears unit negative charge and is coordinated by four electrostatic bonds of strength $\frac{1}{4}$ to four Ca^{2+} ions, thereby achieving local electrical neutrality.

Postulate (*e*) leads to linear coordination for MX_2 arrangements since this is the arrangement in which the Xs are farthest apart. We would expect deviations from this to be found when there are lone or inert electron pairs on M which may exert a stereochemical effect. It would then be more reasonable to regard the lone pair(s) as an additional coordinating group(s) and ask what structure a three-(four)-coordinated arrangement would adopt, as in the Gillespie–Nyholm theory of molecular shape.

For MX_3 systems, the Xs are furthest from each other when in planar disposition (rutile, for example), whilst tetrahedral coordination is the preferred four-coordinate arrangement. (It is helpful to visualize a tetrahedron by recalling that it is defined by two sets of opposite corners of a cube). Note that square-planar coordination involves a higher electrostatic repulsion energy than tetrahedral. Square-planar coordination occurs (mainly) in the structural chemistry of Ni^{II}, Pd^{II}, Pt^{II}, Rh^{I}, and Ir^{I}. Its existence is clearly a problem in valence theory; where square-planar coordination is found in a structure we can be sure that the bonding has a substantial covalent contribution.

Octahedral coordination is preferred for MX_6 arrangements; the trigonal prism arrangement has higher repulsion energy as can be seen on rotating the top half by $2\pi/6$, thus converting it to the octahedral form of MX_6. The coordination about molybdenum in MoS_2 is trigonal prismatic, but this is clearly a highly covalent lattice type.

The Archimedean—or square—antiprism, Figure 4-5, represents the form of lowest energy for MX_8. The anion in $K_3[TaF_8]$ has this structure. However, it is not possible to make a space-filling array by joining identical square antiprisms in the way that one can with cubes, Figure 4-6. This is an example of lattice theory restricting the types of coordination which are possible. Although higher electrostatic repulsive energy is associated with the cube, its symmetry is suitable for lattice formation.

It *is* possible to obtain a lattice of square-antiprismatic units if one 'half' is reduced in size by $1/\sqrt{2}$ so that its diagonal just equals the side of the other square. To do this implies that the central cation is bonded to two different kinds of anion because the two sets of bond lengths cannot be equal. This is exactly what happens in the PbFCl structure (p. 59)! The structure of ThI_4 is also based upon distorted square-antiprismatic coordination.

It is worth noting in passing that many aspects of the considerably greater variety of nearest-neighbour arrangements found in discrete coordination complexes with mono- or poly-dentate ligands can be understood on the basis of ligand–ligand repulsion energies (Kepert, 1972).

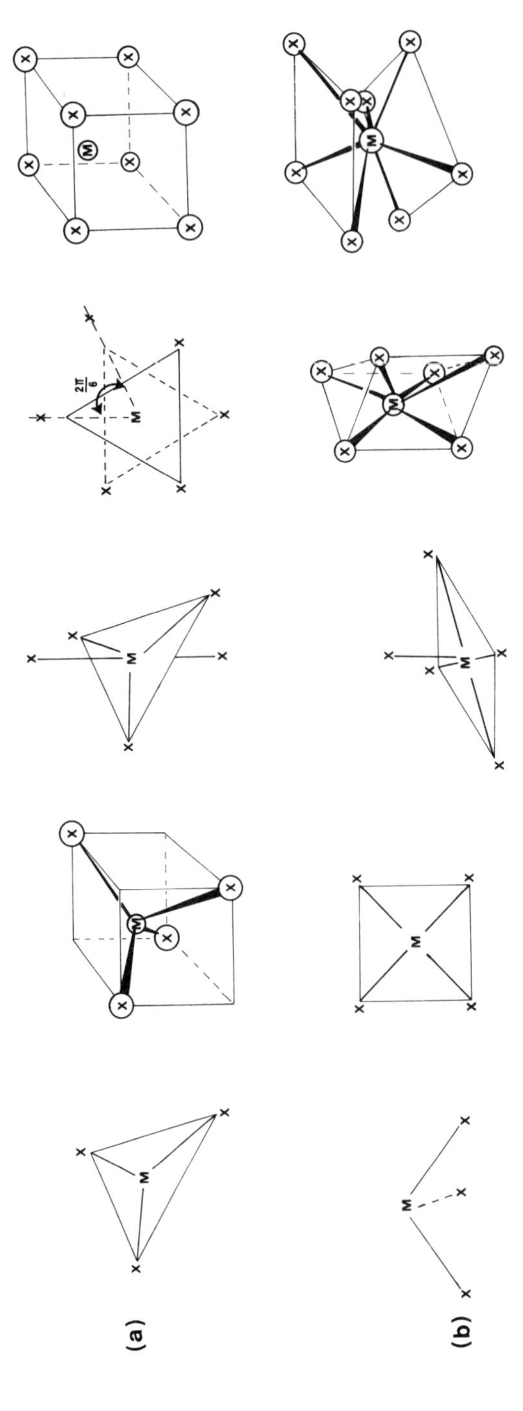

Figure 4-5 MX_n coordination arrangements. Row (a) shows the arrangements which have lowest electrostatic energy. Row (b) shows other common coordination types

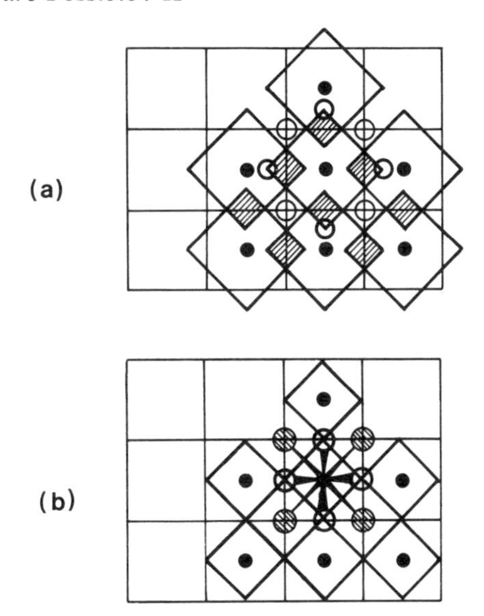

Figure 4-6 (a) Demonstration of the impossibility of forming a layer structure by packing regular Archimedean square antiprisms. (b) Formation of a layer structure (PbFCl) by packing Archimedean prisms which have one face smaller than the other

4.4 RÉSUMÉ

We have begun to form an outline understanding of the relation between crystal structure and bond type. For molecular crystals, packing considerations predominate; when we have worked out the details of packing–lattice theory in Chapter 6 it will be clear that the basic understanding of molecular crystal structure is now very satisfactory. Relative size considerations are evidently of high importance in determining ionic crystal structures and Laves's 'symmetry principle' is given explicit recognition in our discussion of the coordination shapes compatible with minimum potential energy. The over-riding importance of valence theory in determining the *nearest* neighbours of an atom in a covalently bonded solid comes as no surprise, but the crucial importance of topological considerations thereafter is still insufficiently recognized.

BIBLIOGRPAHY

Dunitz, J. D., and L. E. Orgel, *Advances Inorganic Radiochem.*, **2**, 1 (1960).

Kepert, D. L., *Inorg. Chem.*, **11**, 1561 (1972) and references therein.

Pauling, L., *The Nature of the Chemical Bond*, Cornell University Press, Ithaca, 1940.

Wells, A. F., *The Third Dimension in Chemistry*, The Clarendon Press, Oxford, 1956.

CHAPTER 5

Further Factors Affecting Crystal Structure

In preceding chapters some of the more fundamental factors responsible for crystal structure have emerged. We now seek to quantify and extend our understanding, especially in the matter of intermediate bond types which we have deliberately avoided until now. Just to set the scene, it may help to have some specific examples in mind.

Consider the elements in Groups IVb, Vb, and VIb, of the Periodic Table. The upper members (C, Si, P, S) adopt structures in which both the number and the distribution of the nearest neighbours are determined by valence theory, and the more distant neighbours are arranged in conformity either with topological or packing principles. In contrast, the heaviest elements of the same groups (Pb, Bi, Po) are metallic or semimetallic as revealed by a progressive change down the group in properties such as electrical conductivity and optical behaviour, as well as by changes of structure type. Carbon, silicon and germanium all crystallize in the diamond lattice; tin is diamorphic, one form (grey tin) having the familiar diamond lattice, the other (white tin) being more typically metallic both in properties and in having a higher coordination number, see Figure 5-1. It can be considered as derived from the diamond lattice by compression along the *c*-axis. The Group IVb series is completed by lead, a typical metal crystallizing with the cubic closest-packed structure, c.n. = 12. The trend to higher coordination number and more-metallic behaviour is accompanied by a decrease in the directional nature of the bonding and corresponding progressive changes in band structure, commonly

Table 1. The Ratio $\left(\frac{\text{next nearest neighbour distance}}{\text{nearest neighbour distance}}\right)$ in crystals of some elements (after C. S. G. Phillips and R. J. P. Williams, *Inorganic Chemistry*, The Clarendon Press, Oxford, 1965)

P	1·78	S	1·81	Cl	1·65
As	1·33	Se	1·49	Br	1·46
Sb	1·16	Te	1·21	I	1·33
Bi	1·12	Po	1·00		

referred to as 'metallization' or 'dehybridization'. A rather clear indication of dehybridization is given by the ratio of the next-nearest to the nearest neighbour distances, Table 1.

The essential cause of dehybridization (a word which we adopt in preference to metallization as it is more indicative of the changes involved) is the progressive lowering of the energy differences between *s*, *p* and *d* atomic orbitals with increase in principal quantum number, *n*, see Figure 5-2.

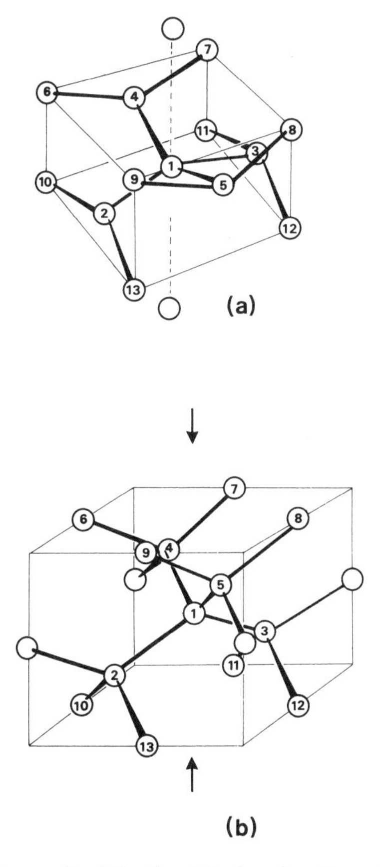

Figure 5-1 The structure of white tin, (a) showing the relation to the diamond structure (b) of grey tin

We now continue our enquiry by first asking whether the predictions of our simple ionic theory of Chapter 4 bear any relation to the facts. This will leave us with a mildly jaundiced view of its applicability, but the succeeding section on bond directionality will restore our confidence in our ability to handle partly ionic crystals.

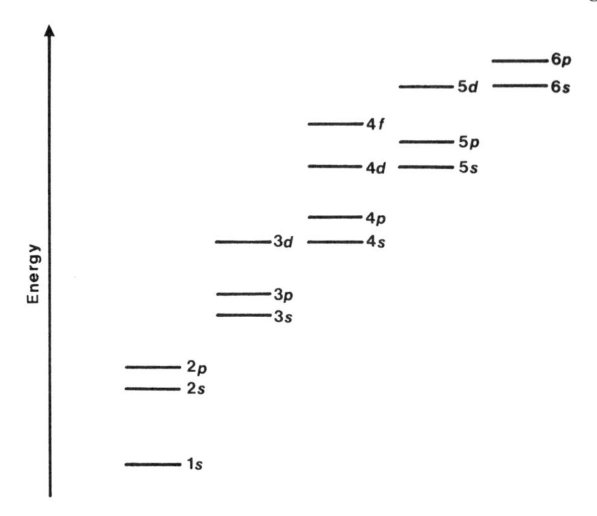

Figure 5-2 Relative energies of atomic orbitals

5.1 THE IONIC MODEL: A CRITICAL APPRAISAL

The term 'ionic model' covers a combination of ideas but two main aspects may be identified. Firstly it is a way of thinking about the crystallographic aspects of mainly ionic structures in terms of hard elastic spheres and of discussing reasons for the adoption of one structure rather than another (e.g. rutile or fluorite). In this its success is moderate only. Secondly, the 'ionic model' provides the basis of simple ways of calculating various properties of crystals, particularly lattice energy. In this it enjoys rather more success than might be expected in view of the above remarks. This appears to be due to certain self-compensatory features of the calculations. Thus, the validity of the ionic model is definitely limited (if we extend its range of application by including so-called polarization effects, it is a matter of semantics whether or not we are still using the ionic model) but its usefulness, especially for discussion of energetics, is quite considerable. In practice therefore, it finds much more favour with those who are concerned with thermodynamic arguments than with those who wish to account for crystal structure.

5.1.1 Energetics and Properties of Ionic Crystals

It is evident from many of the properties of ionic crystals (e.g. their high melting and boiling points) that the cohesive forces are strong. We can investigate the matter further in two broad ways, empirically and theoretically.

The observed internuclear separations in ionic crystals are the result of an equilibrium between the Coulomb attraction of oppositely charged ions, balanced by the repulsive forces associated with the interaction of the outer electron shells of the ions. We may reasonably investigate these forces by seeking physical properties determined by them.

The Attractive Force

The attractive force between two ions A^+, B^- separated by a distance r is given by

$$F = \frac{Z_A Z_B e^2}{r^2}$$

where $Z_A e$ is the charge of the cation, $-Z_B e$ that of the anion. The Coulomb potential energy is then obtained by:

$$V_{AB} = \int_{\infty}^{r} F\,\mathrm{d}r = -\frac{Z_A Z_B e^2}{r}$$

This attractive force can be investigated by looking at processes which cause or are associated with expansion of the crystal, viz., thermal expansion, the melting and boiling points, which we would expect to show a marked inverse dependence upon r. We might also expect that since $F \propto e^2$, the melting points of crystals with multivalent ions would exceed those of univalent ions for the same internuclear separations. This is seen to be true in no uncertain way, Table 2. Hardness also increases considerably with charge.

Table 2. Effect on physical properties of increasing ionic charge

	LiF	MgO	NaF	CaO
M.p., K	850	2770	992	2600
Internuclear separation, Å	2·01	2·11	2·31	2·41
Hardness, Moh's scale (Diamond = 10)	3·3	6·5	3·2	4·5

Both melting and boiling points of the alkali halides decrease with increasing internuclear separation (see Figure 5-3), with the notable exception of the lithium salts which appear to be out of line. Pauling has shown that such irregularities may be understood in terms of the increased anion–anion interaction possible for lithium salts. He corrected the observed melting points by referring them to a set of hypothetical standard crystals which have the same radius sums, $r_+ + r_-$, but a radius ratio of 0·75. These 'corrected' values are much more regular.

The Repulsive Force

The repulsive force increases very rapidly when two ions are brought close together. It is quantum mechanical in nature, being a consequence of the

Heisenberg Uncertainty principle. As the electrons of the ions are compressed into smaller space, defining their position more accurately, their momentum necessarily increases. Born suggested that this repulsion would be inversely proportional to some power of internuclear separation, r^{-n}; n is often referred to as the 'Born exponent'. In more recent calculations a term $(-r/\rho)$ has been employed where ρ is a parameter to be determined, reflecting the form of the radial functions of atomic orbitals.

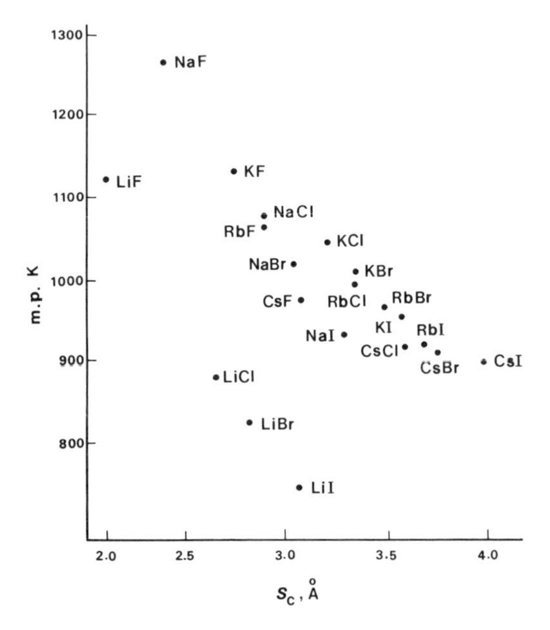

Figure 5-3 Melting points of the alkali halides. Reprinted with special permission. (From Harvey–Porter, *Introduction to Physical Inorganic Chemistry*, Addison–Wesley, Reading (Mass.), 1963)

The compressibility of an ionic crystal will clearly be related to this repulsive force. If a crystal is subjected to a uniform pressure the decrease in volume $-dv$, will be proportional to the pressure change and to the volume of the crystal. Hence,

$$dv = -\beta v \, dp$$

where the proportionality constant β is the volume compressibility. It is found that compressibility decreases with the equilibrium internuclear separation for a series of crystals of the same structure, Figure 5-4, which is what one would expect intuitively on the basis of the ionic model. Compressibility also decreases with increase of ionic charge, being considerably less for MgO than for NaF. Compressibility measurements can be used to determine ρ by the relation

$$\frac{1}{\rho} = \frac{2}{r_e} + \frac{9Cr^3}{Z_A Z_B e^2 A\beta} \tag{1}$$

where A is the Madelung constant (see below) and C is defined in terms of the molar volume, $V = CNr_e^3$, where r_e = equilibrium distance.

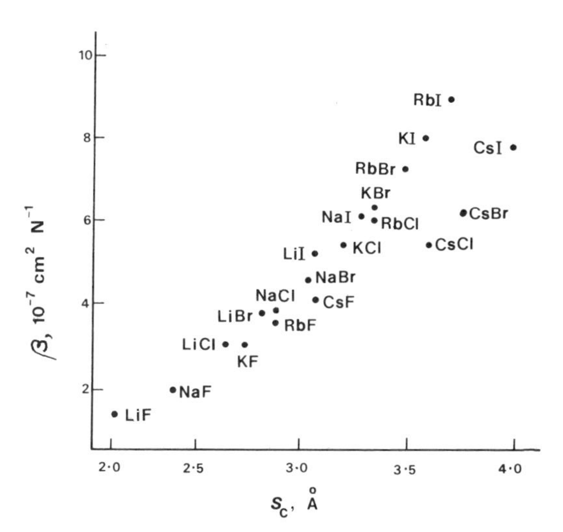

Figure 5-4 Compressibilities of the alkali halides. Reprinted with special permission. (From Harvey–Porter, *Introduction to Physical Inorganic Chemistry*, Addison–Wesley, Reading (Mass.), 1963)

Lattice Energy

Thinking in terms of the ionic model we may consider a crystal to be formed by a hypothetical process from the gas phase. Thus:

$$M^+(g) + X^-(g) \xrightarrow{-UL} MX \text{ (crystal)} \tag{2}$$

where we use monovalent ions for simplicity. By definition the energy given up in the process is the lattice energy, U_L. It may be determined experimentally in a roundabout way; it may also be calculated *a priori.*

(*a*) *Experimental* determination of U_L may be achieved indirectly in terms of other, related, parameters which we *are* able to measure in the laboratory. This illustrates a classic procedure in chemistry: we have stated our problem in terms of energetics (Equation 2). We now use a thermodynamic cycle to restate it in terms of quantities which we can handle. Such cycles were first used by Born and by Haber (1919). In general we may be able to construct several cycles pertinent to a given problem. In practice one is usually easier to relate to experimental quantities than others. For determination of lattice energy the following Born–Haber cycle is convenient.

$$\begin{array}{ccc} M^+(g) + X^-(g) & \xrightarrow{-U_L} & MX\ (\text{crystal}) \\ +I \uparrow \quad -E \uparrow & & \uparrow \Delta H_f^{\ominus} \\ M(g) + X(g) & \xleftarrow{S + \frac{1}{2}D} & M\ (\text{crystal}) + \frac{1}{2}X_2(g) \end{array}$$

where I is the ionization energy given up by the metal; E is the electron affinity of the non-metal and must be supplied; S is the heat of sublimation of the metal, D the dissociation energy (to form *two* atoms of X(g)), both to be supplied. Then, by Hess's law, the lattice energy is given by:

$$U_L = S + \tfrac{1}{2}D + I - E - \Delta H_f^{\ominus} \tag{3}$$

where $\Delta H_f^{\ominus}$ is the heat of formation of the crystal from its elements in their standard states. For accurate calculations, expansion terms and other fine details must be taken into account (see Johnson, 1968) but this formulation is adequate for our purposes. Values are available for all of the terms on the right-hand side of Equation 3, although the electron affinities are by far the most difficult to determine, thus enabling estimation of U_L.

(*b*) *Theoretical.* Combining the attractive and repulsive terms for a pair of ions at their equilibrium separation r_e, the potential energy is given by

$$V_{ij} = -\frac{Z_i Z_j e^2}{r_e} + b \exp\left(-\frac{r_e}{\rho}\right)$$

where b is a constant. The lattice energy is obtained by summing such terms over all pairs of ions. Thus,

$$U_L = \sum_{ij} V_{ij} = -\frac{A Z_i Z_j e^2}{r_e} + B \exp\left(-\frac{r_e}{\rho}\right) \tag{4}$$

where A is known as the Madelung constant and reflects the *geometry* of the lattice; B is a further constant to be determined experimentally. We note that this equation makes the link between lattice energy and properties such as m.p., compressibility, etc., whose dependence we looked at above.

Calculation of A can be illustrated by reference to the sodium chloride lattice. Any one Na^+ ion interacts first with its six nearest Cl^- neighbours giving Coulomb interaction $(6Z^+Z^-e^2)/r_e$. Interaction with the next-nearest neighbours, viz., twelve Na^+ ions at the midpoints of the f.c.c. edges distant $r_e\sqrt{2}$ from the first Na^+, gives a term $-(12Z^+Z^-e^2)/r_e\sqrt{2}$. Continuation of the summation over successively distant sets of neighbours (eight Cl^- at the cube corners, six Na^+ at the centres of neighbouring unit cells, etc.) yields the expression

$$-\frac{e^2}{r_e}\left(6 - \frac{12}{\sqrt{2}} + \frac{8}{\sqrt{3}} - \frac{6}{\sqrt{4}} + \frac{24}{\sqrt{5}} - \cdots\right)$$

for the energy of interaction, where the bracketed part is the Madelung constant, A. *It is a function of the geometry of the crystal only*. Summation of the expression is complicated due to its slow convergence but has been achieved by several methods. Values of Madelung constants for several lattices are given in Table 3, from which it is seen that A takes approximately

Table 3. Madelung constants for some lattices

Structure		A
Rock salt	M^+, X^-	1·74756
Caesium chloride	M^+, X^-	1·76267
Sphaelerite	M^+, X^-	1·63806
Wurtzite	M^+, X^-	1·64132
Fluorite	$M^{2+}, 2X^-$	5·03878
Rutile	$M^{2+}, 2X^-$	4·816
Cadmium iodide	$M^{2+}, 2X^-$	4·383
Corundum	$2M^{3+}, 3X^{2-}$	25·0312

equal (but significantly different) values for AB, AB_2, etc., structures respectively. As the NaCl and CsCl lattices are interconverted in the manner indicated in Figure 2-7, the Madelung constant varies as shown in Figure 5-5.

Since our model relates to the equilibrium positions of the ions, the lattice energy is necessarily a minimum. We state this mathematically by writing $dU_L/dr = 0$. Hence,

$$\frac{dU_L}{dr} = \frac{d}{dr}\left\{-\frac{AZ_iZ_je^2}{r_e} + B\exp\left(-\frac{r_e}{\rho}\right)\right\}$$

$$= \frac{AZ_iZ_je^2}{r_e^2} - \frac{B}{\rho}\exp\left(-\frac{r_e}{\rho}\right) = 0$$

Therefore,

$$B = \frac{AZ_iZ_je^2\rho\exp(r_e/\rho)}{r_e^2} \tag{5}$$

Substituting Equation 5 in 4 and multiplying by Avogadro's number, N_A, we obtain:

$$U_L = -\frac{AZ_iZ_je^2N_A}{r_e}\left(1 - \frac{\rho}{r_e}\right) \tag{6}$$

which is known as the Born–Mayer equation. Clearly, it allows lattice energy to be calculated from a knowledge of (*a*) the geometry of a crystal (and hence r_e and A), (*b*) ρ, which is obtained from compressibility measurements, Equation 1. When using SI units, the Born–Mayer equation takes the form of Equation 7, where ϵ_0 is the permittivity of free space.

$$U_L = -\frac{AZ_iZ_je^2N_A}{4\pi\varepsilon_0 r_e}\left(1 - \frac{\rho}{r_e}\right) \tag{7}$$

Using Equation 7 for the case of NaCl we have: $A = 1{\cdot}74756$, $e = 1{\cdot}602 \times 10^{-19}$ C, $N_A = 6{\cdot}0225 \times 10^{23}$ mol^{-1}, $r_e = 2{\cdot}814 \times 10^{-10}$ m, $\rho = 0{\cdot}345 \times 10^{-10}$ m, and $\varepsilon_0 = 8{\cdot}854 \times 10^{-12}$ F m$^{-\frac{1}{4}}$. Hence $U_L = -757$ kJ mol$^{-\frac{1}{4}}$ ($\equiv -181$ kcal mol^{-1}), which compares well with the Born-cycle value of -770 kJ mol^{-1} ($\equiv -183{\cdot}8$ kcal mol^{-1}). Some further values are given in Table 4 and

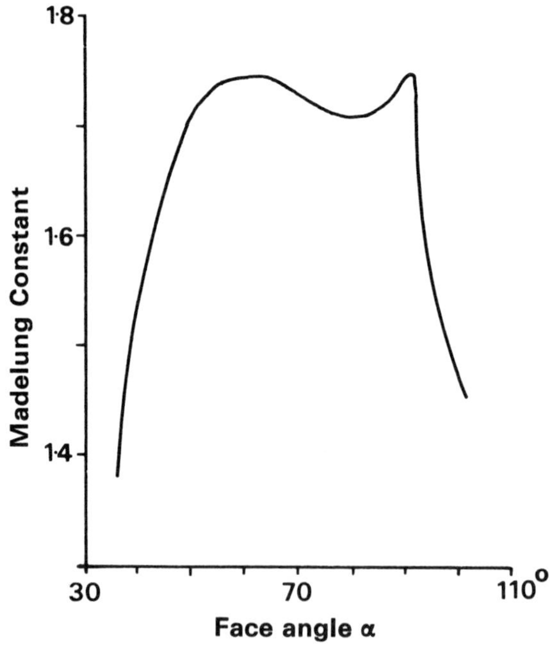

Figure 5-5 Variation of Madelung constant with rhombohedral face angle α (see Figure 2-7). Note maxima at $\alpha = 90°$ (CsCl) and 60° (NaCl). (From H. D. E. Jenkins and T. C. Waddington, *Trans. Faraday Soc.*, **65**, 1231 (1969))

compared with Born-cycle values. The agreement is remarkably good considering the approximations involved. The largest discrepancies are for crystals containing the larger ions, situations in which the strictly ionic model would be expected to be a poorish approximation. Various attempts have been made to improve the fit in these cases by taking into account the effect of

Table 4. Lattice energies of some halides and dihalides, in kJ mol^{-1}

	Structure	Born cycle	Born–Mayer	'Extended' Calculation
LiF	Rock salt	1009·2	999·6	1019·2
LiI	Rock salt	733·9	709·2	736·8
NaF	Rock salt	903·7	894·5	901·2
NaCl	Rock salt	769·0	749·8	767·8
NaBr	Rock salt	736·0	713·4	734·3
NaI	Rock salt	688·3	667·8	687·4
KCl	Rock salt	697·9	682·8	702·5
CsCl	Caesium chloride	641·0	618·0	640·6
CsI	Caesium chloride	587·0	564·4	596·2
MgF_2	Rutile	2908	2915	
CaF_2	Fluorite	2611	2584	
BaF_2	Fluorite	2368	2328	

London and other forces. Thus there is at all temperatures vibrational motion of the lattice which will allow induction of transient dipoles in neighbouring ions resulting in dipole-induced dipole attraction. Clearly this will be greatest for the larger and more polarizable ions: it adds a term $N_A C/r^6$ to the lattice energy, where C is a function of the ionic polarizabilities. However, this is largely nullified by a term of opposite sign which must be added to take account of the zero-point energy of the crystal. These, and other refinements, lead to values very close to the experimental ones; they are referred to as 'extended-calculation' values, Table 4. Note that *this success* in calculating lattice energies *does not validate the ionic model*, but rather illustrates its convenience. In particular we should note an inbuilt self-compensating feature. In calculations of lattice energy we use formal charges (∓ 1, ∓ 2, etc.) but take *experimental* internuclear distances. Now these distances are the net result of the interplay of *all* bonding forces *including* any contributions from van der Waals and covalent bonding, to use language appropriate to the ionic model, and they will have the effect of shortening internuclear distances from those which would have arisen from purely ionic bonding. In conclusion, therefore, the ionic model is of enormous value in chemical energetics (see for example, Sharpe, 1968) but it yields the right answers for reasons which do not always stand up to the closest scrutiny.

5.1.2 Radius ratios (σ)

In Chapter 4 we saw that the relative sizes of the ions in a crystal could be related to preferred coordination arrangements. A predicted correlation between structure and radius ratio, for several arrangements, was summarized in Table 1, Chapter 4. Do the facts bear out these predictions?

One purpose of the section on ionic radii (2.3.2) was to show how difficult it is to decide which radii to use and consequently how cautious must be our acceptance of any conclusions drawn from any theory based upon them. Table 5 gives radius ratios for alkali halides calculated from (*a*) the new experimental radii, and (*b*) Pauling radii. The NaCl structure is expected for crystals with $\sigma = R_{M^+}/R_{X^-}$ having a *minimum* value of 0·414 whilst for $\sigma = 0{\cdot}732$ and above the CsCl structure is more probable.

Table 5. Radius ratios (R_{M^+}/R_{X^-}) for alkali halide crystals

	Li^+	Na^+	K^+	Rb^+	Cs^+
F^-	0·81	1·00	0·78[a]	0·71[a]	0·62[a]
	(0·44)[b]	(0·70)	(0·98)	(0·92)[a]	(0·80)[a]
Cl^-	0·57	0·71	0·91	1·00	0·88[a]
	(0·33)	(0·52)	(0·73)	(0·82)	(0·93)
Br^-	0·52	0·65	0·83	0·90	0·97[a]
	(0·31)	(0·49)	(0·68)	(0·76)	(0·87)
I^-	0·46	0·57	0·73	0·80	0·91
	(0·28)	(0·44)	(0·62)	(0·69)	(0·78)

[a] R_{X^-}/R_{M^+}

[b] Values in parentheses are calculated from Pauling crystal radii; the first value given in each case is calculated from 'corrected' experimental radii of Gourary and Adrian.

Taking the 'corrected' empirical radii first we find that only the seven compounds in the box (continuous line) fit the predictions. Some further values of R_{M^+}/R_{X^-} come out >1 but, since in both NaCl and CsCl structures anions and cations have identical coordination arrangements, there is no reason why we should not also accept suitable R_{X^-}/R_{M^+} values. Thus RbF and CsF also fit the predictions, as do the three compounds with the CsCl structure (·–·–) by similar arguments. But this still leaves eight compounds with radius ratios which suggest that they should adopt the CsCl structure.

Using instead the radius ratios from Pauling radii covers a different set of compounds (––––) but overall the agreement with predictions is no more impressive.

The fluorite (CaF_2) and rutile (TiO_2) structures are also commonly (and fallaciously) taken as typically ionic. They have the metal in cubic eight-coordination and octahedral six-coordination respectively (see pp. 60 and 71) and should therefore have radius ratios within the same ranges as do the CsCl and NaCl structures above. We do not have any 'corrected' empirical radii available for these compounds, but taking F^- as 1·16 Å and using known internuclear distances, the R_M/R_X values of Table 6 are obtained.

For the rutile structure, the Pauling radii-based ratios agree well with predictions whilst those from the experimental radii are only slightly above the value at which a change to eight-coordination might be expected (0·732) and probably no further out than the uncertainty that would be associated in determining the 'corrected' radii. Similar comments apply to the fluorite ratios.

Table 6. Radius ratios R_{M^+}/R_{X^-} for some difluorides with the fluorite and rutile structures

Structure	From 'Experimental' Radii[a]	From Pauling radii (Dunitz and Orgel, 1960)
Rutile		
MnF_2	0·83	0·59
FeF_2	0·81	0·55
CoF_2	0·76	0·53
NiF_2	0·73	0·52
ZnF_2	0·75	0·54
MgF_2	0·72	0·48
Fluorite		
CaF_2	1·14	0·87
SrF_2	1·28	0·97
BaF_2	1·43	1·13
CdF_2	1·11	0·84
HgF_2	1·18	0·92

[a] These have not, in fact, been determined experimentally, see text p. 35.

Whatever the exact interpretation to be placed upon the meaning of the various types of radii, and hence on the derived radius ratios, it is quite clear that the radius ratios in Table 6 fall into two distinctly separated sets associated with the rutile and fluorite structures respectively. This is particularly well emphasized for dioxides by comparing the list of compounds adopting the rutile structure (Section 3.6) with those having the fluorite form (Table 5, Chapter 3); the larger metal cations clearly opt for eight-coordination.

5.1.3 Some Cautionary Thoughts on Radius Ratios

Despite the very real difficulties associated with choosing meaningful ionic radii and, consequently, of estimating radius ratios the predictions of the ionic model bear some qualitative relation to reality; but they are not to be taken too literally in their quantitative predictions. It is most desirable that the 'corrected' empirical radii derived by Gourary and Adrian for the heavier alkali halides be checked by experiment. Until this is done some slight doubt must remain about the applicability of radius ratios based upon them.

It is undoubtedly true that the larger cations are found in situations of high rather than low coordination number; this is a phenomenological conclusion and as such is perfectly valid, although it also receives semi-quantitative support from radius-ratio considerations.

We must now consider two problems thrown up by the comparison of radius-ratio predictions with the facts. (*a*) When they apparently work well, do they do so for the right reasons? (*b*) How do we account for the retention of a structure of lower coordination number (e.g. NaCl) when the radius ratio of the compound implies that one of higher coordination (e.g. CsCl) should be adopted?

(*a*) Radius ratio predictions seem to work particularly well for oxides and fluorides adopting the rutile (TiO_2) and fluorite (CaF_2) structures but the reasons are, at best, suspect. Consider rutile itself. If we agree to describe it as an *ionic* crystal (and we must if we are to expect an 'ionic' model to work for us!) it consists of Ti^{4+} and O^{2-} ions. But electrical gradients of this magnitude are absolutely out of the question, ions of charge $>1e$ being improbable. Indeed, its very structure suggests that it is not wholly ionic. When relating the rutile structure to a hypothetical close-packed precursor (p. 72) we considered the process as one of distortion from a condition of unfavourable electrostatic potential to a more stable variant, the extra stability being achieved by semi-covalent bond formation between the elements which can be considered in terms of pre-band formation states sp^2(O) and d^2sp^3(Ti) respectively. (We consider the electronic structure of the rutile lattice in detail in Section 5.5.6). It follows, therefore, that in estimating a radius ratio for rutile we should not adopt an *ionic* radius for oxygen. The currently quoted values of ionic radius for O^{2-} are ca 1·40 Å (Table 4, Chapter 2) but these are almost certainly far too high; the experimentally determined radius for oxygen in MgO is 1·09 Å (Table 5, Chapter 2) whilst covalently bonded oxygen is reliably accorded a value of 0·81 Å (Table 8, Chapter 2). If we must assign oxygen a radius in rutile then a value somewhere between the ionic and covalent values would seem appropriate, say 0·95 Å, corresponding to $\sigma = 0{\cdot}95$ (since $d(\text{Ti—O}) = 1{\cdot}96$ Å) a value significantly larger than usually quoted and one that implies adoption of the fluorite structure. Clearly, the six-coordinate environment of titanium in rutile is stabilized by a covalent-bond contribution. The eventual switch to the eight-coordinate fluorite structure of the heavier-metal oxides and fluorides is probably to be seen as much in terms of the tendency to dehybridization as in terms of relative size.

It follows from this discussion that it is easy to be lulled into a false sense of security and to accept with satisfaction predictions of radius-ratio calculations which are invalid by reason of neglect of covalent overlap and consequent variation of the radii that should be used in such considerations.

(*b*) One rather notable feature of the rock-salt structure is its retention by compounds for which radius ratio rules indicate that the CsCl lattice is probable. We have already indicated an ingredient of the answer to this problem when discussing rutile above.

The NaCl lattice is ideal for formation of covalent bonds by overlap of p_x-, p_y- and p_z-orbitals which have lobes aligned with the orthogonal A–B directions. But why not satisfy *both* radius ratio and orbital overlap requirements in the heavier alkali halides by adopting the CsCl structure? Two points should be held in mind in answering this question: (*a*) if we are talking about satisfying a covalent bonding requirement we are, by definition, outside the terms of reference of the ionic model and hence of radius ratio predictions although, if the degree of covalent bonding is not large then radius ratio predictions might well be a tolerable guide; (*b*) covalent bonding requirements can be satisfied much better in the rock salt lattice than in that of CsCl. Here's why.

In the 6:6-coordinate rock-salt structure the three orthogonal *p*-orbitals can be considered as lying along the vectors joining any one cation to its six nearest neighbours. Extensive overlap (i.e. band formation) can therefore take place, and be further reinforced by admixture of *s* character. But in the CsCl structure there are *eight* nearest neighbours and it is impossible to share six *p*-orbital lobes equally between them. We *can* form overlap schemes with similar orbitals on the set of six *next-nearest* neighbours but due to the longer distance involved this would be less efficient and the bond scheme would be of an entirely different nature as it would involve A—A or B—B contact alone and not A—B.

We conclude that retention of the NaCl structure by compounds such as RbCl, for which radius ratios predict the CsCl structure, is due primarily to a small covalent-bond contribution which is more favourably accommodated by the former structure. Nevertheless, with the larger ions the tendency towards dehybridization (i.e. loss of bond directionality) coupled with the need to make more efficient use of volume (see Section 5.1.4) makes the difference in energy between the two structures exceedingly small. This is reflected in the transformation of RbCl to the CsCl structure at a very low excess pressure, ca 5 kbar, a point we return to in the next section.

5.1.4 Space-filling Considerations

We may obtain further insight into the importance of relative atomic sizes of components in crystals in determining structure type by considering the following relationships due to Laves and to Parthé (1961).

Treating atoms or ions as hard spheres with definite radii r_i, r_j, etc., the volume occupied by them in a solid is

$$\frac{4\pi}{3}\sum_i a_i r_i^3$$

where a_i is the number of atoms of types 1, 2, etc., per unit cell. We then define a space-filling factor, ϕ, by

$$\phi = \frac{(4\pi/3)\sum_i a_i r_i^3}{V}$$

where V = volume of unit cell. For simple AB structures the unit cell edge, l can be expressed in terms of radii r_A, r_B for the special cases of A—A, B—B and A—B contact. Thus, consider the NaCl structure,

$$\phi = \frac{(4\pi/3) \times 4(r_{Na}^3 + r_{Cl}^3)}{l^3}$$

For Na—Na contacts $4r_{Na} = l\sqrt{2}$ and

$$\phi = \frac{(16\pi/3)\,(r_{Na}^3 + r_{Cl}^3)}{16\sqrt{2}\,r_{Na}^3} = \frac{\pi}{3\sqrt{2}}\,\frac{\sigma^3 + 1}{\sigma^3} \tag{8}$$

where $\sigma = r_{Na}/r_{Cl}$. Similarly for Na—Cl contacts, $r_{Na} + r_{Cl} = l/2$ and hence

$$\phi = \frac{2\pi}{3}\,\frac{\sigma^3 + 1}{(\sigma + 1)^3} \tag{9}$$

whilst for Cl—Cl contacts, $l = 2\sqrt{2}\,r_{Cl}$ and

$$\phi = \frac{\pi}{3\sqrt{2}}\,(\sigma^3 + 1) \tag{10}$$

These functions are plotted in Figure 5-6 which shows the envelope of those parts of the three ϕ functions which have the lowest value at any given value of σ. Starting from low σ the space filling of NaCl increases in conformity with Equation 10 with only anions in mutual contact. At the discontinuity $\sigma = 0{\cdot}414$ each chlorine is in contact with both chlorine and sodium and Equation 9 takes over until the second discontinuity at $\sigma = 2{\cdot}42$, at which Equation 8 obtains, representing only sodium–sodium contact. The discontinuities are, of course, the lower limiting radius-ratio values of Table 1, Chapter 4.

The Equations 8 to 10 have the general forms:

$$\phi = K_{A-A} \frac{a\sigma^3 + b}{\sigma^3} \quad \text{for A—A contacts} \tag{11}$$

$$\phi = K_{A-B} \frac{a\sigma^3 + b}{(\sigma + 1)^3} \quad \text{for A—B contacts} \tag{12}$$

$$\phi = K_{B-B}(a\sigma^3 + b) \quad \text{for B—B contacts} \tag{13}$$

where a and b are the numbers of atoms of types A and B in the unit cell. It is clear from Equations 11 to 13 that for a given composition (e.g. $a = b$) all the ϕ vs σ curves will be of the same shape but shifted relative to each other along the ϕ-axis due to geometrical differences represented by K_{A-A}, etc.

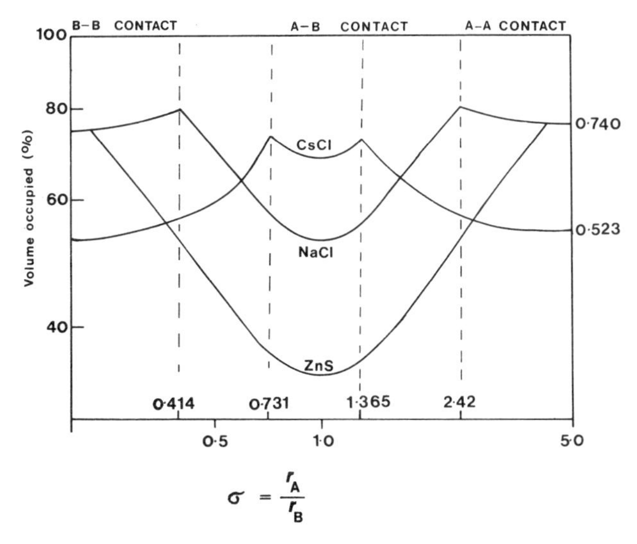

Figure 5-6 Parthé space-filling curves for the CsCl, NaCl and ZnS lattices. σ = radius ratio

Curves for NaCl, CsCl and ZnS (both wurtzite and zinc blende) structures are shown correctly scaled. Since these equations depend upon σ, two structures of the same formula type and the same coordination about each ion will have identical ϕ vs σ functions. Thus, the zinc-blende and wurtzite (for $c/a = 1{\cdot}63$†) curves are coincident.

† ϕ is also dependent upon axial ratio for non-cubic structures. However, in view of the implications of variation in axial ratio in terms of band theory, it is doubtful whether ϕ vs axial ratio discussions have much meaning.

The inflections at 0·225 (ZnS), 0·414 (NaCl) and 0·732 (CsCl) show the known critical radius ratios *below* which each structure will be unstable due to anion–anion (B—B) contact. There is a large range of σ, 0·414 to 2·42, for the rock-salt structure until it becomes unstable again, this time due to cation–cation repulsion. In contrast the CsCl structure has a rather limited range of geometrical stability (0·732 to 1·365) for *ionic* compounds. What *is* rather effectively demonstrated by these curves is that for $\sigma > 0{\cdot}59$ the CsCl structure offers more efficient use of space than the rock salt structure and might therefore be expected to be widely adopted if space filling were the only consideration. Of course, for the CsCl structure a σ value less than 0·732 requires Cl—Cl contact and this is clearly not allowed for an ionic structure. However, A—A contact *is* permissible in metals and alloys, all of which suggests why the CsCl structure is so widely adopted by intermetallic compounds, but is very rare for ionic ones.

In Section 5.1.2 we did not consider radius ratios for tetrahedrally coordinated lattices (such as zinc blende) because their highly covalent nature deprives any such procedure of meaning: such structures are governed by principles laid out in Section 4.2. Since the space-filling formulae depend only upon unit cell edge and σ we can still use them in a general way for covalent structures in this sense; for any realistic radius (covalent or ionic) accorded to zinc and sulphur the value of σ is such that ϕ is not more than 40%. In other words, tetrahedrally coordinated structures make highly inefficient use of space, a point which further emphasizes the transcendental importance of valence and topological principles in determining them.

5.1.5 Phase Transitions

Many phase transitions can be understood basically in terms of space-filling considerations. RbCl is such a case: at −190°C it transforms from the rock salt to the CsCl structure. Note that it has $\sigma = 1{\cdot}00$ (Table 5) and is therefore at what we might term a 'point of maximum disadvantage' with respect to the CsCl structure (see Figure 5-6) since at this σ value little more than 50% of space is filled using the rock salt structure. The transformation can also be effected by application of a lowish pressure (ca 5 kbar). All the alkali halides other than lithium salts can be transformed from the rock salt to the CsCl structure with transition pressures that decrease from ca 10–20 kbar (sodium and potassium salts) to ca 5 kbar (rubidium salts). There seems to be a rough correlation between these transitions, their pressures (which are often very imprecisely located due to hysteresis), and space-filling considerations.

It is probable that space-filling considerations also enter into phase transitions in which the low-pressure form has a tetrahedral structure. This will only occur if the bond directionality is borderline so that a small change in temperature or pressure can swing the balance in favour of a 6:6- rather than a 4:4-coordinate form. The stable form of AgI at room pressure has the zinc blende structure, $d(\text{Ag–I}) = 2{\cdot}80$ Å, but at 3·7 kbar it transforms to the rock salt structure, $d(\text{Ag–I}) = 0{\cdot}03$ Å, which can be understood in terms of the relative ϕ vs σ curves, Figure 5-6. The increase in internuclear separation accompanying the transformation does not necessarily imply *less* efficient use of space but, rather, indicates that re-hybridization has altered the volume distribution of electron density.

5.2 BOND DIRECTIONALITY

In Section 5.1 we saw that the 'ionic model' enjoys some success in handling the energetics of crystals, most notably those which adopt NaCl, CsCl, rutile or fluorite, structures and their close relatives. Its success in predicting coordination arrangements is moderate only. Even in these fairly ionic materials we recognized the importance of some covalent interaction. At another extreme of bond type we encounter highly covalent bonding. How do we handle intermediate bond types; indeed, how do we recognize them? There are also many questions which ionic theory cannot begin to answer, for example, the adoption of the $CdCl_2$ rather than the CdI_2 structure. Let us spell it out: *ionic theory is a good starting point for getting some general guidance on the relative importance of factors such as size and coordination arrangement and is very important in energetics, but for anything beyond this we must use the concepts and language of modern valence theory and talk in terms of orbital overlap and band structure. Ionic theory has had a good run (well over 50 years), it is still heavily over-emphasized; so far as detailed considerations of crystal structure are considered it is time it was interred.*

Some guidance in relation to bond type is clearly given by the positions in the Periodic Table of the atoms involved, see Figure 5-7. In addition changes in colour, departures of observed internuclear distances from those expected from ionic radii, variations in solubility or heat of formation can all give some insight, whilst direct determination of some measure of orbital overlap can be had from appropriate physical techniques such as nuclear quadrupole or electron spin resonance.

As an example, consider the silver halides. The radii of Gourary and Adrian (Table 6, Chapter 2) correctly reproduce internuclear separations in the alkali halides to within about one per cent. Using these radii for halide ions and subtracting them from the experimental internuclear distances in silver halides (NaCl structure) reveals that the 'apparent' ionic radius of Ag^+ is highly anion dependent, Table 7. The calculated lattice energies also show a progressive departure from experimental values. This trend is emphasized by the fact that AgI adopts the zinc blende (4:4) structure; in this

Table 7. 'Apparent' ionic radius of Ag^+ and lattice energy of silver halides

	d(Ag-X), Å	$R(X^-)$, Å	'Apparent' $r(Ag^+)$, Å	Lattice energy at 0°K[a] Born cycle	'Extended' calculation
AgF	2·46	1·16	1·30	954	921
AgCl	2·77	1·64	1·13	904	833
AgBr	2·89	1·80	1·09	895	816

[a] kJ mol^{-1}.

compound the degree of covalent bonding has become just sufficient to determine the type of coordination. The colours of the silver halides also change from white (AgCl) to yellow (AgI). The radii of B-metal cations in general show a pronounced anion dependence, as in the silver halides, accompanied by increasingly poor lattice energy fits.

All this is very well but we really need a more general approach to the topic. Preferably it should be based upon quantities which can be understood theoretically so that the theory can be used in a predictive capacity.

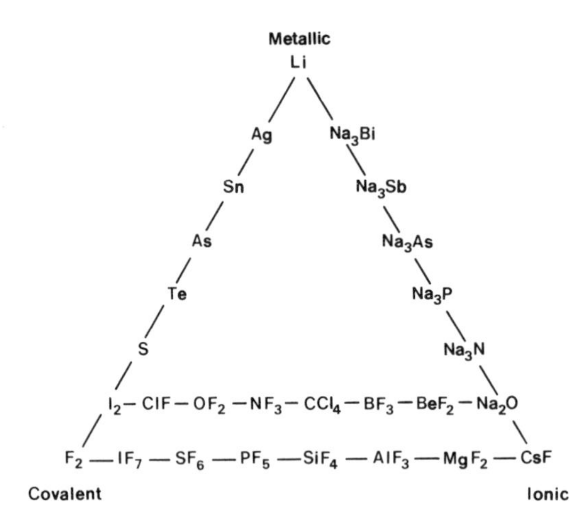

Figure 5-7 Schematic indication of variation of bond type. (From J. A. A. Ketelaar, *Chemical Constitution*, Elsevier, Amsterdam, 1960)

5.2.1 Choice of Parameters to Indicate Directionality of Bonding

One parameter with an obvious (to a chemist, at least!) bearing on bond character is electronegativity, χ. Compounds formed from elements of widely differing electronegativity are more highly ionic than those for which $\chi_A - \chi_B = \Delta\chi$ is small. Thus, in the series with principal quantum number three, NaCl, MgS, AlP and Si, the electrical conduction mechanism changes gradually from ionic to semiconduction as $\Delta\chi$ changes from 2·1 to 0. If $\Delta\chi$ is small the resulting solid may be either highly covalent (e.g. SiC with $\Delta\chi = 0{\cdot}76$) or metallic.

Electronegativity was defined by Pauling as 'the power of an atom *in a molecule* to attract electrons to itself'. It is a simple and valuable concept but one that has resisted attempts to quantify it. As defined, electronegativity does not have a fixed value for any one element but varies with oxidation state, the nature of the bonding to other atoms, coordination number, etc. Moreover, the established scales relate to molecules and not crystals. For a discussion of the definition and relations of various electronegativity scales the reader is referred elsewhere (e.g. Cotton and Wilkinson, 1972).

Pauling and others have made heroic attempts to relate bond ionicity to electronegativity. The results are never of general applicability and deviations can be very large. Although this is an unsatisfactory situation we can nevertheless make excellent progress in understanding crystal-structure types even with this rather blunt instrument.

Although we can undoubtedly make some sort of correlation between bond type and electronegativity, the precision of our description would be greatly improved by concurrent use of a second parameter. Dehlinger (1955) has shown that a good measure of bond *directionality* is given by n, the principal quantum number of the valence shell of an atom. For compounds, an average principal quantum number defined by

$$\bar{n} = \sum_i c_i n_i / \sum_i c_i$$

has been introduced by Mooser and Pearson (1959) where n_i is the principal quantum number and c_i the number per formula unit of atoms of the ith kind. With increase of n or $\bar{n}$, atomic orbitals become larger, more diffuse and less strongly directional. For $n = 2$ (the carbon row) s–p hybridization yields highly directional bonds of considerable strength. As n increases, d- and f-orbitals become comparable in energy with s- and p-orbitals, so that symmetry-allowed combinations of them appear. The essential feature is a pronounced falling off of bond directionality with n. The process is called 'metallization' or 'dehybridization'. We have already seen its effect on crystal structures of Group IVb elements (p. 88).

Mooser–Pearson plots

Diagrams of $\bar{n}$ versus $\Delta\chi$ were first published by Mooser and Pearson in 1959. So far as I know they have only been mentioned subsequently in one textbook and were then accorded the status of an interesting curiosity. In fact they warrant a place of high prominence in solid-state studies, especially now that they have been given precise theoretical backing (see Section 5.2.2).

Figure 5-8 shows an $\bar{n}$ vs $\Delta\chi$ plot for compounds AB which adopt the rock-salt or one of three types of tetrahedrally coordinate structure. There is a beautiful, clean separation of the 6:6- from the 4:4-coordinate structures with only a few discrepancies. The arrow indicates the direction of increase of bond directionality. It follows from this that compounds which adopt the tetrahedral structures do so because of the directional character of their bonding. Those occurring near the borderline might reasonably be expected to be diamorphic.

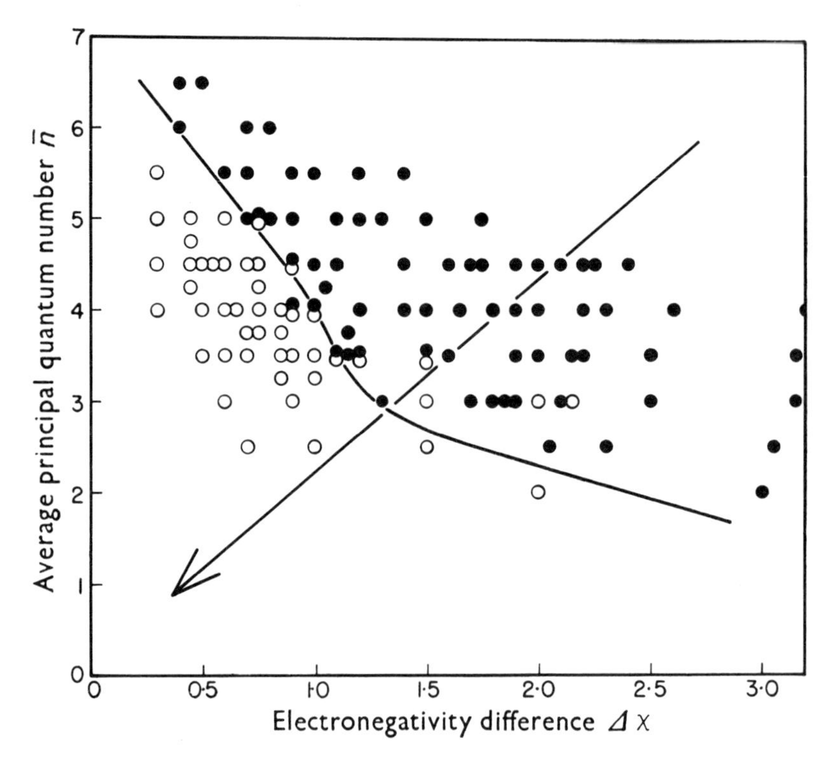

Figure 5-8 Mooser–Pearson diagram for AX structures with tetrahedral (○) or octahedral (●) coordination. (This, and the other Mooser–Pearson plots in this chapter are reproduced with permission from *Acta Cryst.*, **12**, 1015 (1959))

The existence of a sharp borderline in Figure 5-8 suggests the idea of a *critical ionicity* which determines whether a given AB compound will have the rock salt or a tetrahedral structure. The curious shape of the borderline shows that any mathematical function found to fit it would require several free parameters (i.e. variable constants). This is not too surprising in view of the problems associated with defining electronegativities and, for that matter, $\bar{n}$ is not the most delicate of parameters. Is there any way of developing this very promising approach?

5.2.2 Phillips and van Vechten Dielectric Electronegativities

Very recently Phillips and van Vechten (1969; 1971) have discussed the properties of a set of AB crystals (some 70 compounds) of the type $A^N B^{8-N}$, where N is a group of the Periodic Table, which have a total of eight valence electrons between A and B. Using two parameters for which relations to $\bar{n}$ and $\Delta\chi$ can be demonstrated (see below) a remarkable straight-line separation of 6:6- from 4:4-coordinated structures was obtained. Although this work is

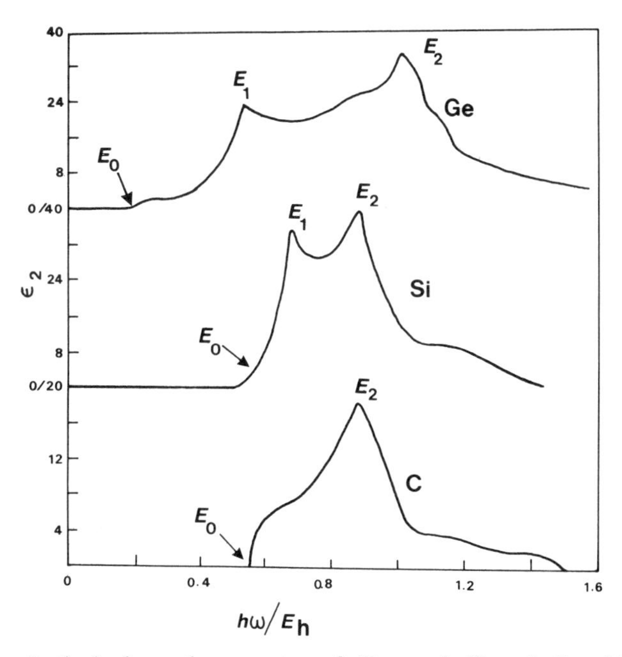

Figure 5-9 Optical absorption spectra of diamond, Si and Ge. (From J. C. Phillips, *Rev. Mod. Phys.*, **42**, 317 (1970))

at present restricted to $A^N B^{8-N}$ compounds it is clearly pregnant with possibilities and implications for future delineation, description and understanding of crystal structures and their properties in terms of quantum-mechanically defined quantities.

Consider, first, a crystal with the diamond lattice (C, Si or Ge). In molecules the electronic energy levels can be studied by means of absorption spectra, the frequencies of light absorption being directly related to the energy separation of the molecular orbitals by $E_i = h\nu_i$. In an entirely analogous manner, the optical absorption spectra of crystals are directly related to the separations of the energy levels within them, although now we must talk in terms of transitions between bands rather than molecular orbitals. Absorption spectra of diamond, silicon and germanium, are shown in Figure 5-9 and represent various transitions from the filled valence band to empty conduction band levels. Happily, for our purposes, we do not need to go into the detailed interpretation of these spectra because a simplifying assumption (due to Penn, 1962) is made that all these transition energies can be replaced by a single average energy, E_g. This is the 'average band gap', the average width (in energy units) of the forbidden gap between the filled valence and the empty conduction band levels. The suggestion is that the structure of a crystal is to be understood in terms of the relative energies of the bands (constructed by orbital overlap, along the lines described in Section 2.2). Rather than be faced with the problem of conjuring with the many separate transition energies represented by optical spectra such as those of Figure 5-9, a single average energy gap, E_g, is employed in the theory. For diamond this is a tolerable approximation but it becomes progressively poorer as the Group is descended; in particular, a peak E_1 develops at lower energy, and the spread of the spectrum increases. Nevertheless E_g is a parameter firmly related to a spectroscopic observable of fundamental meaning and importance.

All this applies to the elements. Spectra similar to those of Figure 5-9 are shown by all tetrahedrally coordinated crystals, even relatively ionic ones. Hence, for any given heteropolar crystal, $A^N B^{8-N}$ an equivalent average energy gap, E_g, is likewise determined. Because there is now an ionic contribution to the bonding due to the differing electronegativities of A and B we write

$$E_g = E_h + iC$$

where E_h is a homopolar energy gap and C represents the charge transfer between A and B and can be termed the 'ionic energy'. It is convenient to work with

$$E_g^2 = E_g E_g^* = E_h^2 + C^2 \tag{14}$$

Consider an example involving three isoelectronic atom cores. For germanium $E_h = 4{\cdot}3$ eV whilst E_g for ZnSe $= 7{\cdot}0$ eV (from experiment). Hence from Equation 14, $C = 5{\cdot}5$ eV.

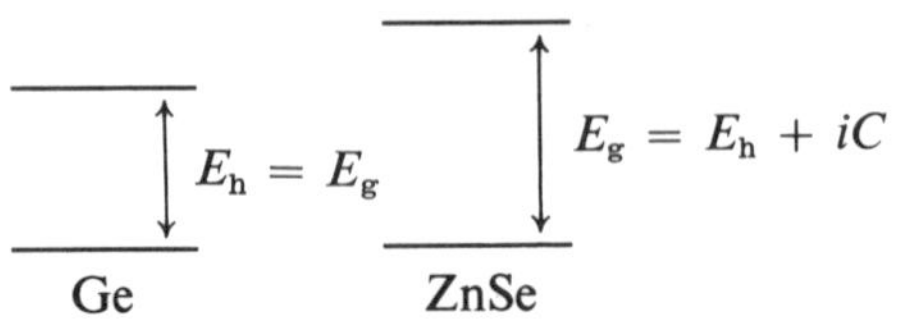

$E_h = E_g$

$E_g = E_h + iC$

Ge ZnSe

Table 8. Dielectrically defined *s–p* electronegativities (Phillips, 1969)

Li 1·00	Be 1·50	B 2·0	C 2·50	N 3·00	O 3·50	F 4·00
Na 0·72	Mg 0·95	Al 1·18	Si 1·41	P 1·64	S 1·87	Cl 2·10
Cu 0·79	Zn 0·91	Ga 1·13	Ge 1·35	As 1·57	Se 1·79	Br 2·01
Ag 0·57	Cd 0·83	In 0·99	Sn 1·15	Sb 1·31	Te 1·47	I 1·63
Au 0·64	Hg 0·79	Tl 0·94	Pb 1·09	Bi 1·24		

Since C appears directly as a measure of the charge transfer energy in a heteropolar crystal bond it forms the natural basis for a scale of dielectrically defined electronegativities. Note carefully though, that in distinction from the commonly used definitions of χ, C measures the *relative* electronegativity of A and B, and that it does so specifically for sp^3-hybridized atoms. For discussing crystal structure, C is used directly, but such values *have* been apportioned between atoms forming heteropolar bonds with the results shown in Table 8. We stress that these dielectric electronegativities, χ_D, refer

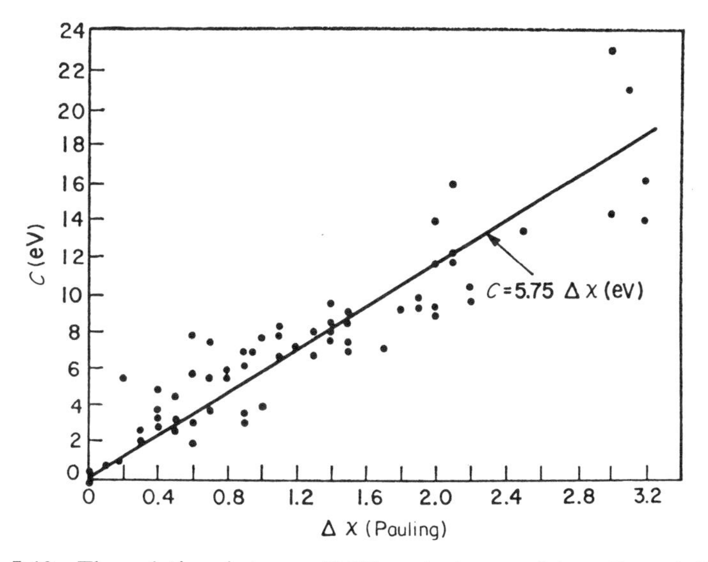

Figure 5-10 The relation between Phillips electronegativity, C, and Pauling electronegativity, χ. (From J. C. Phillips, *Rev. Mod. Phys.*, **42**, 317 (1970))

specifically to sp^3-hybridized atoms *in crystals*. Nevertheless there is some cause for preferring them to the usual Pauling or Allred–Rochow electronegativities even in molecular situations. As Phillips has pointed out, $\chi_D(Cl) < \chi_D(N)$ whereas these values are reversed on both Pauling and Mulliken (though not on Allred–Rochow) scales. It is recalled that N—H hydrogen bonds are stronger than Cl—H bonds. A plot of C values against Pauling electronegativity $\chi_A - \chi_B$ for $A^N B^{8-N}$ crystals, Figure 5-10, shows considerable scatter but the clear general trend is given by

$$C = 5{\cdot}75\Delta\chi$$

where C is in eV.

Dielectric Definition of Ionicity

A convenient scale of ionicity, f_i, would have values from zero (when $C = 0$) to unity. Such a scale can be defined in terms of E_h and C by treating

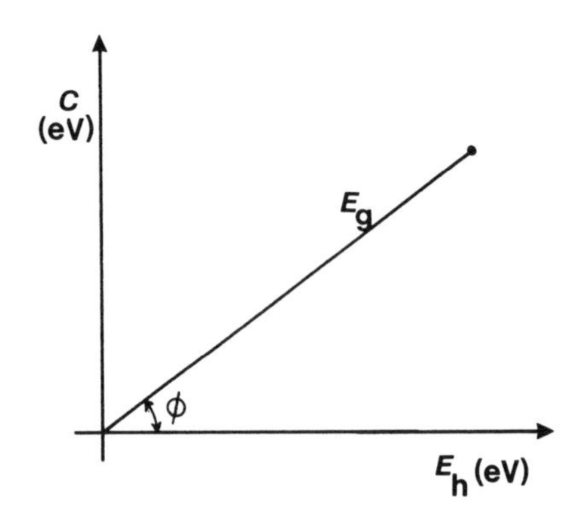

Figure 5-11 Definition of a scale of ionicity

them as cartesian coordinates (Figure 5-11). Transformation to polar coordinates casts E_g as the polar radius, a definition clearly suggested by Equation 15. Then

$$f_i = \sin^2 \phi = C^2/E_g^2 \tag{15}$$

which scales between zero and unity. An analogous covalent bond fraction, f_c, is likewise defined as

$$f_c = \cos^2 \phi = E_h^2/E_g^2 \tag{16}$$

so that $f_i + f_c = 1$.

5.2.3 E_h versus C Diagrams and their Relation to Mooser–Pearon Plots

Figure 5-12 shows an E_h vs C plot for about seventy $A^N B^{8-N}$ crystals. The result is truly magnificent! There is an absolutely clean separation between

6:6- and 4:4-coordinated structures with *no* errors. In contrast to the analogous Mooser–Pearson plot, Figure 5-8, the border is a straight line which even passes through the origin! The slope of the line corresponds to a 'critical ionicity' $F_i = 0{\cdot}785 \pm 0{\cdot}010$. That is, for any value of $f_i(AB) < F_i$

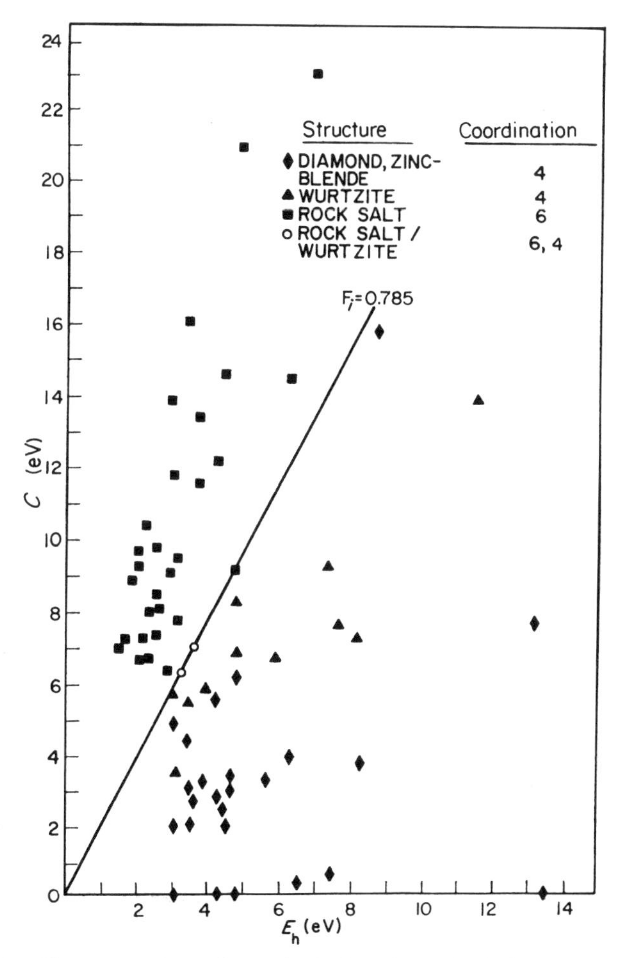

Figure 5-12 The separation of tetrahedral and octahedral structures using the spectroscopically defined covalent (E_h) and ionic (C) energy gaps. (J. C. Phillips, *Rev. Mod. Phys.*, **42**, 317 (1970))

one of the four-coordinated structures is adopted whilst for $f_i(AB) > F_i$ the structure is of the rock salt or CsCl type. Ionicity values for specific compounds are collected in Table 9 and are seen to follow the commonly accepted trends. Thus, compounds with large cations and small anions are the most ionic, typically RbF and SrO.

Table 9. Phillips–van Vechten ionicities, f_i, (on a scale $0 \rightarrow 1$) for some $A^N B^{8-N}$ crystals[a]

Group Ia and Ib Halides						
	Li	Na	K	Rb	Cu	Ag
F	0·915	0·946	0·955	0·960	0·766[b]	0·894
Cl	0·903	0·935	0·953	0·955	0·746	0·856
Br	0·899	0·934	0·952	0·957	0·735	0·850
I	0·890	0·927	0·950	0·951	0·692	0·770

Group IIa Chalcogenides			
	Mg	Ca	Sr
O	0·841	0·913	0·926
S	0·786	0·902	0·914
Se	0·790	0·900	0·917
Te	W[c]	0·894	0·903

[a] Critical ionicity for this series, $F_i = 0{\cdot}785 \pm 0{\cdot}010$.
[b] Compounds in the box have the zinc blende structure; all others adopt the rock salt structure.
[c] Wurtzite structure.

It is of interest to consider the relationships between E_h vs C, and $\bar{n}$ vs $\Delta\chi$ plots, and to seek reasons for the somewhat arthritic borderlines in the latter. The ionic energy C is clearly the equivalent of $\Delta\chi$ (Pauling, Allred–Rochow, etc.), although we recognize that $\Delta\chi$ is not a very reliable measure of ionic character. It can be shown that $E_h \propto a^{-2{\cdot}5}$ where a is the cubic lattice constant (i.e. the length of the unit cell edge). a Changes with principal quantum number, n, but not smoothly. Thus, the percent changes in a are:

$$n = 2 \xrightarrow{50\%} 3 \xrightarrow{4\%} 4 \xrightarrow{15\%} 5$$

The relation is even more non-linear for compounds with A and B from different rows of the Periodic Table. Phillips points out that MgTe, GaAs and CdS, all have $\bar{n} = 4$ but that the bond lengths are 2·76, 2·44 and 2·53 Å, respectively. In other words, for these three compounds use of $\bar{n}$ masks a true variation in E_h in excess of 30%!

We conclude this section by emphasizing that the critical factor determining crystal structure adoption by $A^N B^{8-N}$ compounds is *ionicity* and not relative electronegativity taken alone. The Phillips–van Vechten ionicity, f_i, is quantum-mechanically defined and related directly to observable and measurable properties of crystals. Now that the interpretation of the Mooser–Pearson plots has been supported and clarified we can proceed to use them for groups of compounds not yet treated on the basis of the quantum-mechanical approach with the assurance that we are obtaining fruitful and meaningful information.

5.2.4 Discussion of AB Compounds on the Basis of E_h vs C, and $\bar{n}$ vs $\Delta\chi$, Plots

In the E_h vs C diagram of Figure 5-12 there are regions below the critical-ionicity line which correspond to wurtzite and zinc blende structures respectively. The delineation is not as sharp as that between 6:6- and 4:4-coordinate

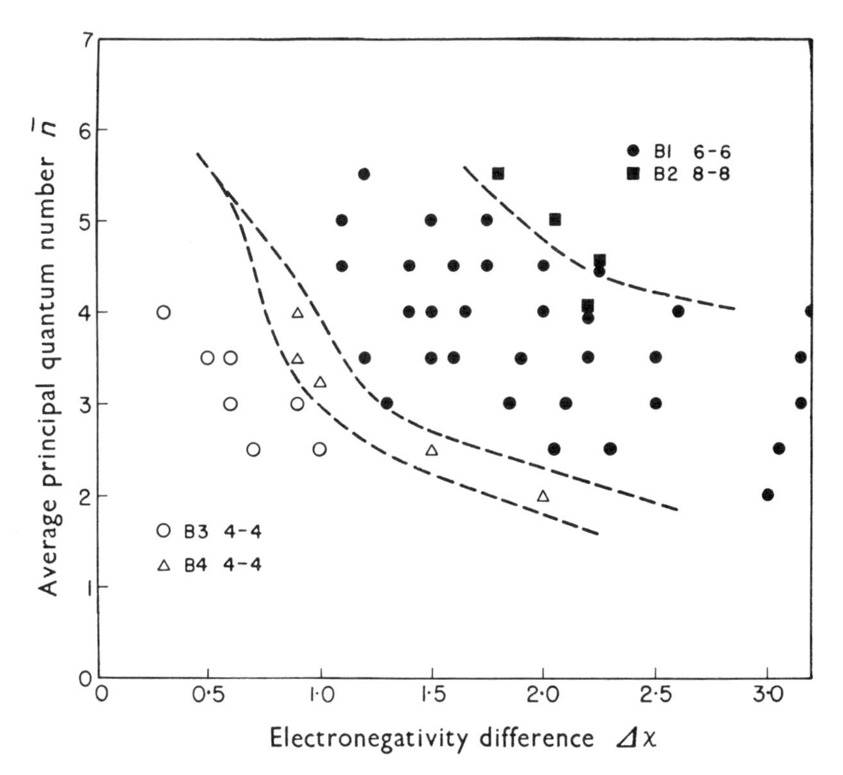

Figure 5-13 Mooser–Pearson diagram for AX compounds containing A-Group cations. B1 = NaCl; B2 = CsCl; B3 = zinc blende; B4 = wurtzite

structures but it is clear that the order of increasing ionicity is

$$\text{zinc blende} \longrightarrow \text{wurtzite} \longrightarrow \text{rock salt}$$

which is also the order of increasing Madelung constant (Table 3). Equivalent features are found in the $\bar{n}$ vs $\Delta\chi$ plots shown in Figures 5–13 and 5-14. For the two sets containing respectively A- and B-Group cations we see the full progression of structure types with increasing ionicity and Madelung constant.

$$\text{Zinc blende} \longrightarrow \text{wurtzite} \longrightarrow \text{rock salt} \longrightarrow \text{CsCl}.$$

The predominantly covalent behaviour of B-group cations in their solid compounds is clear from the relative population densities of the domains in each figure.

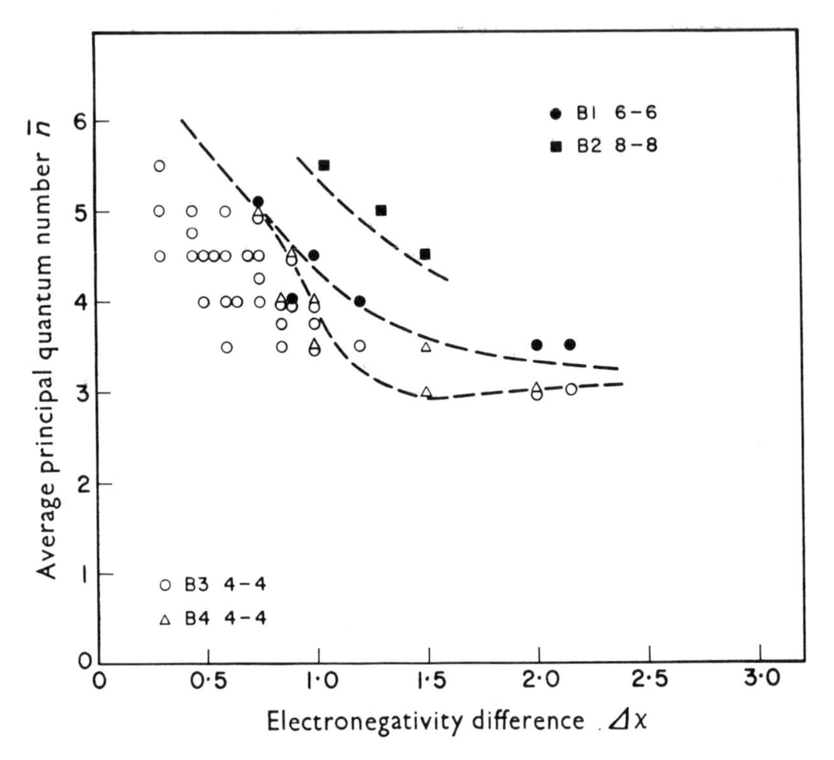

Figure 5-14 Mooser–Pearson diagrams for AX compounds containing B-Group cations. The numbers by the labels indicate coordination type. Key to structure types as in Figure 5-13

In each diagram the narrow domain associated with the wurtzite structure suggests that the factors determining its adoption are rather delicately balanced. The majority of compounds with this structure also exist in either the zinc blende or rock salt structures (and occasionally in all three, e.g. MnS), if not under normal conditions, then with a little persuasion (thermal or hydrostatic). It appears to be favoured by oxides and nitrides in which the high electronegativity of the non-metal assures a higher f_i value than for phosphides, arsenides, etc. (see Table 6, Chapter 3) that adopt the zinc blende form. **In summary**, the wurtzite structure is the more ionic of the two but not by a large margin.

A number of well-known changes in structure type within a Group are readily explained by knowledge of critical ionicity. Thus, for the silver halides, f_i drops from AgF (0·894) to AgBr (0·850) and is only 0·770 for AgI which therefore adopts the zinc blende structure. In this case it is the electronic structure of iodine which swings the balance. The cuprous halides *all* have

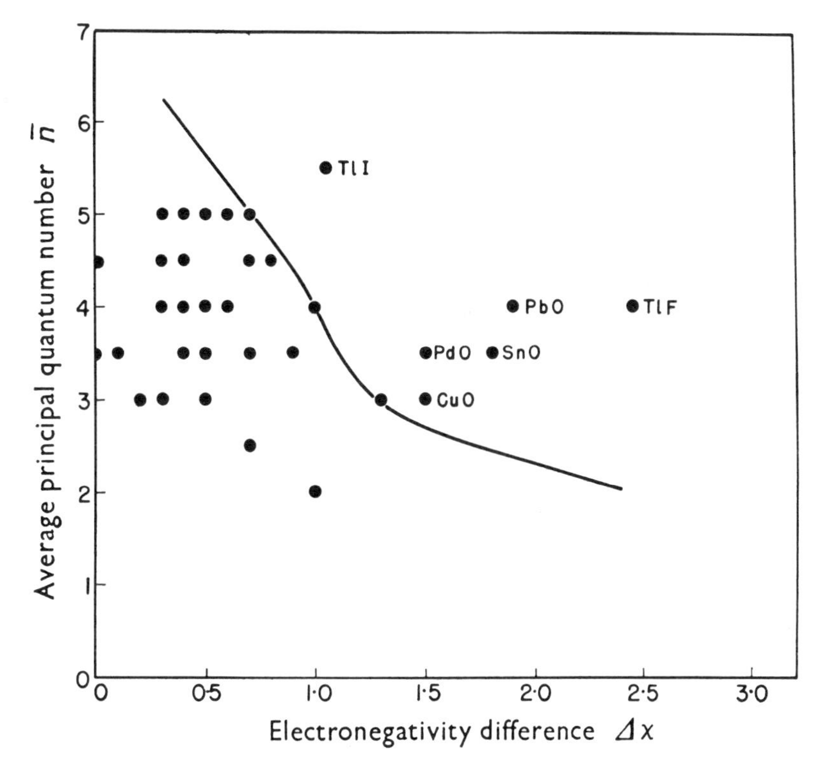

Figure 5-15 Mooser–Pearson diagram for some AX compounds with less common structures

the zinc blende structure because of the more directional bonding associated with a lighter element. The switch from 4:4- in ZnO to 6:6-coordination in CdO is understood similarly, while the chain structure of HgO has a special explanation (Section 5.4.3).

Distorted and Rare AB Structures

There are very many compounds that crystallize in distorted variants of the basic structure types we have so far considered, and others which are the sole or nearly the sole representatives of their type. They mostly have quite strong directional bonding as indicated by the distribution of points in Figure 5-15 relative to the borderline, which has been transferred from Figure 5-8. This is only to be expected as many of these unusual structures are formed to accommodate special electron configurations (such as low spin d^8, or d^{10}), and these configurations are commonly associated with directed bonding. We consider this topic in detail in Section 5.4 where we also describe some of the relevant structures.

Of the points on the 'ionic' or non-directional bonding side of the borderline TlF, PbO and SnO, can be regarded crudely as containing non-spherical ions due to the 'inert' pair which becomes prominent in the chemistry of these elements. They all have structures that are distorted variants of rock salt. Although they lie on the 'ionic' side of the borderline we emphasize that their position is not inconsistent with a degree of directed bonding—as in the alkali halides. The implication is that in them the directed bonding is not so strong as to *determine* the overall structure; instead a compromise has been reached in which the stereochemical effect of the 'inert' pair results in distortion of the highly symmetric parent structure by a process that can be considered as *ionic in origin* though the resultant structure may well be more convincingly described in terms of a specific band or hybridization scheme.

5.2.5 Anion-rich Compounds AX_2 and AX_3

$\bar{n}$ vs $\Delta\chi$ plots for AX_2 and AX_3 structures are shown in Figures 5-16 and 5-17. They include data for several structure types which we have not described (e.g. $PbCl_2$, AlF_3, LaF_3, BiF_3, etc.) if only because a line has to be drawn somewhere. The reader is referred to Naray-Szabo (1969) or Wells (1962) for details. The implications can be appreciated without specific knowledge of the structures. In both figures, and also in one (not reproduced here) for A_2X_3 structures (Al_2O_3, La_2O_3, Mn_2O_3, Zn_3P_2), the domains are separated by borderlines which, remarkably, delineate identical areas of $\bar{n}$ vs $\Delta\chi$ space. And it *is* remarkable! The occurrence of the same borderlines for AX_2, A_2X_3 and AX_3, structures strongly suggests that the same critical

Table 10. Structural details for AX_2 and AX_3 crystals (after Mooser and Pearson, 1959)

	AX_2	AX_3	Close-packed array made of:
Cubic	CaF_2	BiF_3	Cations → dense structures
Hexagonal	—	LaF_3	
Cubic	$CdCl_2$	$CrCl_3$	Anions → open structures
Hexagonal	CdI_2	BiI_3, AlF_3	

ionicities govern the switch from one structure type to another: from fluorite to rutile, from LaF_3 to AlF_3, etc. As for AX compounds there is a trend from highly ionic behaviour with high c.n. at the upper right (La^{3+} is (5 + 6)-coordinate in LaF_3; Pb^{2+} nine-coordinate in $PbCl_2$; Ca^{2+} eight-coordinate in CaF_2) through intermediate values (six-coordinate cations in $CdCl_2$ and CdI_2 lattices) to low values at the lower left, where directed bonding determines the structure and very low coordination arrangements occur (two and four in polymorphs of SiO_2, for example).

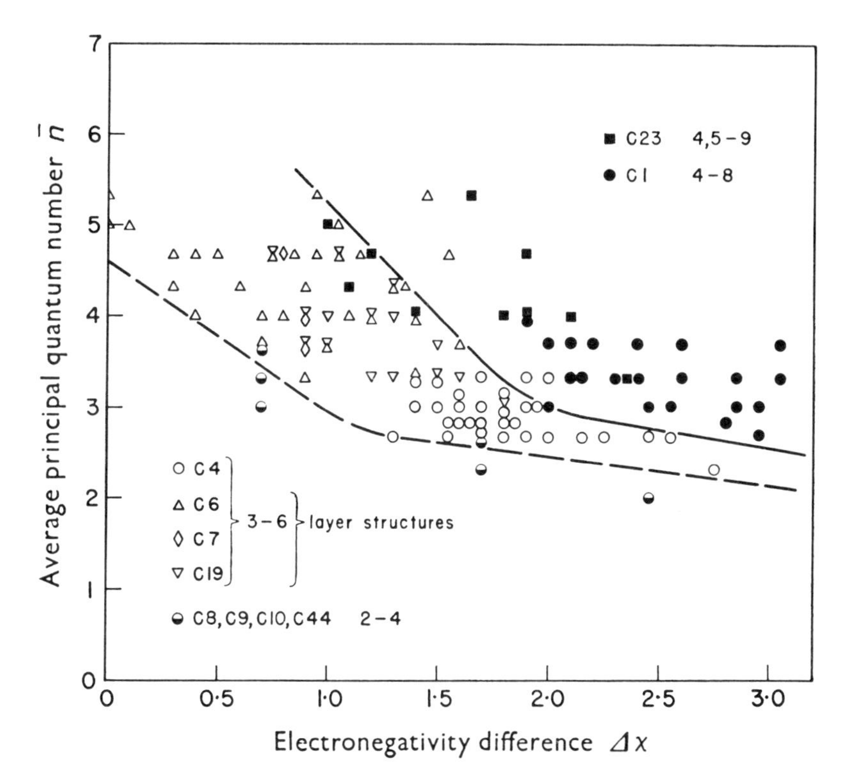

Figure 5-16 Mooser–Pearson plot for compounds AX_2.C1 = fluorite; C23 = $PbCl_2$ type; C4 = rutile, TiO_2; C6 = CdI_2; C7 = MoS_2; C19 = $CdCl_2$; C8, 9, 10 and 44 are less common structures with 2–4 coordination (e.g. SiO_2 polymorphs)

In conformity with these trends, Mooser and Pearson have pointed out that the highly ionic materials form dense structures, in which efficient use of space is made, based upon close-packed cation arrays. The rather more covalent compounds have anion-based close-packed lattices which make much less efficient use of volume. The situation is summarized in Table 10.

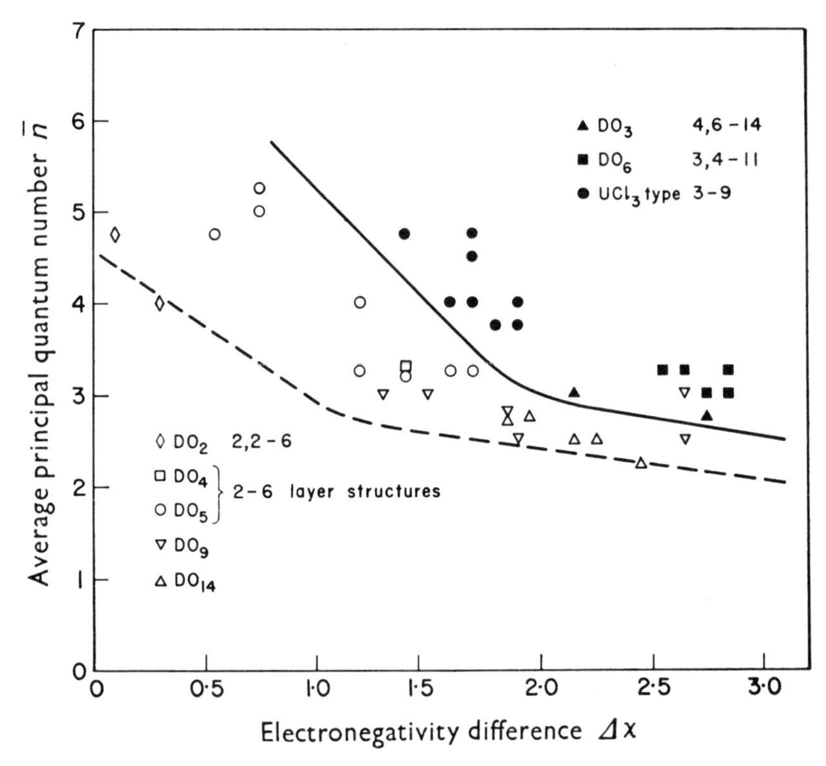

Figure 5-17 Mooser–Pearson plot for compounds AX_3. DO_3 = BiF_3 type; DO_6 = LaF_3; DO_5 = BiI_3; DO_9 = ReO_3; DO_{14} = AlF_3

5.2.6 Physical Properties and Degree of Ionicity

For A^NB^{8-N} *structures*, although there is a clean separation of 6:6- from 4:4-coordinate structures at F_i = 78%, there is good evidence that for some *properties* of these crystals the transition from typically covalent to ionic behaviour comes at much higher ionicity, nearer 94% in fact. To go far into this topic would lead us into solid-state physics which is already well served with accounts at all levels, but consider one example. The refractive index of the alkali halides increases with pressure whereas that of compounds with diamond, zinc blende and wurtzite structures *de*creases. MgO has the rock salt structure but its refractive index also *de*creases under pressure; it has a Phillips ionicity of 84%. From this, and much other evidence, it now appears that so far as properties are concerned, full ionic behaviour is only reached in the potassium and rubidium halides. The behaviour of lithium and sodium halides is intermediate between ionic and covalent (van Vechten and Phillips, 1970).

Prior to development of the Phillips ionicity scale, it was understood that certain vibrational properties of crystals are related to ionicity. Consider two atoms or ions in a binary crystal vibrating relative to each other. Due to the displacement of charge accompanying vibration an oscillating electric dipole moment, $\boldsymbol{p}$, given by

$$\boldsymbol{p} = e^*\boldsymbol{\chi} \tag{17}$$

will be present, where $\boldsymbol{\chi}$ is a vector describing relative displacement of the atoms and e^* is known as the 'effective charge'. It can be shown that† vibrations in crystals fall into two classes: (*a*) 'acoustic' modes which are of low frequency and correspond to in-phase motion of atoms in neighbouring unit cells; (*b*) 'optical' modes, which are of higher frequency and, as the name suggests, are usually infrared- or Raman-active or both. There are three acoustic modes for each unit cell corresponding to wave propagation along three orthogonal directions, and $3N$–3 optical modes, where N is the number of atoms in the unit cell.

Optical branch modes are further split in the more ionic or dipolar crystals by long-range electrostatic forces, into vibrations that are parallel or transverse to the direction of light propagation in the crystal. The resulting vibrations have slightly different frequencies (say from 1 or 2 cm^{-1} up to 100 cm^{-1} or more) and are referred to as longitudinal-optic (LO) and transverse-optic (TO) modes respectively. A schematic illustration is given in Figure 5-18. For cubic crystals their frequencies, ω_{LO} and ω_{TO}, are related by the Lyddane–Sachs–Teller formula

$$\left(\frac{\omega_{\text{LO}}}{\omega_{\text{TO}}}\right)^2 = \frac{\varepsilon_{\text{S}}}{\varepsilon_{\text{O}}}$$

where ε_{s} = static dielectric constant at very low frequency (below ω_{LO} and ω_{TO}) and ε_{o} = dielectric constant at optical frequencies above these modes. This relation clearly suggests itself as a measure of ionicity as ε_{s} and ε_{o} differ chiefly in being measures of dielectric constant below and above frequencies at which ionic motion can follow the electric field. The interrelation of these quantities is shown schematically in Figure 5-18. Rewriting Equation 17 for TO and LO modes respectively we have

$$\boldsymbol{p}_{\text{T}} = e^*_{\text{T}}\boldsymbol{\chi}$$

$$\boldsymbol{p}_{\text{L}} = e^*_{\text{L}}\boldsymbol{\chi} = \frac{e^*_{\text{T}}}{\varepsilon_{\text{o}}}\boldsymbol{\chi}$$

† A euphemism indicating that we have dodged a large chunk of physics. For more information the reader is referred to C. Kittel, *Introduction to Solid State Physics*, Wiley, New York, 1953; or to many other sources.

It can then be shown that for a diatomic unit cell

$$\omega_{\rm LO}^2 - \omega_{\rm TO}^2 = \frac{4\pi\varepsilon_0}{MV}(e_{\rm L}^*)^2 = \frac{4\pi}{MV}\frac{(e_{\rm T}^*)^2}{\varepsilon_0} \tag{18}$$

where M = reduced mass of the vibrating atoms, and V = volume of the unit cell.

For the same set of $A^N B^{8-N}$ crystals for which the f_i scale was defined $(e_T^*)^2/\varepsilon_0$ is very nearly linear in f_i although there is a fair amount of scatter. The function $\omega_{\rm LO}^2 - \omega_{\rm TO}^2$ is undoubtedly related, and can be taken as a good

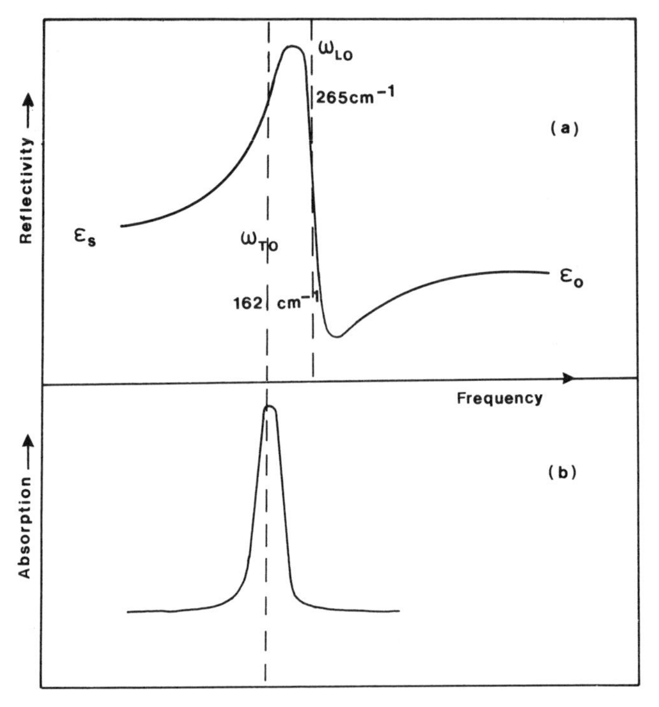

Figure 5-18 Schematic illustration of relations between $\omega_{\rm TO}$ and $\omega_{\rm LO}$ frequencies for NaCl and the far-infrared reflectance spectrum (a), where ε_s = static dielectric constant, ε_0 = dielectric constant at optical frequencies. The absorption spectrum, (b), can be computed from the reflectance spectrum.

relative guide, to ionicity. The literature contains a good many broadly equivalent types of plot in which $\omega_{\rm TO}^2/\omega_{\rm LO}^2$ or $\omega_{\rm LO}^2 - \omega_{\rm TO}^2$ is so used (see Tateno, 1970, and references therein). Whilst these were understood in a general way, the link with f_i is now made and should be exploited for other classes of crystal.

5.2.7 Résumé

As a result of the work of Mooser and Pearson (1959) and Phillips and van Vechten (1969) a good semi-quantitative and, in many instances quantitative, understanding has emerged of the relationships between ionicity, or bond directionality, and crystal structure for many of the structures adopted by binary compounds. Although there is a great deal still to be done in this field, the way in which it is likely to develop is now clear. We have to hand an excellent means of accounting for the structures of materials of intermediate bond types and can even use our knowledge in a predictive capacity with a fair chance of success.

But we have not finished yet! At this point in our investigation we have successfully sorted structures into groups in order of the influence of bond directionality. The next step is to describe that directionality in terms of orbital overlap and band theory, and to account for some of the subtler features which have so far resisted us (e.g. adoption of the $CdCl_2$ as opposed to the CdI_2 structure). We do this in a descriptive way beginning in Section 5.4 and continuing in Chapter 7.

5.3 THERMODYNAMICS AND CRYSTAL STRUCTURE

In our investigation thus far we have attempted to understand the existence of stable structures in terms of parameters such as size, ionicity, topology, etc. We have also come across compounds that hover on the borderline between two or more structures, and others that transform under application of pressure or temperature. An alternative description of stability is given in thermodynamic terms by saying that, under given conditions, the most stable structure is the one of lowest Gibbs free energy, $G = H - TS$ where H is the enthalpy and S the entropy.† This statement is subject to a major qualification: even if a material has a structure of *higher* free energy than an alternative structure which it can in principle adopt, it will not transform unless there is a suitable mechanism. The point is well illustrated by diamond which is unstable with respect to graphite at room temperature and pressure by virtue of a difference in free energy of 2·89 kJ mol^{-1}. Nevertheless, the resistance of diamond to spontaneous change is such that it is regarded as a good long-term investment!

The purpose of this section is to outline relations between structural and thermodynamic approaches to crystal stability. We shall only skim the surface of a very complex subject, especially in relation to non-stoichiometric phases.

† The word entropy was invented by R. Clausius; it comes from the Greek word *trope* meaning transformation and he deliberately added the prefix *en* to convey a connection with energy. (D. Tabor, *Gases, Liquids and Solids*, Penguin Books Ltd, Harmondsworth, 1969).

For an excellent account of thermodynamic properties of solids in relation to reactivity the reader is referred to Johnson (1968).

Thermodynamics is a language which enables us to state a problem (in this case by requiring structures of lowest Gibbs free energy) and then restate it in terms of various other parameters (such as enthalpy and entropy) which, if well chosen, are susceptible to experimental investigation. Ultimately it leaves us holding these bulk quantities which we can only understand if we make further steps of a theoretical nature by introducing ideas extra to thermodynamics, e.g. the relation of trends in redox potentials in solution to ionization potentials.

For solids, the enthalpy term is determined by the cohesive forces in the crystal. In metals, the largest contribution is from nearest-neighbour interactions, but on the ionic model—which works remarkably well for many problems in energetics—the more extensive interactions described by the Madelung constant must be taken into account. The lattice energy of an (assumed) ionic crystal, U_L, determines the enthalpy term since $\Delta H_f^\ominus = U_L + nRT$. U_L can be calculated as shown in Section 5.1.1. With increasing covalency in a series U_L becomes a progressively poorer approximation to $\Delta H_f^\ominus - RT$. Indeed, differences in $\Delta G_L^\ominus$ in a related series are associated with variations in covalent contribution, i.e. in ionicity. Since a good quantitative understanding of ionicity is available for the $A^N B^{8-N}$ compounds, the question naturally arises as to whether heats of formation of these compounds can be calculated from a knowledge of their band-gap properties as defined in Section 5.2.2. This has, in fact, been achieved very recently by Phillips and van Vechten (1970) and represents a most important advance, although it should be noted that empirical relations between band gap and heat of formation have been known for a long time.

It has been possible to handle lattice energies in ionic crystals for many years; we now have a fundamental approach to covalent and semi-ionic crystal energies based upon spectroscopic observables of basic meaning. Admittedly this applies only to the $A^N B^{8-N}$ set at present, but there is real hope that more general extensions will be forthcoming. We now outline this method.

5.3.1 Calculation of Heats of Formation of $A^N B^{8-N}$ Compounds

It will be evident from our discussion of ionicities in Section 5.2.2 that in $A^N B^{8-N}$ crystals, bonding is viewed in terms of a covalent fraction, f_c, due to orbital overlap, and an ionic fraction $f_i = 1 - f_c$ arising for the case of heteropolar bonds. Further, according to the rows of the Periodic Table from which the combining elements are drawn, there will be more or less dehybridization (or metallization) thus reducing the electron density involved in covalent bond formation and resulting, in the limit, in formation of a metallic structure. Both of these features must be accommodated in calculating ΔH_f.

Phillips and van Vechten's formula for ΔH_f is

$$\Delta H_f(AB) = \Delta H_0 \left(\frac{a_{Ge}}{a_{AB}}\right)^s Df_i(AB) \qquad (19)$$

with

$$D = 1 - b\left(\frac{E_2}{\bar{E}}\right)^2 \qquad (20)$$

D being a measure of dehybridization. a_{Ge} and a_{AB} are the lattice constants (i.e. unit cell edges) of germanium and crystal AB respectively; $f_i(AB)$ is defined as shown in Section 5.2.2; ΔH_0 and s are scaling factors; $\bar{E} = \frac{1}{2}(E_0 + E_1)$ where E_0, E_1 and E_2 are spectroscopic energies shown in Figure 5-9; b is related to E_0, E_1 and E_2, as described below.

Before looking at the effectiveness of this means of computing heats of formation, we need to amplify the definitions of the quantities in Equation 19, as this leads to a fuller appreciation of the basis of the calculation. However, the reader who wishes to skip this can do so while remembering that the data fed into Equation 19 are (a) spectroscopic energies, (b) experimental lattice constants. In other words, this equation accounts for heats of formation directly in terms of the fundamental electronic band structure of the related group of $A^N B^{8-N}$ crystals.

The Quantities of Equation 19

The approximation upon which the Phillips ionicity concept is based is that all the various transitions from valence to conduction bands in an $A^N B^{8-N}$ crystal can be replaced by a single average energy gap, E_g.

Consider now what happens with increase in atomic number down Group IVb; the increasing tendency towards metallic properties (i.e. dehybridization) develops as a conduction band of *s*-character shifts to lower energy (leaving others at high energy) and begins to hybridize with the valence bands. As the conduction band is lowered, so is the energy gap between it and the valence band, a process revealed in the spectra by development of absorption at *lower* energy than E_g; this is the E_1 peak of Figure 5-9. It is associated with transitions from a valence band of bonding *p*-character to an antibonding band of *s* character. E_2 represents the main valence to conduction band transition to which E_g is an approximation. The increasing difference between E_g and E_2 is an indication of the progressively poorer validity of the one-gap approximation with atomic number. The spectrum of germanium also shows a weak absorption, E_0, below E_1, and a trace of it can be detected in silicon and diamond: this has been interpreted as a measure of the *smallest* direct valence-conduction band gap in the crystal. Whether or not this is correct is not important; E_0 goes along with E_1 so both are used as a measure of the degree of dehybridization by defining $\bar{E} = \frac{1}{2}(E_0 + E_1)$. In essence then, dehybridization occurs due to lowering of *s* relative to *p* atomic orbital energies on descending the Group. The extent of valence-conduction band mixing is reflected in the physical size of the unit cell which varies as the trend towards a metallic structure develops. It is therefore assumed that $\Delta H_f \propto a^{-s}$ where a is the cubic-lattice constant and s is a parameter to be determined.

In order to scale the dehybridization process we make reference to tin which is balanced (at ca 270K and one bar) on the diamond (α-tin) $\leftrightarrow$ metallic (β-tin) transition and assume that values of $\bar{E}$ and E_2 close to those of tin represent essentially complete dehybridization. If the ratio $E_2/\bar{E}$ (which increases down the Group) is taken as a measure of dehybridization of the wave function, the effect on crystal energy should be proportional to $(E_2/\bar{E})^2$. We thus arrive at a measure of the energy difference between the tetrahedral AB structure and the metals, $D = 1-b(E_2/\bar{E})^2$, where b is expected to take a value close to that of $(\bar{E}/E_2)^2$ for α-tin. Finally, ΔH_0 and s in Equation 19 are fixed empirically by adjusting for best fit with ΔH_f values of BN, BeO and ZnO.

Discussion of Equation 19

Values of ΔH_f(AB) computed with Equation 19 are accurate to about 10%. Rather more impressive is the success in fitting very small (InSb, InAs, GaSb) as well as much larger heats of formation for crystals with very similar lattice constants. The basic reason for this impressive fit is the explicit inclusion of parameters taking care of dehybridization. Some values are given in Table 11. Finally, we note that if $\bar{E}$ and E_2 take values such that $D \approx 0$,

Table 11. Heats of formation for some $A^N B^{8-N}$ crystals (J. C. Phillips and J. A. van Vechten, *Phys. Rev. B*, **2**, 2147 (1970))

	ΔH (Equation 4)[a]	ΔH (Experimental)[a]
BN	266·1	254·4
AlP	92·9	166·5
GaAs	68·2	71·1
InSb	37·7	30·5
ZnS	176·1	205·9
CdS	149·8	161·9
GaSb	39·7	41·8
ZnSe	166·5	163·2
ZnTe	108·4	117·6
InAs	48·1	58·6

[a] in kJ mol^{-1}.
$s = 3$; $\Delta H_0 = 300$ kJ mol^{-1}; $b = 0{\cdot}0467 = (\bar{E}/E_2)^2$ for grey tin.

Equation 20, then a tetrahedral structure will not be stable. This may explain the occurrence of TlSb and TlBi in the rock salt lattice despite their ionicities which are $\ll$78%. If this is indeed true, it represents a most important exception to the bond-directionality criterion for structure adoption.

5.3.2 The *TS* Term

The influence of the *TS* term on the free energy of solids is felt mainly through temperature. With increase of temperature there is an increase in entropy due, usually, to greater thermal motion of atoms and the consequent possibility of disorder, and sometimes also a change in electronic or magnetic interactions. If there is another structure type of lower free energy that can accommodate the higher entropy term there will be a phase transition, provided that a suitable mechanism exists.

The well-known α(grey) $\rightleftharpoons$ β(white) tin transformation is such a case. β-Tin is the stable form at 13°C or above, but the transformation is rapid only at -50°C or below in the absence of a catalyst. (The conversion, which occurs naturally in very cold climates, is apparently known as 'tin pest'!) The $\beta \rightarrow \alpha$ transition is accompanied by energy evolution of 2·1 kJ mol^{-1}. The entropy term dominates the heat term so that overall the β-form is the more stable at higher temperatures because it accommodates the increased thermal motion more readily than does the α (diamond) lattice with its greater binding energy.

5.3.3 The Graphite–Diamond Transition

To emphasize the importance of kinetic considerations in phase transitions, we briefly consider the graphite-to-diamond transition. Diamond is unstable with respect to graphite at one atmosphere and room temperature, having a positive free-energy difference of 2·88 kJ mol^{-1} which increases to 10·05 kJ mol^{-1} at ca 1200°C at which temperature the transition begins to proceed at an observable rate. This, of course, is a reaction of academic interest only and more people wish to accomplish the reverse transition, graphite to diamond. Since $(\partial G/\partial P)_T = V$, application of sufficient pressure should bring the desired transition within reach. Figure 5-19 shows the relation between diamond and graphite as a function of *P* and *T*. However, even under conditions of very high pressure and temperature (which requires elaborate equipment) the transformation to diamond does not take place. Since the thermodynamic conditions have been met we conclude that kinetic reasons prevent it.

The theory of reaction rates shows that

$$\log(\text{rate}) = \text{const} - \frac{\Delta V^* P}{RT}$$

where ΔV^* is the molar difference in volume between diamond and the transition state and is known to be at least $+10$ ml. The frustrating situation therefore obtains that although the graphite–diamond transition becomes more thermodynamically allowed with increase of pressure the *rate* of the reaction actually decreases and apparently outweighs the thermodynamic requirement.

To cut a very long and involved story short, diamond was synthesized from graphite at very high pressure and temperature in 1954 using catalysts of, for example, nickel or iron compounds. Synthetic diamonds are usually

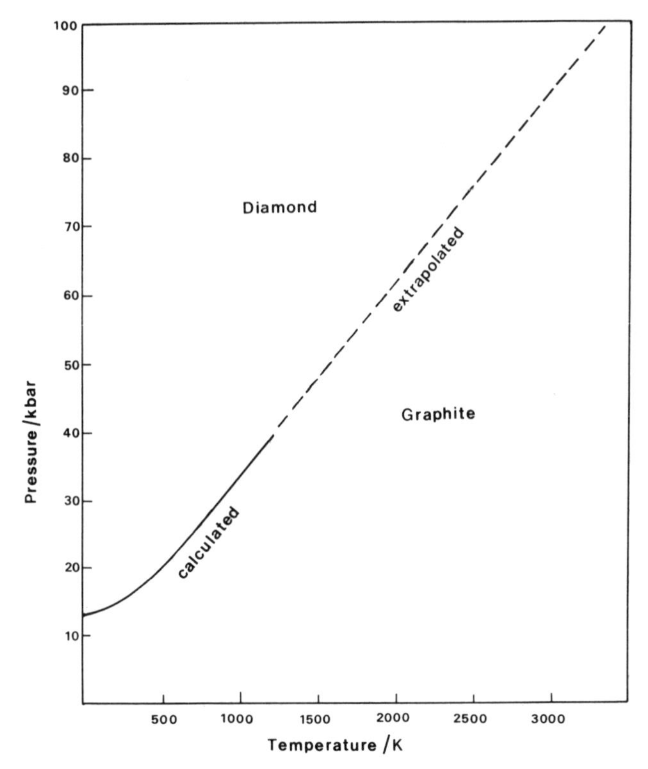

Figure 5-19 Part of the carbon phase diagram. (From H. J. Hall, *J. Chem. Educ.*, **38**, 484 (1961))

very small, often only a few tens of micrometres in size, but quite acceptable for industrial use. Within the last few months (at the time of writing) the first report has appeared of the synthesis (from graphite) of diamonds up to 5 mm in diameter; when cut in 'brilliant' form they were indistinguishable from gem-cut natural diamonds, all of which raises an extremely interesting situation for those in the diamond trade!

5.3.4 Phase Diagrams and their Importance in Understanding Crystal Structure

We live in a world so constructed that we take one bar pressure and ca 290 K as our reference point. Implicit in our whole enquiry into structures of crystals is the assumption that we mean those forms stable at n.t.p., unless otherwise stated. However, very often quite modest changes of temperature or pressure are sufficient to induce transition to another structure and, if our universal reference point happened to be somewhat different from n.t.p., we should find ourselves seeking to explain a partly different set of observations. It is therefore clear that we cannot claim to understand the structure of a solid unless we can also say why it does or does not adopt certain other structures that are in principle available to it (assuming that kinetic restrictions can be circumvented).

The relations between phases in the $T - P$ domain are represented on phase diagrams. If enough were known about the quantitative relations between crystal structure and electron distribution it would be possible to make *a priori* calculations of the stabilities of various polymorphs, but we are not yet in that blissful state. It is therefore of the highest interest to have to hand any means of predicting phase diagrams because, among other things, it leads

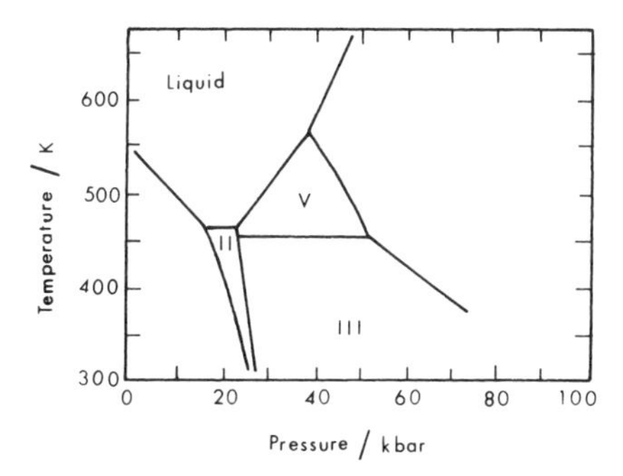

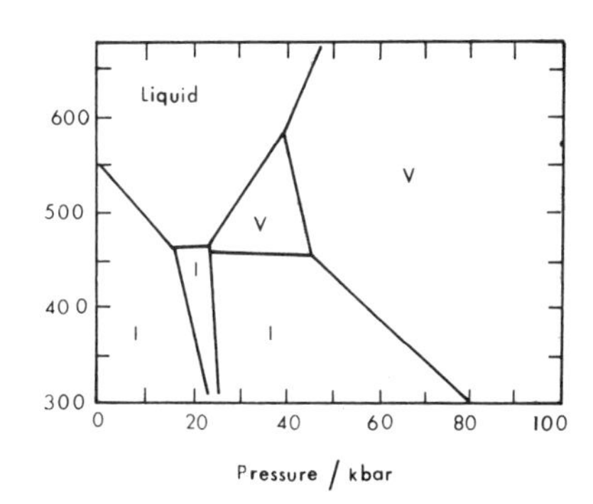

Figure 5-20 The phase diagram for bismuth: (a) observed, (b) calculated. (From L. Kaufman and H. Bernstein, *Computer Calculation of Phase Diagrams*, Academic Press, New York, 1970)

to suggestions for further experiment and reveals problems of relative stability which may not have been recognized before.

Recently some remarkable calculations of phase diagrams have been made by Kaufman and coworkers based upon a piece of work which has lain around in the literature for a long time. Van Laar (1908) proposed that multiphase equilibria could be described by specifying, for each solid phase, the heat of fusion, its melting point, and its heat of mixing with a phase adjacent to it on the phase diagram, an ideal entropy of mixing being assumed. Unfortunately all these quantities are not always accessible to experiment and, in any case, part of the object of the exercise is to *predict* regions of phase stability. To summarize a complex situation—the interested reader is referred to Kaufman and Bernstein (1970) for details—estimates of the unavailable quantities have been made for many phases by using heats of mixing and various experimental data on known phase transitions. The elaborate thermodynamic equations are solved by computer and the phase diagram printed out.

The work is limited, so far, to elemental metals and to their binary and ternary alloys. Figure 5-20 shows the computed phase diagram for bismuth and compares it with the experimental one. The agreement is remarkable. To give but one other example, a new, high-pressure, phase of thallium was predicted and its existence later confirmed by experiment, see Figure 5-21. The extraordinary extent to which this method has been developed is emphasized by the fact that enthalpy differences between the three phases of thallium (b.c.c., h.c.p. and f.c.c.) are only 7·1 to 0·7 kJ mol^{-1}.

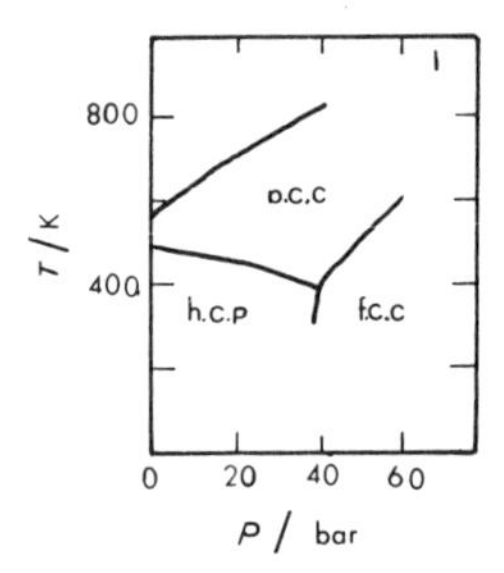

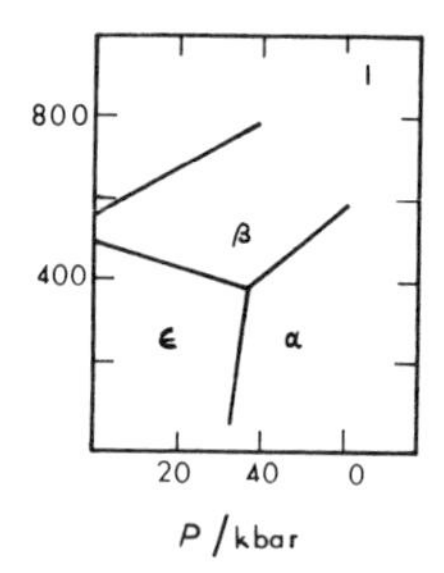

Figure 5-21 The phase diagram for thallium: (a) observed, (b) calculated. (This and Figure 5-20, are from L. Kaufman and H. Bernstein, *Computer Calculation of Phase Diagrams*, Academic Press, New York, 1970)

5.4 ELECTRON CONFIGURATION

In Section 1.6 some very basic points about the relation between electron configuration and type of structure formed were noted. We now look in some detail at the structural consequences of four main types of special electron configuration.

5.4.1 The Inert Pair

A prominent feature of the chemistry of the B-metals is their tendency to adopt an $N - 2$ valence state, where N is the Group number, in addition to the expected Group valence. The effect is especially pronounced for the heavier elements Tl, Pb and Bi, and is of importance for their congeners In, Sn and Sb, as well as for Ga. It is associated with a tendency for the *ns* pair of valence electrons to remain associated with the metal nucleus rather than take part in covalent bonding, and is usually referred to as the 'inert'-pair effect. Coming after the *d*-block metals, the B-metals have firmly held *ns*- and *np*-electrons, and the *ns* are the more firmly held of the two due to their greater penetration.

It should be understood that their inertness is due to their *s* character, not because they are paired. Although the inert pair is usually thought of as occupying a 5*s*- or 6*s*-orbital it does, in fact, commonly appear with directional character and must then be in hybrid orbitals. In that case, it can and does have pronounced stereochemical effects, somewhat analogous to those of

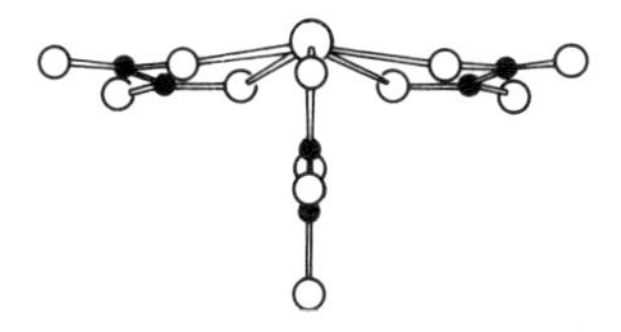

○ Sb ○ O ● C **Figure 5-22** The Structure of $[Sb(oxalate)_3]^{3-}$

'lone' pairs in molecular situations. Although we are concerned principally with lattice compounds, we note that stereochemical effects due to the inert pair are present also in molecular situations. In $K_3[Sb(oxalate)_3]$, instead of the simple octahedral MO_6 coordination arrangement found in analogous complexes such as $K_3[Al(oxalate)_3]$, a highly distorted structure is adopted

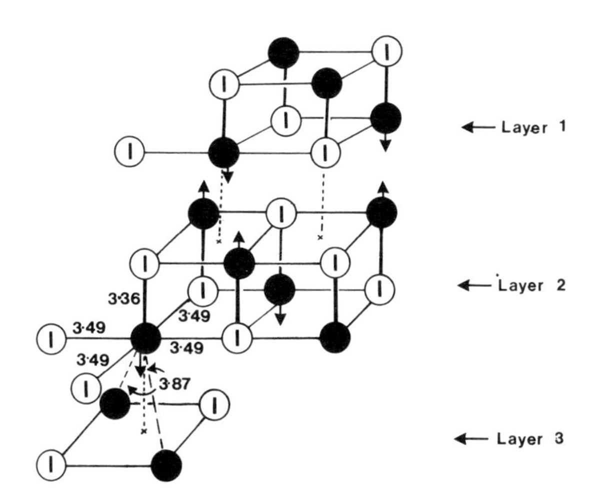

Figure 5-23 The structure of thallous iodide, TlI

by the anion, Figure 5-22. In contrast, the complex halides $[SbX_6]^{3-}$ and $[TeX_6]^{2-}$ show no stereochemical evidence of the inert pair, which must therefore be in a pure *s*-type orbital.

We consider a small group of compounds as typical of inert-pair effects. Thallium iodide, **TlI**, has the structure shown in Figure 5-23. It is a layer structure composed of pairs of thallium and iodine atoms so close (3·36 Å) that they can almost be regarded as molecules. Each Tl atom has four other near neighbours in a square plane at 3·49 Å with the sixth octahedral position occupied by an inert pair in an *sp* hybrid orbital, indicated by arrows in the figure. In order to accommodate these pairs the double layers are offset relative to each other so that the electron pairs are directed between the atoms in the next layer. In contrast the unstable red form of TlI has the CsI structure with a Tl—I distance of 3·64 Å; the vapour-state bond length of the molecule TlI is 2·87 Å. From this it is evident that in TlI there is a high degree of covalent bonding between Tl and its nearest iodine neighbour to which it is bound by the other lobe of the *sp* hybrid. Bonding to the four iodines at 3·49 Å is achieved by overlap of filled *p*-orbitals on iodine with empty *p*-orbitals on the metal.

The stereochemical effect of the inert pair in **tin** and **lead** compounds shows some interesting variations. Consider first the structures of their chalcogenides, ME.

E	O	S	Se	Te
Sn	T	D	D	R
Pb	T + S	R	R	R

T = tetragonal layer structure; S = special structure; R = rock salt; D = orthorhombically distorted rock salt structure.

There is a clear preference for a high-symmetry structure (NaCl) shown by the heavier element combinations; this is to be understood in terms of the trend towards metallic structures down the Group. The inert pair does not exert a stereochemical effect in this group of solids. In a Mooser-Pearson plot they fall in the directed-bonding domain rather than the ionic area which might have been suggested by their coordination. Krebs has described the bonding in PbS in terms of overlap of the two 6*p*-electrons on lead and four 3*p*-electrons on sulphur, a total of six valence electrons per atom pair. Since this *p*-type band is not filled, electronic conduction is expected. Although PbS behaves in some respects as an intermetallic compound (lustrous black appearance, low water solubility) its electronic conduction is down in the semiconductor range because the high charge on sulphur and its low principal quantum number cause strong localization of electrons on it. PbTe and SnTe can really be described as alloys; it is therefore no surprise that substitution can take place leading to $Pb_{1-x}Cd_xTe$, $Pb_{1-y}In_yTe$ (and similarly for SnTe) where x and y can take values of up to 0·2 and 0·35 respectively. In these alloys *p*-electrons on Cd or In clearly take the place of those of lead, and full electronic conductivity is developed.

In SnO and the red form of PbO the stereochemical effect of the inert pair is fully developed, see Figure 5-24, and a layer structure results in which the directed non-bonding pairs are accommodated in much the same way as in TlI.

It differs from the TlI structure because of the need to accommodate tetrahedral bonds to oxygen. The SnO structure may reasonably be regarded as equivalent to the CsCl lattice distorted tetragonally by the directed inert pair. Both SnO and PbO fall in a domain of the Mooser–Pearson plots which suggests fairly non-directional bonding, lending some support to the idea

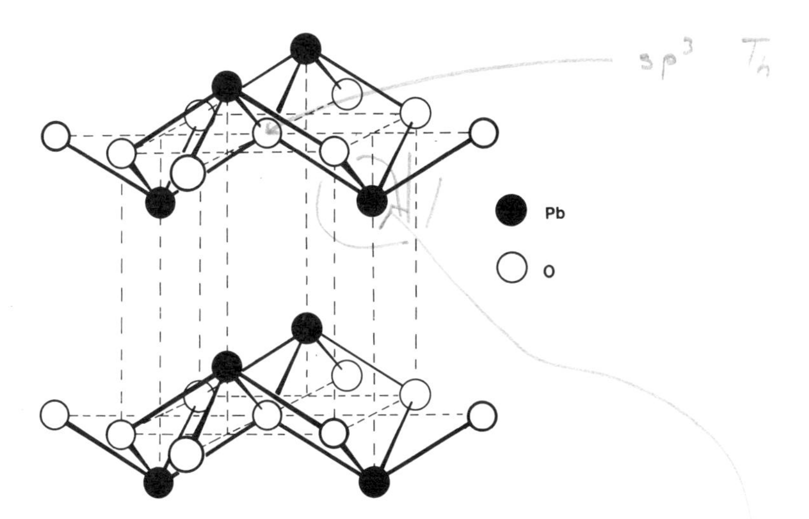

Figure 5-24 The structure of red PbO. (From A. F. Wells, *Structural Inorganic Chemistry*, The Clarendon Press, Oxford, 1962)

that we can regard this structure, crudely, as formed to accommodate unsymmetrical ions. Oxygen is sp^3 hybridized, having four tetrahedrally disposed metal neighbours. The strong directional effect of the inert pair is achieved by hybridization with either d_z^2 or p_z, or an admixture of both. Bonds to the four oxygen neighbours of each metal atom are formed by use of *d–p* hybrids.†

Yellow PbO has a chain structure, Figure 5-25, which represents an alternative and almost equally favourable way of satisfying the atomic bonding requirements. As the O—Pb—O angle is 90°, bonding is via two $6p$-orbitals on lead with oxygen sp^2 hybridized (Pb—O—Pb angle = 120°), although hybridization with d-orbitals could also occur without altering bond angles.

The SnS structure can now be seen as intermediate between those of SnO and SnTe; it is usually described as distorted rock salt, Figure 5-26, although

† In terms of group theory, the requirement for Pb—O σ bonding is $A_1 + B_1 + E$. Available orbitals are of symmetry $A_1(s, p_z, d_z^2)$, $B_1(d_{x^2-y^2})$ and $E(p_{x,y}; d_{xy,yz})$.

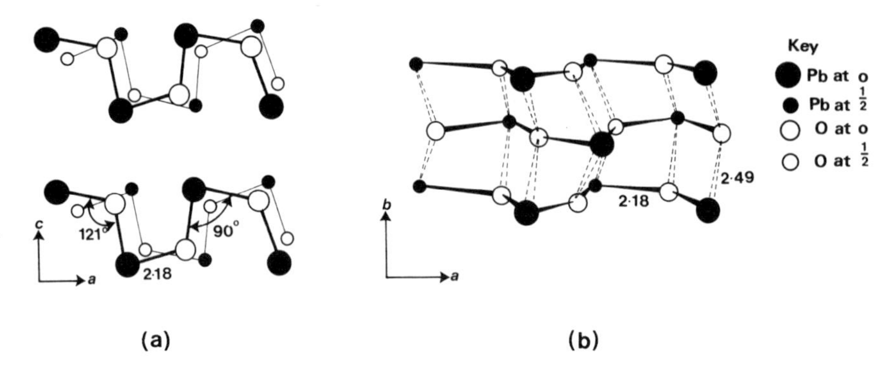

Figure 5-25 The yellow PbO structure. Infinite chains parallel to the *a*-axis form layers as shown in (b). These layers are stacked as shown in (a). Note that the interlayer contact is between lead atoms only: oxygen is the 'filling' in the lead 'sandwich' (Dickens (1965))

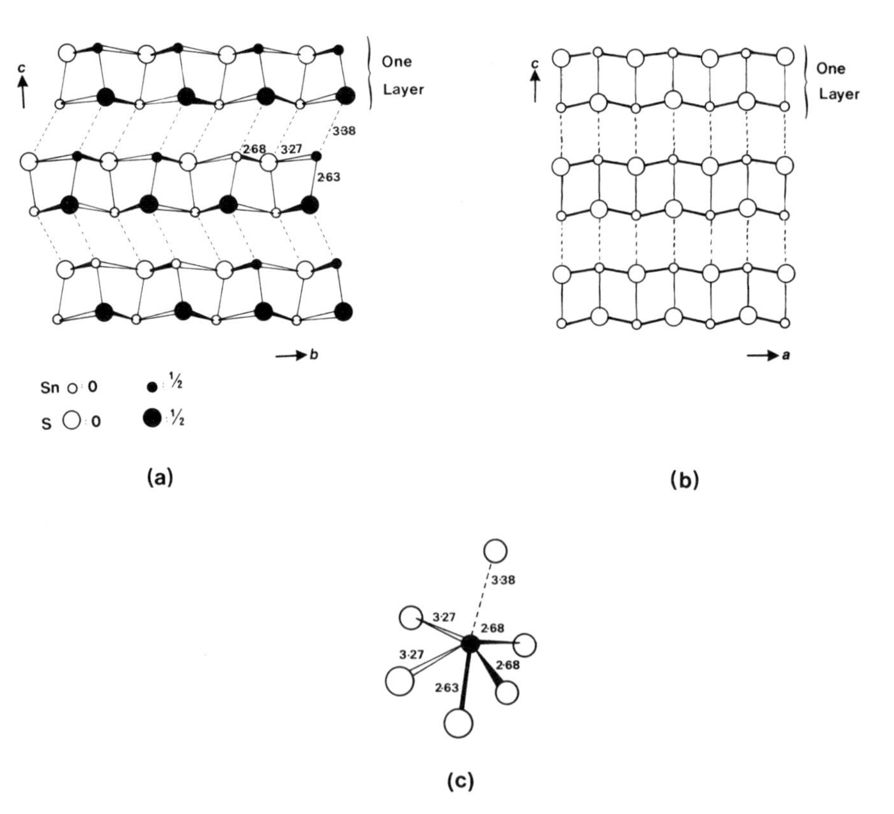

Figure 5-26 The structure of SnS. (a) and (b) show projections on the (*bc*) and (*ac*) planes respectively. (c) shows the coordination about one tin atom

it is more readily visualized by noting a formal relation to the black-phosphorus layer structure with alternate atoms replaced by tin or sulphur (see Figure 7-8) although the layer stacking sequence is different. The coordination sphere of tin is highly asymmetrical. Pure *p* overlap as in PbS is not achieved, and neither is sp^3 hybridization at S. The observed structure is to be seen as a compromise, the inert pair achieving *some* directional character by *p* and/or *d* hybridization, whilst Sn—S bonds achieve sufficient directionality to place the structure in the directed-bonding domain of Mooser–Pearson diagrams.

Many further cases of the stereochemically active inert pair are to be found. For example, $SnCl_2$, which has a chain structure not unlike that of yellow PbO; $K_2SnCl_4.H_2O$ in which pyramidal $SnCl_3^-$ ions with a directed inert pair are found; $K_2[SbF_5]$ which has a square-pyramidal anion; Bi_2O_3 of which one polymorph has Bi coordinated to six oxygens at the corners of a *cube* leaving two diagonally opposed corners 'unoccupied'.

5.4.2 d^0 Ions

Of all ions, those with configuration d^0 should exhibit the simplest, most classical, structural behaviour. This is very largely borne out in practice by the mainly ionic lattice compounds of the Group Ia and IIa metals, compounds of high symmetry and coordination number which can be understood principally in terms of the packing of ions and space filling. With the substitution of polyatomic for monatomic anions, related distorted structures are formed (see Section 3.5.5) which can still be understood along much the same lines. The object of this short section is to draw together some recent data which suggest that, under some conditions, these typical A-metals also show a tendency to ion-pair formation, and can occur in unusual coordination environments; see Truter (1970).

A structure of particular interest in this connection is $\mathbf{CaCO_3.6H_2O}$. Contrary to reasonable expectation (see Section 6.3.2) it is not composed of $Ca(H_2O)_6^{2+}$ and CO_3^{2-} but of discrete ion pairs $Ca^{2+}CO_3^{2-}$ solvated by a sheath of eighteen water molecules. Calcium is eight-coordinate having a bidentate anion and six water molecules around it as shown in Figure 5-27. $MgSO_4.4H_2O$ consists of two Mg^{2+} ions bridged by two sulphate groups, the double ion-pair complex being similarly solvated.

Complexes related to biological systems. Further examples of ion pairing are known in which a complex multidentate ligand is used to solvate the pair. Apart from their intrinsic interest, these complexes are of importance in connection with biological systems: the ability to complex Na^+ rather than K^+, Ca^{2+} rather than Sr^{2+}, is relevant to ion balance in cells and to radiation protection.

A cyclic polyether rejoicing in the name 'dibenzo-18-crown-6' (2, 3, 11, 12-dibenzo-1, 4, 7, 10, 13, 16-hexaoxocyclooctadeca-2, 11-diene to the organic

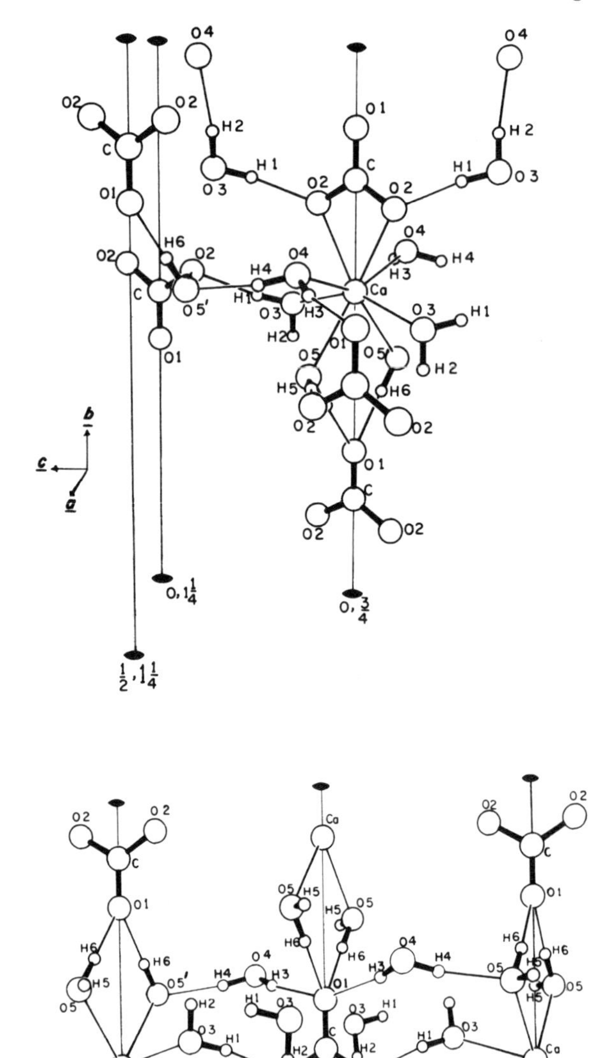

Figure 5-27 The structure of $CaCO_3.6H_2O$. The top view shows the coordination at Ca^{2+}; the lower one details the environment of CO_3^{2-}. (Dickens and Brown (1970))

chemist) forms a series of complexes with many different metals. In $(Rb_{0\cdot55}Na_{0\cdot45})NCS, 7{\cdot}5L$ where L is the above polyether, Rb^+NCS^- ion pairs are found solvated by the ligand which has altered its shape in order to wrap around the Rb^+ ion and coordinate with all six oxygens, Figure 5-28. With NaBr the same ligand forms a complex which also contains water. In it sodium is found in two different environments. (*a*) As in Figure 5-28 an ion pair Na^+Br^- is complexed by the six ether oxygens and a water molecule is coordinated to Na^+ opposite to Br^-. (*b*) Six ether oxygens coordinate as in Figure 5-28 but now two water molecules occupy the positions above and below, with a free bromide in the lattice.

A further solvated ion pair, $Na^+ClO_4^-$, is found in a complex with bis-[N,N′-ethylenebis(salicylideneiminato)copper(II)], Figure 5-29, which thereby renders sodium perchlorate soluble in xylene–ethanol.

Coordination arrangements that are electrostatically preferred for isolated coordination polyhedra but which are not usually found in binary solids may be adopted or approached in complexes. $Cs^+[Y(CF_3COCHCOCF_3)_4]^-$ (both Cs^+ and Y^{3+} are d^0 ions) contains Y^{3+} with the eight oxygen donor atoms disposed at the corners of a dodecahedron (*not* a cube), an arrangement which is more favourable than cubic electrostatically and allows more efficient packing. Cs^+ is coordinated to eight fluorine atoms, six at 3·50, two at 3·75 Å. On the other hand, the antibiotic macrotetrolide known as nonactin, disposes eight oxygen atoms around K^+ at the corners of a *cube* as shown schematically below. In this case, the possible orientations of the ligand must

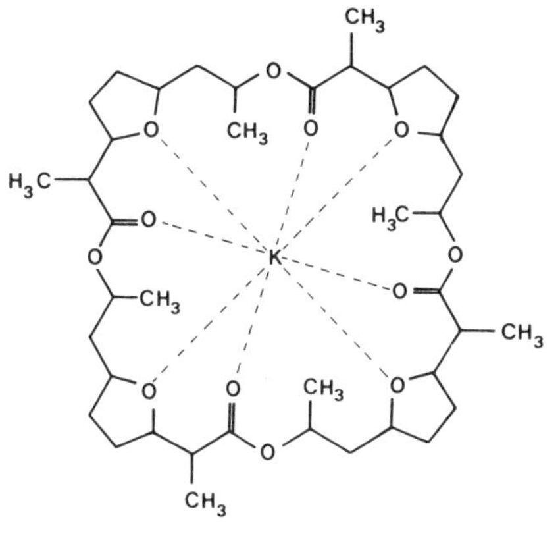

be responsible for an electrostatically unfavoured arrangement. There are other complexes in which quite unsymmetrical arrangements are found, so that a range of cation–oxygen distances is formed.

Taking a more complex example, it has recently been shown that the enzyme thermolysin, chiefly notable for its high thermal stability, owes that stability to the presence of two Ca^{2+} ions which are deep in the interior of the structure. They form strong electrostatic bonds with oxygens of carboxyl and carbonyl groups, thereby strengthening the protein structure and making it less susceptible to denaturing.

We see that the coordination behaviour of d^0 ions is not as simple as would appear at first sight. Even the simple principle that the larger cations have more neighbours seems to take a knock occasionally, as in the salts of N-3 oxy 5-bromo-6-methyl uracil. The potassium salt has ten oxygen neighbours some of which are shared, but the rubidium cation has only *six* (unshared). To restore confidence in our understanding of some of the basic factors in d^0-ion crystal chemistry we must in all fairness declaim that the

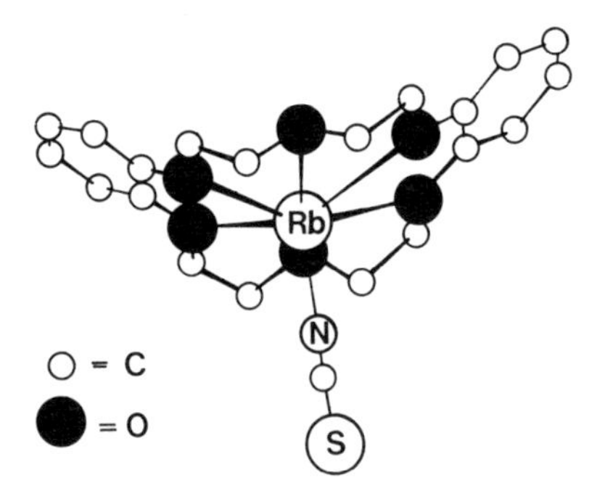

Figure 5-28 Structure of RbNCS, dibenzo-18-crown-6, showing the ligand with the six oxygen atoms co-planar and nearly equidistant from rubidium. (Truter (1970))

special coordination arrangements mentioned above are closely associated with the special stereochemical properties of the ligands, including possible ways in which they can stabilize themselves in certain forms via hydrogen bonding. (This applies especially to polypeptides.) On the other hand, we would give formation of ion pairs some prominence as probably being one of the regular features of d^0-ion chemistry. Of course, the sceptic could always point out that given a sodium ion complexed as in Figure 5-28, and faced with the problem of how to fit its accompanying anion into a crystal, where on earth would you expect to find it other than in the position of highest positive potential!

Having raised the idea of ion pairs in solids for situations in which it makes some sense, i.e. when the pair can be 'solvated' by water or a fancy ligand, let us at least ask whether a tendency towards pair formation is evident in binary compounds; the **hydroxides of Group Ia** are particularly instructive and show a range of structures which are worth considering.

LiOH adopts the anti-type of the layer structure of red PbO, Figure 5-24, accommodating the requirements of both lithium and hydroxyl. Lithium is commonly found in tetrahedral coordination because covalent overlap is more effectively achieved through use of sp^3 hybrids than through overlap of the unhybridized *p*-orbitals which must be used in high-symmetry octahedral coordination (as in lithium halides). Lithium does not, repeat *not*, occur in tetrahedral coordination due to its 'small size and the corresponding radius ratio requirement'; the radii of Li^+ and O^{2-} are very nearly equal if we accept the 'experimental' radii, Table 5, Chapter 2, and even if we insist on being traditional in our choice of radii (Table 6, Chapter 2) values of Li^+/O^{2-} well above the octahedral coordination minimum of 0·414 are still obtained. Thus,

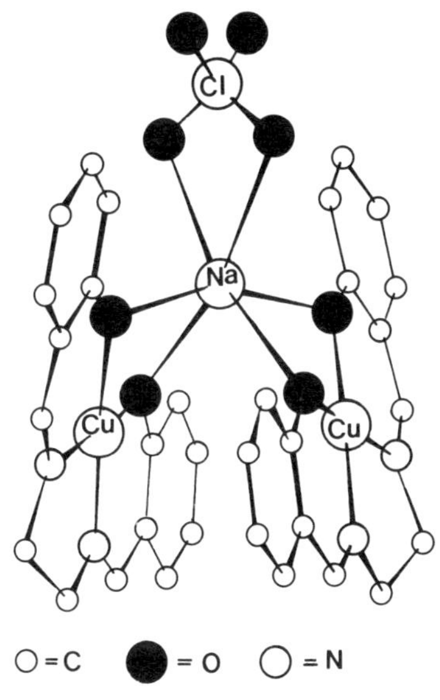

Figure 5-29 The $NaClO_4$ complex with bis-[N,N-ethylenebis(salicylidene-iminato)Cu^{II}]. (Truter (1970))

the preference of lithium for tetrahedral coordination is associated with the necessity of placing it in an environment most favourable to covalent bonding. In its halides, LiX, the available tetrahedrally coordinate AB structures (wurtzite and zinc blende) require bonding more highly directional than lithium can supply because the Phillips ionicity of its halides is $> 78\%$. These halides therefore adopt the rock salt structure in which the covalency is taken care of by pure unhybridized *p*-orbital overlap. However, in Li_2O the different atomic proportions allow selection of a structure in which lithium can be tetrahedrally coordinated using sp^3 hybrids, and oxygen can adopt a higher coordination number (8), viz. the anti-fluorite structure (p. 60). Just how oxygen can be eight-coordinate and bond covalently to lithium requires

a little comment; we follow Krebs's description of Be_2C, transposing it for our purpose.

A multicentre molecular-orbital scheme is constructed using (*a*) the *s*- and *p*-orbitals of oxygen together with (*b*) one sp^3-hybrid function from each of the eight lithium atoms that surround oxygen cubically (designated ψ_1 to ψ_8). Four functions can then be written as follows,

$$\begin{aligned}
\psi_1 &= 2^{-1/2}N_0'\psi_{0,s} + N_{Li}8^{-1/2}(\psi_1 + \psi_2 + \psi_3 + \psi_4 + \psi_5 + \psi_6 + \psi_7 + \psi_8) \\
\psi_2 &= 2^{-1/2}N_0\psi_{0,p_x} + N_{Li}8^{-1/2}(\psi_1 + \psi_2 + \psi_3 + \psi_4 - \psi_5 - \psi_6 - \psi_7 - \psi_8) \\
\psi_3 &= 2^{-1/2}N_0\psi_{0,p_y} + N_{Li}8^{-1/2}(-\psi_1 + \psi_2 + \psi_3 - \psi_4 - \psi_5 + \psi_6 + \psi_7 - \psi_8) \\
\psi_4 &= 2^{-1/2}N_0\psi_{0,p_z} + N_{Li}8^{-1/2}(\psi_1 + \psi_2 - \psi_3 - \psi_4 + \psi_5 + \psi_6 - \psi_7 - \psi_8)
\end{aligned}$$

which can accept eight valence electrons (six from oxygen, one quarter from each lithium).

While we are considering anti-fluorite, it is worth accounting for the bonding in **fluorite**. Although compounds adopting the fluorite structure are usually highly ionic there is nevertheless some covalent bonding; fluorine being tetrahedrally coordinate uses sp^3 hybrids whilst calcium bonds to the cubically disposed fluorines via d^3s hybrids, Figure 5-30, which are directed

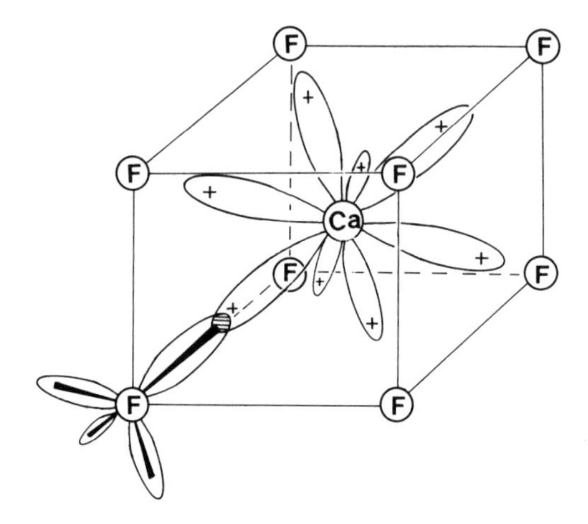

Figure 5-30 The bonding in fluorite. Fluorine uses sp^3 hybrids and calcium d^3s (see also Figure 8-7). Only one sp^3 hybrid is shown, for clarity

at the corners of the cube where the *d*-orbitals are of the t_{2g} set (the e_g-orbitals are directed to the centres of the cube faces). In CaF_2 itself, the covalent contribution is small, which is to be expected in view of the rather high energy of *d*-orbitals on calcium.

Returning to LiOH, the reasons for adoption of the red PbO structure become clearer. The hydrogens are accommodated between layers (there is no hydrogen bonding), as were the inert pairs of PbO, and lithium uses sp^3 hybrids. Oxygen has available s, p_x and p_y; it is probable that p_x and p_y are used for sideways overlap with the sp^3 functions with a non-bonding pair in an s/p_z hybrid which gives it some slight directional effect, the other lobe of the function being used for bonding to hydrogen.

NaOH, at high temperature, has the rock salt structure in which, presumably, thermal motion gives OH^- ions an effectively spherical appearance. At room temperature it has the TlI layer structure (p. 131). Why? A distorted rock salt structure would seem more reasonable, like that of NaCN or CaC_2 in which the linear anions are aligned with one axis. It is possible that in the TlI form of NaOH there is a tendency to ion-pair formation and that the packing of linear ion pairs in this form yields a lower free energy than would the CaC_2 structure because the latter would have associated with it a higher entropy term due to random orientation of OH^- ions parallel and antiparallel to the c-axis. We emphasize the curious coordination around sodium in NaOH; i.e. an actahedral arrangement with the sixth ligand missing but replaced by two next-nearest neighbours at much larger distances (3·70 Å)—not a typical d^0 ionic environment. The five Na—O distances in NaOH are said to be equal, but this is an early structure and needs reinvestigation.

KOH and RbOH are isostructural but different from NaOH, although they all have the rock salt structure at higher temperature. In KOH, potassium has six nearest oxygen neighbours but the structure is considerably distorted in such a way as to allow formation of a zig-zag hydrogen-bonded chain, closely similar to that of HF, Figure 5-31 (Hamilton and Ibers, 1968).

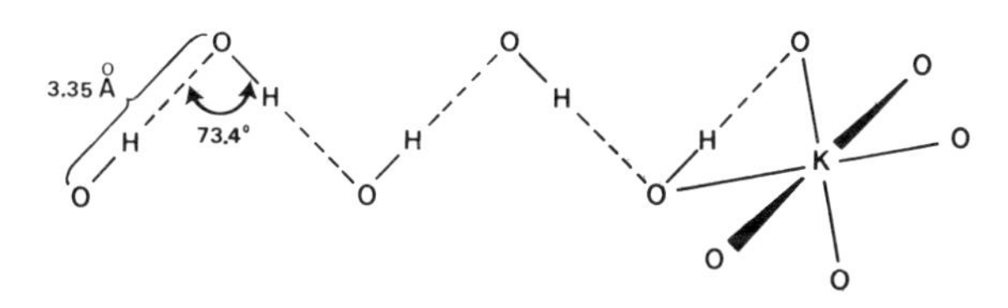

Figure 5-31 The H-bonding scheme in solid KOH. (From Hamilton and Ibers. *Hydrogen Bonding in Solids*, copyright © 1968 by W. A. Benjamin, Inc., Menlo Park (Calif.))

The hydrogen bonding in this case is certainly weak (compared with that in hydrates, for example) but it apparently contributes sufficient Madelung energy to stabilize this structure in preference to that of TlI.

In summary, we see that although the crystal chemistry of d^0 ions is apparently simple, this appearance is deceptive and a full account of their structural behaviour requires a surprising range of concepts.

5.4.3 d^{10} Systems

The Group Ib metals in their +1 oxidation state and the Group IIb metals in their +2 state are all accorded the formal electron configuration $(n - 1)d^{10}$, having been relieved of their only valence electrons, those in *ns*-orbitals. Since the *d* shell is filled and spherically symmetric it is reasonable to expect that the structural chemistry of Cu^{I}, Ag^{I}, Au^{I} and of Zn^{II}, Cd^{II} and Hg^{II} would be straightforward. In fact these d^{10} systems are chiefly notable for showing a pronounced tendency towards linear two-coordination which becomes dominant for Au^{I} and Hg^{II}, just those elements for which a tendency towards metallic structures and high coordination number might be predicted. The key to understanding these effects is to realize that the *ns*-orbital energies are close to those of $(n - 1)d$ and *np*, allowing the possibility of forming strongly directional *s*/*p* or *s*/*d* hybrids. Although the *d* shell is completed at the end of each transition series, the ionization potentials are not so high that it cannot be reopened to d^9. The strength of the covalent bonds so created evidently stabilizes structures in forms of lower energy than can be achieved by dehybridized bonding of a semimetallic nature. The ultimate reason for this behaviour lies in the strength with which *ns*-electrons are held in these atoms (due to orbital penetration) and, to this extent, a common cause underlies the structural features of both these elements and those which exhibit inert-pair effects.

We illustrate the occurrence of these d^{10} systems in linear coordination by considering a selection of the simpler compounds, although the same principles apply throughout their structural chemistry (which is highly complex in the case of mercury; for further detail see Grdenič, 1965).

The halides (Cl, Br, I) of Cu^{I}, Ag^{I}, Zn^{II} and Cd^{II}, all adopt structures in which the metal is in regular octahedral or tetrahedral coordination; see Table 12. HgX_2 (X = Cl, Br, I) forms crystals with linear X—Hg—X molecules which have bond lengths close to those found for the same species in the vapour phase. For HgI_2 the molecular crystal structure is unstable with respect to red HgI_2 which has a layer structure (p. 242) with the metal tetrahedrally coordinated. AuI forms a chain, Au—I—Au—I—Au— with I—Au—I angles of 180° and Au—I—Au angles 72°. AuCN and AgSCN also occur as chains with the metal in linear two-coordination. Thus,

Au—CN—Au—CN— and Ag—S—C—N—Ag—S—C—N (N—Ag—N angle 103·8°)

The oxides show perhaps the most varied behaviour. Cu_2O and Ag_2O are isostructural; the structure is shown in Figure 5-32. Note that there are two

Table 12. Structures of some compounds of d^{10} systems

Tetrahedral (wurtzite or zinc blende)	CuCl, CuBr, CuI, AgI, ZnO, ZnS, CdS
Octahedral (rock salt)	AgCl, AgBr, CdO
Chains with M in two-coordination	AuI, AuCN, AgSCN, HgO, HgS

interpenetrating lattices in it with no primary bonds between them†. The basic reason for this (Wells, 1962) is that the length of the —O—Cu—O— unit leads to a very open structure in which there is room for two such lattices. A valuable comparison can be made here with cristobalite, SiO_2; the —Si—O—Si— unit length is much shorter than that of the corresponding O—Cu—O unit in Cu_2O and, although space is inefficiently occupied, there is not room for a further lattice.

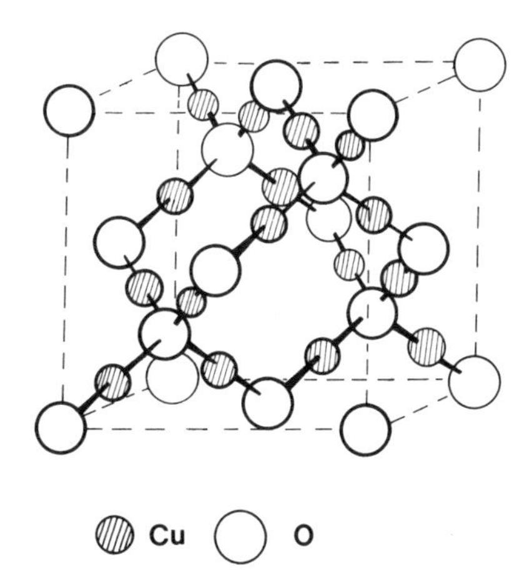

Figure 5-32 The cuprite (Cu_2O) structure. (A. F. Wells, *Structural Inorganic Chemistry*, The Clarendon Press, Oxford, 1962.) In the crystal *two* such structures interpenetrate, without forming primary bonds to each other

In Cu_2O, oxygen (sp^3) is tetrahedrally coordinated to four copper atoms which each have two oxygen nearest neighbours. The *anti*-cuprite form is adopted by $Zn(CN)_2$ and $Cd(CN)_2$, which then has the *metal* in tetrahedral coordination. In Group IIb the oxide structures vary from wurtzite (ZnO) to rock salt (CdO) to two specialized structures for HgO, characterized by the presence of spiral chains. One form of HgO is isostructural with the stable form of HgS which occurs as a bright-red mineral, cinnabarite, in which infinite chains spiral along the *c*-axis. Mercury is linearly coordinated to two

† The record appears to be held by $[Cu\{NC(CH_2)_3CN\}_2]NO_3$ which has no less than *six* interpenetrating lattices.

sulphur atoms and the Hg—S—Hg angle is 105° (not far off tetrahedral); Figure 5-33.

From this brief survey it is clear that there is a pronounced tendency towards two-coordination for these d^{10} systems, the most marked examples being Au^{I} and Hg^{II}. As indicated above, an explanation is sought in terms of hybridization of *ns* with *np* and $(n-1)d$ orbitals (Orgel, 1958). Consider

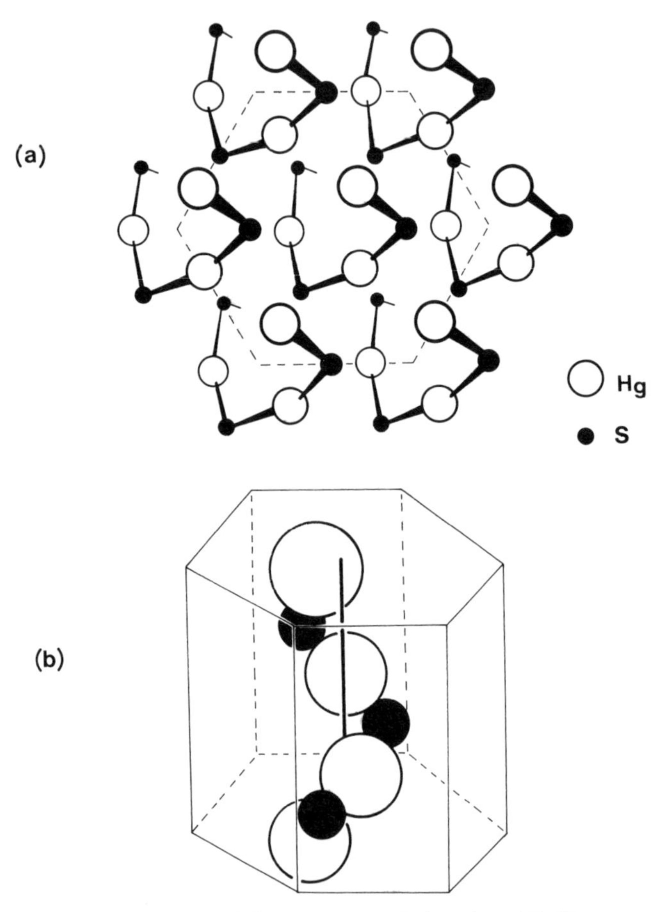

Figure 5-33 The structure of cinnabar (HgS)

formation of a d_{z^2}–s hybrid (Figure 5-34) giving linear combinations $\psi_+ = 2^{-1/2}(d_{z^2} + s)$ and $\psi_- = 2^{-1/2}(d_{z^2} - s)$. Further admixture of p_z allows construction of two hybrid orbitals $2^{-1/2}(\psi_+ \pm p_z)$, also shown in the figure, which are ideal for formation of the two co-linear covalent σ bonds by overlap with orbitals on ligands. The extent of d/p mixing doubtless varies from metal to metal but the basic explanation is believed to hold.

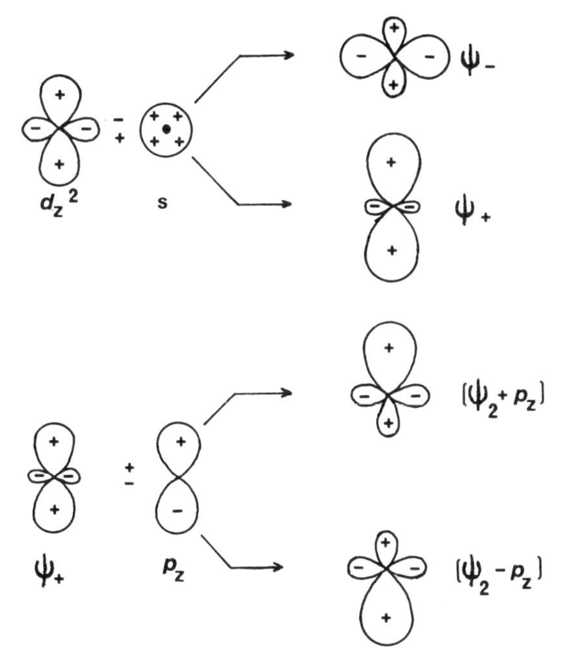

Figure 5-34 Formation of hybrid orbitals suitable for linear bonding in d^{10} systems. In each sketch, the z-axis is vertical and the actual orbital is generated by rotating about z. (After Cotton and Wilkinson (1972))

5.4.4 d^{1-9} Systems

d-Electron configuration profoundly affects the type of structure adopted by a solid, including possible distorted variant standard structures, and a wide range of physical properties. Of these, electrical conductivity and magnetic properties are two of the most important indicators of the nature of the bonding; indeed the band scheme (or schemes) compatible with a particular structure type, and magnetic ordering (where it exists), probably exerts a stabilizing influence in favour of adoption or retention of a structure. We find as a general rule that binary transition metal compounds commonly adopt structures that are compatible with some flexibility as shown, for example, in axial ratios. The rutile and NiAs structures are classic examples of this effect, both encompassing phases with a surprising variety of physical characteristics. Details, and often quite major features, of *d*-electron effects in solids continue to be a topic of substantial research, both theoretical and experimental, much of it of great complexity. We shall only take a few fairly clearcut cases, but hope to include in them a sufficient variety so that the interested reader can go on to more detailed accounts with some basic understanding.

d^n *Configurations*

In order to understand the effect of *d*-electrons on structure and properties, we first need to establish some basic results of crystal field (CF) theory. We do this in a descriptive and qualitative manner whilst emphasizing that formal and more complete treatments will be found in the bibliography (Cotton and Wilkinson, 1972; Kettle, 1971) indeed we only delve into this aspect of standard valence theory here for completeness and for the convenience of the reader who may not have the principal results at his or her finger-tips. CF theory is based upon a purely ionic model and cannot therefore yield quantitatively satisfying results. Its qualitative predictions are quite adequate for our present needs. The modifications necessary to accommodate covalent interaction and achieve quantitatively correct results may be found via the bibliography.

In a free gaseous transition-metal atom or ion the five *d*-orbitals all have the same energy, i.e. they are degenerate. Consider now the effect of bringing six anionic ligands (ions such as F^-, Cl^-, etc., or polar molecules such as NH_3, H_2O, PR_3, etc.) from infinity to an equilibrium bonding distance from the metal ion. From the shapes of the *d*-orbitals and the orientation of their lobes it is evident that there will be substantial electrostatic repulsion between electrons in d_{z^2} and $d_{x^2-y^2}$ orbitals and the charged ligands. In other words, they will be destabilized with respect to their free-ion energies. The three orbitals, d_{xy}, d_{yz} and d_{zx} will also be destabilized but not so much as the other two, as they point between the approaching ligands. The net result is a splitting into two sets known as t_{2g} (or d_ε) and e_g (or d_γ) respectively.†

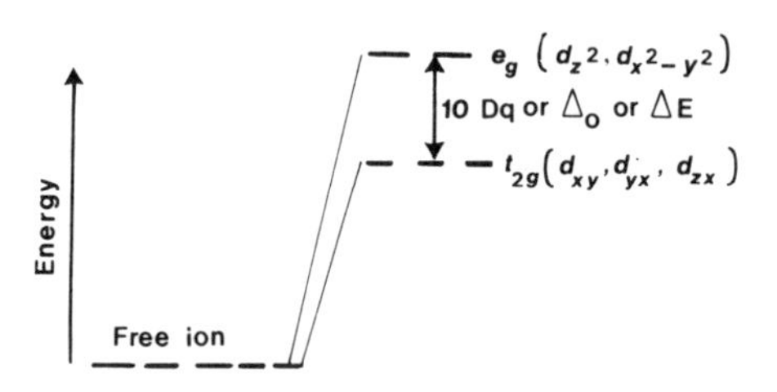

The splitting is similar, though smaller, for tetrahedral coordination ($\Delta_{tet} = \frac{4}{9}\Delta_{oct}$) and there is a crossover of *t* and *e*-symmetry labels with the '*g*' being dropped for tetrahedral geometry since it has no inversion centre. These relations and several others of importance are shown in Figure 5-35. Crystal field theory does not give good *quantitative* agreement with observations relating to relative orbital energies. The magnitude of Δ_o is strongly dependent upon the amount of covalent overlap in the M-ligand bonds.

† These are symmetry labels. *t* indicates a triply degenerate state, *e* one that is doubly degenerate, whilst *g* stands for the German *gerade* (≡even) showing that the *d*-orbitals are symmetric with respect to inversion through their centre points. Cf. *p*-orbitals are *ungerade* (≡uneven) due to their different phase properties.

From the Pauli exclusion principle, two electrons in the same orbital must have antiparallel spins. One of Hund's rules requires that given a set of orbitals of equivalent energy the spin configuration of lowest energy is that with maximum multiplicity; in plain English, the orbitals will be singly

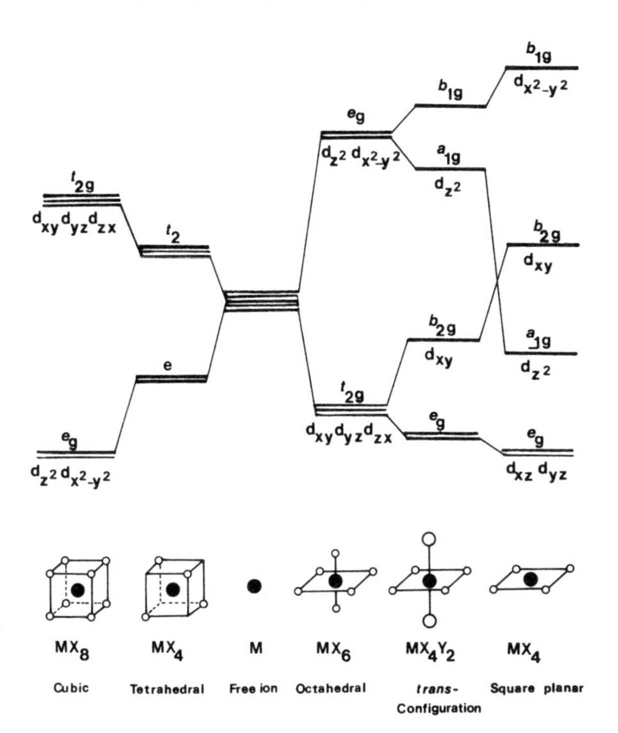

Figure 5-35 Term splitting of a *d*-electron in situations of various symmetry types. (From *Fundamentals of Inorganic Crystal Chemistry* by H. Krebs. © 1968 McGraw-Hill Book Company (UK) Ltd. Used with permission)

occupied with electrons having parallel spins. These rules govern the filling of the *d*-orbital manifolds of Figure 5-35 just as they do for other situations. Consider the octahedral field case for d^3.

Energy ↑ — — e_g — —

Δ

↑ ↑ ↑ t_{2g} ⇅ ↑ –

(*a*) Allowed (*b*) Not allowed

Electrons will enter the t_{2g}-orbitals first until the d^3 configuration shown in (*a*) is attained. A situation such as (*b*) violates Hund's rule. A fourth *d*-electron can *either* enter one of the pair of e_g-orbitals *or* it can spin-pair as shown in (*d*) below, depending upon the size of Δ compared with the spin-pairing energy P.

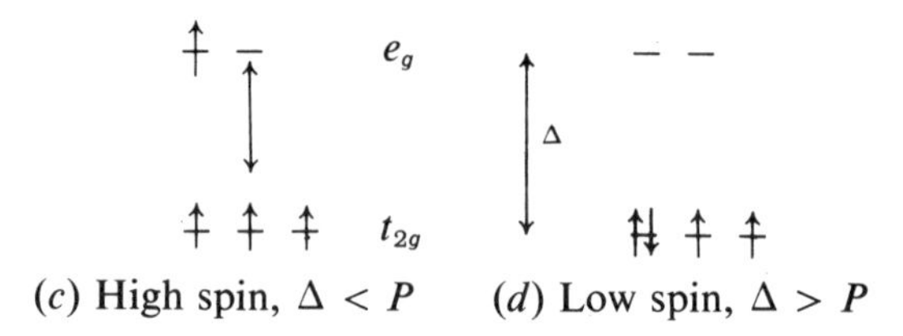

(*c*) High spin, $\Delta < P$ (*d*) Low spin, $\Delta > P$

The various possibilities are summarized in Table 13. For a tetrahedral field the splitting, Δ_t, is so much smaller than for the octahedral case that low-spin behaviour is never observed because Δ_t is always less than P.

The size of Δ varies with (*a*) the ligands, (*b*) the metal, the most important factor being the row in which it occurs. Thus, Δ increases considerably from the 3*d* to the 4*d* to the 5*d* series. For ligands of high field strength, Δ is higher than in the converse situation. High-field ligands (typically CN^-) therefore generally give rise to low-spin complexes in the 3*d* series. For the 4*d* and 5*d* series Δ is so large that low-spin behaviour invariably occurs. Δ can be measured directly from observation of electronic spectra. For details see references given in the bibliography.

Crystal Field Stabilization Energy

The relative stabilities of the various possible *d*-electron configurations may be listed conveniently by reference to the weighted mean energy levels shown below.

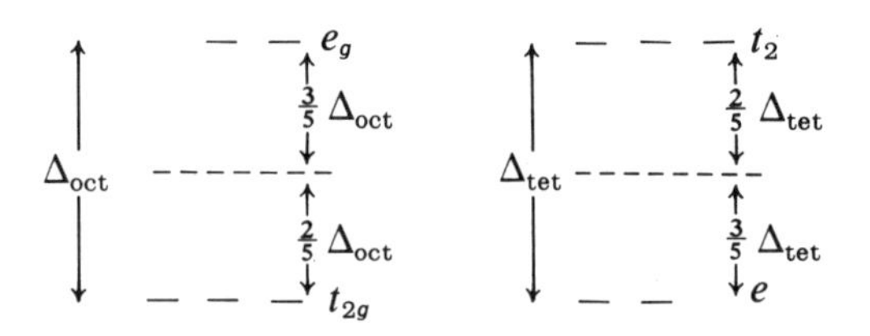

The crystal field stabilization energies (c.f.s.e.) are then given by

$$\xi = 4n_{t_{2g}} - 6n_{e_g}$$

in terms of $\Delta/10$ units, where $n_{t_{2g}}$ and n_{e_g} are respectively the numbers of electrons in t_{2g} and e_g levels; see Table 14.

The effect of c.f.s.e. is seen in the systematic changes observed in a variety of properties of transition-metal compounds. Figure 5-36 shows the bicuspidal variation of lattice energy (from thermodynamic cycles) with *d*-electron configuration. We note that all of these values are above a line determined by

Table 13. Spin configurations for *d*-electrons in an octahedral field

	High spin					Low spin				
	t_{2g}			e_g		t_{2g}			e_g	
d^1	↑					↑				
d^2	↑	↑				↑	↑			
d^3	↑	↑	↑			↑	↑	↑		
d^4	↑	↑	↑	↑		↑↓	↑	↑		
d^5	↑	↑	↑	↑	↑	↑↓	↑↓	↑		
d^6	↑↓	↑	↑	↑	↑	↑↓	↑↓	↑↓		
d^7	↑↓	↑↓	↑	↑	↑	↑↓	↑↓	↑↓	↑	
d^8	↑↓	↑↓	↑↓	↑	↑	↑↓	↑↓	↑↓	↑↓	
d^9	↑↓	↑↓	↑↓	↑↓	↑	↑↓	↑↓	↑↓	↑↓	↑

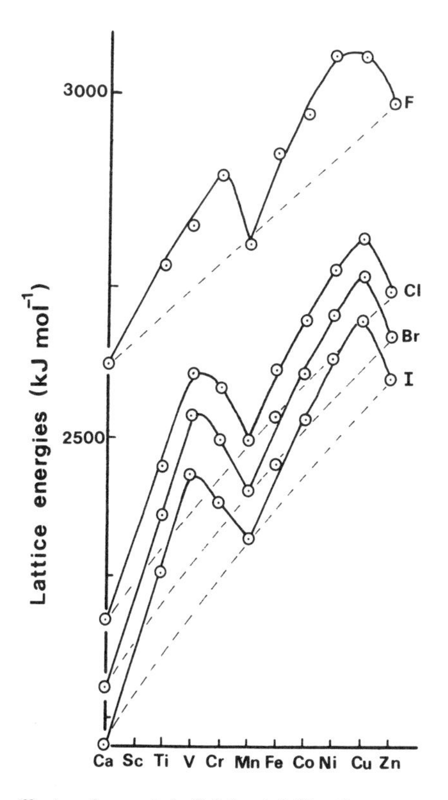

Figure 5-36 The effect of crystal field stabilization energy upon the lattice energies of transition-metal dihalides. (After Waddington, *Adv. Inorganic & Radiochem*, **1**, Academic Press (1959))

Table 14. Crystal field stabilization energy (ξ) in units of $\Delta/10$ for d-orbitals in octahedral and tetrahedral fields

Examples	No. of d-electrons	Tetrahedral[a]	Octahedral[b] High spin	Octahedral[b] Low spin	$\xi_{high\ spin} - \xi_{tet}$[c]
Ti^{3+}	1	6	4	4	1·3
Ti^{2+}, V^{3+}	2	12	8	8	2·7
V^{2+}, Cr^{3+}	3	8	12	12	8·4
Cr^{2+}, Mn^{3+}	4	4	6	16	4·2
Mn^{2+}, Fe^{3+}	5	0	0	20	0
Fe^{2+}, Co^{3+}	6	6	4	24	1·3
Co^{2+}	7	12	8	18	2·7
Ni^{2+}	8	8	12	12	8·4
Cu^{2+}	9	4	6	6	4·2
Zn^{2+}	10	0	0	0	0

[a] $6n_e - 4n_{t2}$ in units $\Delta_{tet}/10$.
[b] $4n_{t_{2g}} - 6n_{e_g}$ in units $\Delta_{oct}/10$.
[c] Taking $4\Delta_{oct} = 9\,\Delta_{tet}$

$Ca^{2+}(d^0)$, $Mn^{2+}(d^5$ high spin) and $Zn^{2+}(d^{10})$, none of which have any c.f.s.e. Furthermore, the values on each half of the curve rise and fall broadly in agreement with predictions of c.f.s.e: all refer to high-spin situations.

Similar variation with d-electron configuration is shown by heats of hydration and ligation of divalent transition-metal ions, and by their ionic radii.

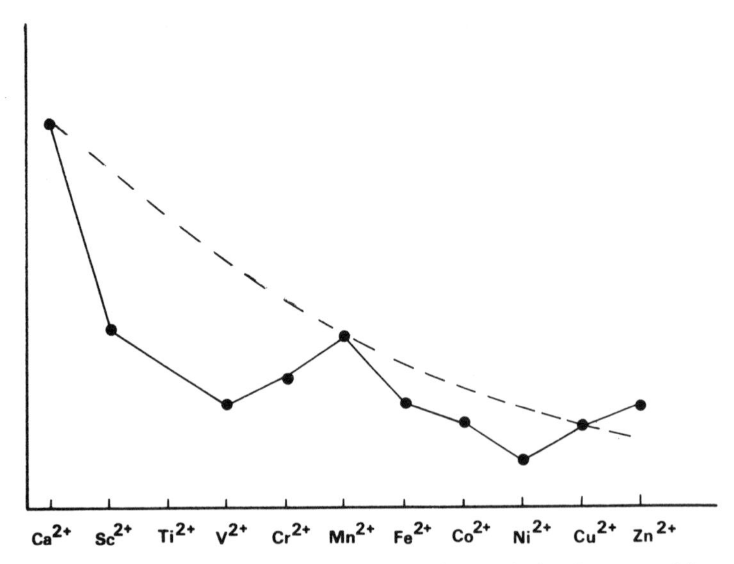

Figure 5-37 Relative ionic radii of divalent ions of the first transition series. (Cotton and Wilkinson, 1972)

Ionic radii, like lattice energies, all depart from a line determined by Ca^{2+}, Mn^{2+} and Zn^{2+}; Figure 5-37. The magnitude of the effect is well emphasized by V^{II} which shows a contraction of ca 25% with respect to the value it would have had in the absence of c.f.s.e. (i.e. as estimated by its distance from the above line). For transition-metal ions which can have both high- and low-spin states two different ionic sizes have been assigned. For example, the sizes for Fe^{II}, d^6, differ by 0·16 Å.

5.5 EXPLANATION OF VARIOUS STRUCTURAL CONSEQUENCES OF *d*-ELECTRON CONFIGURATION

The structural consequences of *d*-electron configuration are many, varied, and profound. We consider below several situations which can only be understood in terms of *d*-electron effects, each chosen to illustrate a different aspect.

5.5.1 Spinels

In our description of the spinel group of structures (p. 68) we noted that the distribution of A^{2+} between octahedral and tetrahedral sites could be described in terms of a parameter γ. For a 'normal' spinel AB_2O_4 all of the A^{2+} ions are in tetrahedral sites and $\gamma = 0$. For 'inverse' spinels with $\gamma = 1$ all A^{2+} ions are in octahedral sites. In fact, a full explanation of the many compounds with the spinel structure or a relative of it is a complex matter: Greenwood (1968) has given an excellent account. Here we are mainly concerned to highlight the principal *d*-electron effects.

For spinels of our type (*a*), i.e. A^{2+}, B^{3+}, it has been shown that if Madelung constant only is considered, almost all 'normal' spinels have larger lattice energy than corresponding 'inverse' ones, whilst for $A^{4+}(B^{2+})_2$ spinels the 'inverse' structure is the more stable. The existence of 'inverse' variants of the $A^{2+}(B^{3+})_2$ variety therefore suggests that c.f.s.e. of transition-metal ions swings the balance in those cases. The crystal field strength of O^{2-} is low enough to allow high-spin configuration for most first-row ions. Table 14 shows (right-hand column) that for $d^5(Mn^{2+}, Fe^{3+})$ and $d^{10}(Zn^{2+}, Ga^{3+})$ ions there is no c.f.s.e. to be gained in either octahedral or tetrahedral coordination for A^{2+}: these spinels are found to be partly inverted, e.g. ferrites such as $ZnFe_2O_4$. On the other hand, Cr^{3+} and Mn^{3+} have very high excess octahedral c.f.s.e. and therefore occur in 'normal' spinels $(A^{2+})_{tet}\{(B^{3+})_2\}_{oct}O_4$ in which they are the B^{3+} ions, and the A^{2+} are required to be in tetrahedral sites, e.g. $(Cd^{2+})_{tet}\{(Cr^{3+})_2\}_{oct}O_4$; this *d*-electron preference for the 'normal' arrangement reinforces the Madelung preference. However, for divalent ions with configuration $d^6(Fe^{2+})$, $d^7(Co^{2+})$, $d^8(Ni^{2+})$ the additional octahedral c.f.s.e. prevails over the Madelung requirement, placing these A^{2+} ions in octahedral sites, hence creating an 'inverse' spinel $(B^{3+})_{tet}(A^{2+}B^{3+})_{oct}O_4$; typical examples are $CoFe_2O_4$, $NiGa_2O_4$.

As final examples we take the important special cases Fe_3O_4 (the mineral 'magnetite') and Co_3O_4. These contain both M^{2+} and M^{3+} ions and are referred to as 'valency-disordered' spinels. We noted above that $Fe^{2+}(d^6)$ has a strong preference for octahedral coordination but that $Fe^{3+}(d^5)$ gains no c.f.s.e. in either this or tetrahedral surroundings. Magnetite is therefore an 'inverse' spinel $(Fe^{3+})_{tet}(Fe^{2+}Fe^{3+})_{oct}O_4$. In contrast $Co^{2+}(d^7)$ is 'normal', $(CO^{2+})_{tet}\{(Co^{3+})_2\}_{oct}O_4$; in this case $Co^{3+}(d^6)$ is low-spin and therefore has very high c.f.s.e. in octahedral coordination, which more than compensates for placing Co^{2+} in tetrahedral sites.

5.5.2 Jahn–Teller Distortions

Consider the d^9 octahedral configuration. Written in full it reads $(t_{2g})^6(e_g)^3$ but even this is not the whole story. The e_g set consists of the orbitals $d_{x^2-y^2}$ and d_{z^2} so that we can write *either* $(t_{2g})^6(d^2_{x^2-y^2}, d^1_{z^2})$ *or* $(t_{2g})^6(d^1_{x^2-y^2}, d^2_{z^2})$ because there is no energy difference between the two e_g orbitals (i.e. they are degenerate). However, the *effect* of placing a d^9 ion in an octahedral environment would differ profoundly according to which of the two electron configurations were used. Single occupation of d_z^2 is equivalent to creating a spherical ion with a hole top and bottom since there is less electron density along the $\pm z$ directions than there is along the x and y directions associated with the doubly occupied $d_{x^2-y^2}$. Conversely, the arrangement $(d_{x^2-y^2})^1(d_{z^2})^2$ concentrates charge along the z-axis whilst allowing a deficit (relative to a d^{10} ion) along x and y axes. Thus, if surrounded by six octahedrally disposed anionic ligands, for the configuration $(d_{x^2-y^2})^2(d_{z^2})^1$ the two ligands along the z-axis would be able to approach the metal more closely than those along x and y axes. Conversely, for $(d_{x^2-y^2})^1(d_{z^2})^2$ the four xy-plane ligands would take up equilibrium positions closer to the metal than would the z-axis ligands. In any one compound one or other of these situations pertains (there is never a mixture of the two) although the reasons for adoption of one rather than the other are obscure. What has happened is that an electron configuration that was degenerate in the isolated ion has been forced to become non-degenerate due to its surroundings. This is known as a Jahn–Teller distortion: their theorem states that 'any non-linear system in a degenerate electronic state will distort so as to remove the degeneracy'. The driving force for this distortion is that there is a net energy stabilization. For an octahedral system this amounts to $\frac{1}{2}E_1$ where E_1 is defined in the scheme below which shows the effect of a small tetragonal distortion (i.e. elongation, or compression, along one axis) where E_1 and $E_2 \gg \Delta_{oct}$.

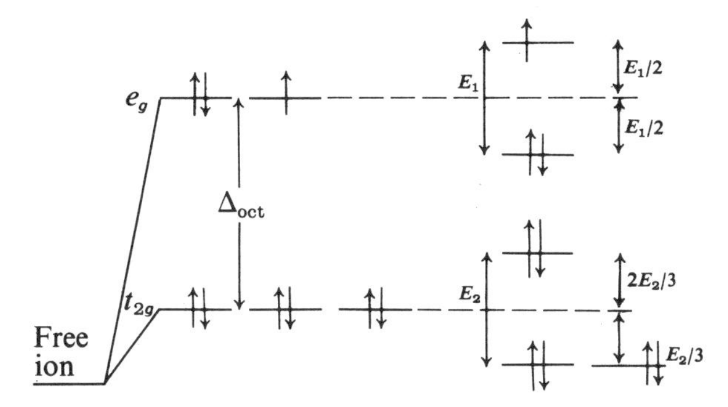

E_1 and E_2 can take any values although they are always small in comparison with Δ_{oct}. An important feature of Jahn–Teller distortions is that a centre of inversion is maintained.

Although we have introduced the Jahn–Teller effect in terms of the d^9 configuration, it occurs for other orbitally-degenerate situations, of which high-spin d^4 (i.e. $(t_{2g})^3(e_g)^1$) and low-spin d^7 (i.e. $(t_{2g})^6(e_g)^1$) are the most important from a structural viewpoint. Jahn–Teller distortions can also be associated with the t_{2g} set (as in $(t_{2g})^1$ where the electron could be considered as initially occupying any one of the set of three orbitals) but these are always very small as the t_{2g} orbitals are not involved in σ bonding.

We limit our discussion of examples to some $Cr^{II}(d^4)$, $Mn^{III}(d^4)$ and $Cu^{II}(d^9)$ compounds. Although difluorides of all of the other first-row transition metals adopt the normal rutile structure, that of CrF_2 and CuF_2 is distorted to allow formation of four short and two longer bonds, equivalent to the configuration $(t_{2g})^6(d_{x^2-y^2})^1(d_{z^2})^2$. The distances are:

	4 fluorines at	2 fluorines at
CrF_2	2·00 Å	2·43 Å
CuF_2	1·93 Å	2·27 Å

Examples of the two short and four long bond distortion are rather rare, two of them being K_2CuF_4 (K_2NiF_4 structure, see p. 228) which has two fluorines at 1·95 Å and four at 2·05 Å; $KCrF_3$ has a distorted perovskite structure (p. 65) with two fluorines at 2·00 Å and four at 2·14 Å distant from the metal. $KCuF_3$ is isostructural with it.

When ions such as Cu^{2+} and Mn^{3+} occur in spinels the lattice is altered to accommodate them. Thus $Mn_3O_4 \equiv (Mn^{2+})_{tet}\{(Mn^{3+})_2\}_{oct}O_4$ (i.e. a 'normal' spinel) is tetragonally distorted by the d^4 configuration of the ions in octahedral sites.

There are many examples of Jahn–Teller distortions in Cu^{II} chemistry. Although the resultant structures are often unique they can always be related to parent structures; the importance of these relationships is the theme of the rest of this section.

$CuCl_2$ might have been expected to adopt the $CdCl_2$ structure like all the other first-row dichlorides (Mn to Zn): in this it would have regular octahedral coordination which is unstable for d^9 but the Jahn–Teller requirement can be accommodated by a tetragonal distortion of each ($CuCl_6$) unit. This is equivalent to spreading apart sections within each 'sandwich' of the $CdCl_2$ layer structure as shown in Figure 5-38. Each copper atom retains two neighbouring chlorines from the chlorine layer forming the top of the sandwich

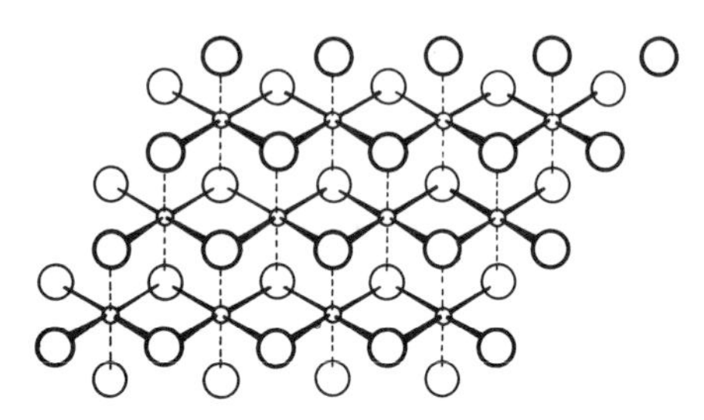

Figure 5-38 Illustration of the distortion of the $CdCl_2$ layer (Figure 3-18b) to give the $CuCl_2$ structure. The dashed lines represent the long bonds (2.95Å), and the smaller circles are copper atoms

and two from the lower layer thereby forming slices of planar structure with the planes of the slices slantwise to that of the original sandwich. The net result of this distortion is to give copper four short (2·30 Å) and two long 2·95 Å) bonds to chlorine, with the short bonds so arranged that copper has square-planar coordination (as it must in a tetragonal distortion of an octahedron) and with the square planes linked to form planar chains:

Cl Cl Cl
Cu Cu Cu
Cl Cl Cl

Wells (1962) describes the $CuCl_2$ structure in terms of chains of the above type packed so as to give copper(II) its preferred 4 + 2 bonding, and compares it with the structure of $PdCl_2$ (p. 192) which has chains of the same type: this comparsion has gained wide currency but is highly misleading. The packing of chains in $PdCl_2$ is of the type classically associated with molecular crystals (this theme is developed in Chapter 6); the chains are planar because the d^8 configuration of Pd^{II} requires square-planar coordination. $CuCl_2$ has planar chains of the same structure for a completely different reason—the occurrence of a centrosymmetric Jahn–Teller distortion. Any comparison between $CuCl_2$ and $PdCl_2$ is of formal significance only; in view of its misleading nature it is best avoided because the relation of $CuCl_2$ to the $CdCl_2$ structure of the other first-row dihalides makes more sense chemically.

First-row transition metals also form compounds such as $CsNiCl_3$ (described on p. 64) in which chains $(MX_3^-)_n$ are present. Cu^{II} forms an analogous compound, $CsCuCl_3$, but again the need to accommodate the Jahn–Teller requirement leads to a characteristic tetragonal distortion giving copper four short (2·30 Å) and two long (2·65 Å) bonds. Because the parent structure is a chain of octahedrally coordinated cations the result of the distortion is also a chain, see Figure 5-39, but because of the two considerably different

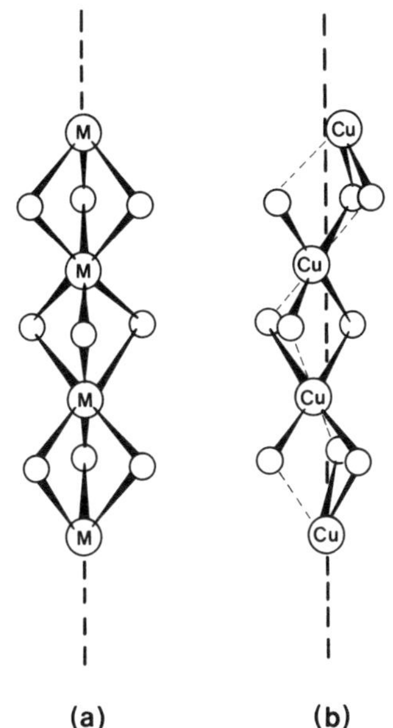

Figure 5-39 Comparison of the structures of the $(MCl_3^-)_n$ chains present in (a) $CsNiCl_3$, (b) $CsCuCl_3$

bond lengths now present it can also be described as formed from square-planar $(CuCl_4)$ units, bridged by common chlorines, which spiral slowly around the chain axis with a hexameric repeat unit $Cu_6Cl_{18}^{6-}$.

```
               Cl      Cl        Cl      Cl
                 \    /            \    /
                  Cu                 Cu
                 /    \            /    \
   \Cl         Cl      Cl        Cl      Cl
       \      /          \      /          \
        Cu                 Cu
       /    \            /    \
    Cl       Cl        Cl      Cl
```

There are many compounds of the type CuX_2L_2 (X = Cl, Br, I; L = donor ligand such as H_2O, amine, etc.) which form structures having the characteristic 4 + 2 coordination. These are commonly described as chain structures and this is helpful in that it emphasizes the natural relation to analogous compounds of other first-row metals. Thus, $CoCl_2.2H_2O$ has the chain structure shown in Figure 5-40a; this is simply distorted to a 4 + 2 form in $CuCl_2.2H_2O$ in which two of the four short bonds are from chlorines and two from oxygens.

$CoCl_2.2H_2O$	4Cl at 2·53 Å	$2H_2O$ at 1·93 Å
$CuCl_2.2H_2O$	2Cl at 2·28 Å	$2H_2O$ at 1·93 Å
	2Cl at 2·95 Å	

$CuCl_2.2H_2O$ may also be described as *trans*-$[CuCl_2(H_2O)_2]$ molecules packed so as to form chains, although the cause of its particular structure is better emphasized by reference to the cobalt compound.

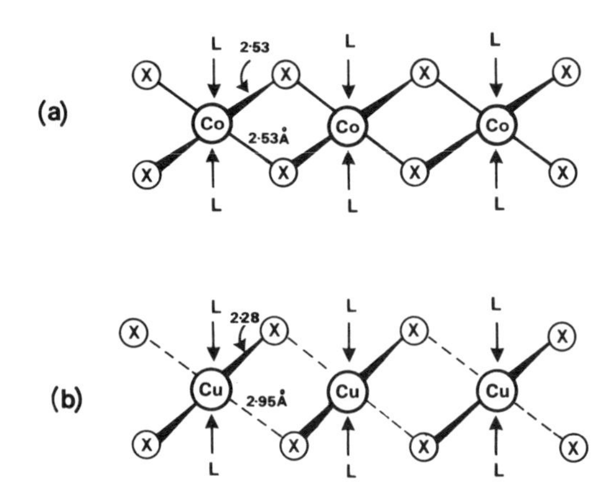

Figure 5-40 Comparison of the structures of (a) $CoCl_2.2H_2O$, (b) $CuCl_2.2H_2O$. L = H_2O

As a final example, consider $K_2CuCl_4.2H_2O$ which may be formally compared with K_2PtCl_6 (anti-fluorite structure). As in $CuCl_2.2H_2O$, 4 + 2 coordination is satisfied by 2Cl at 2·29 Å, $2H_2O$ at 2·01 Å and 2Cl at 2·93 Å. Although formal analogy with K_2PtCl_6 is helpful, an equally good description is of $CuCl_2.2H_2O$ molecules in a lattice of K^+ and Cl^-.

We have given a sufficient number of examples in this section to illustrate the principles. Further detail is to be found in Wells's excellent review of copper structural chemistry (1962).

5.5.3 Transition-metal Oxides, Carbides and Nitrides

Due to the oxidation states available, a range of oxides is formed by transition metals varying from divalent monoxides to molecular MO_4(M = Ru, Os) in which the metal has a formal oxidation state of eight. We restrict our discussion to mono- and di-oxides, MO and MO_2. Structurally these are rather simple: the monoxides all have the rock salt structure and, although rather more variation is exhibited by the dioxides, nearly all have the rutile structure or a close relative of it. The interest lies in the extraordinary range of physical properties shown by these deceptively simple materials. A hint of their complexity is given by the range of their colours: TiO (black), VO (grey), CrO (black), MnO (green), FeO (black), CoO (grey), NiO (green or yellow), CuO (black). A full study of these aspects would take us into main-stream solid-state physics; however, we can outline some of the more important conduction and magnetic properties in a descriptive manner and in so doing illustrate relations between *d*-electron configuration on the one hand and structure and properties on the other. The most basic principle underlying the variation in properties with change of metal is that *d*-electron ionization potentials increase across each transition series due to the poor shielding effect of *d*-electrons, and there is an accompanying orbital contraction. Thus, at Ti *d*-orbitals are large and well extended but as we move along the 3*d* series to Ni they become progressively more localized on the metal.

5.5.4 Monoxides, MO

Consider first the oxides, MO.† TiO and VO exhibit electrical conductivity in the metallic range. This is readily understood in terms of t_{2g} *d*-orbital overlap throughout the lattice. Figure 5-41 shows their orientation relative to the

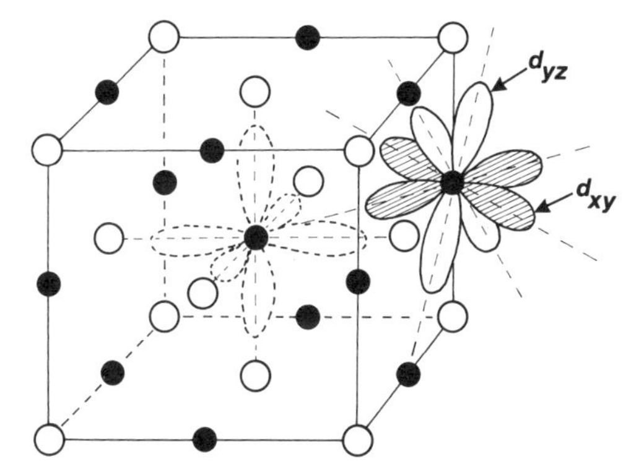

Figure 5-41 The NaCl lattice, showing the orbitals used for overlap in transition metal oxides, MO

lattice whilst Figure 5-42 indicates the related description in terms of band theory. The filled 2*p* band is populated by electrons mainly associated with oxygen and there are higher (antibonding) bands not populated. The e_q *d*-levels are non-bonding and unoccupied and can be represented by an energy

† These oxides are non-stoichiometric and are further considered in Chapter 9.

level rather than a band, but the cooperative overlap of the t_{2g} set must be described as a band. Since this is only one-third filled (Ti^{2+} is d^2) metallic conductivity arises, ca 10^3 ohm^{-1} cm^{-1}.

As we progress across the 3*d* series, nuclear attraction of the *d*-electrons causes them to be more firmly held and reduces the extent of overlap throughout the crystal. In pure NiO the *d*-electrons are effectively localized on nickel: we must now talk of t_{2g} *levels* rather than a band, see Figure 5-42. Consequently metallic conduction is impossible; the very low observed electrical conductivity of ca 10^{-14} ohm^{-1} cm^{-1} is due to hopping of electrons from filled to empty *levels*. In fact, this is mainly due to interaction with levels of a small proportion of Ni^{3+} present duc to partial nickel deficiency in the material which, strictly speaking, should be written $Ni_{1-x}O$. Between the extremes of delocalization and localization of t_{2g}-electrons represented by TiO and NiO there is a transition which can be described as a narrowing of

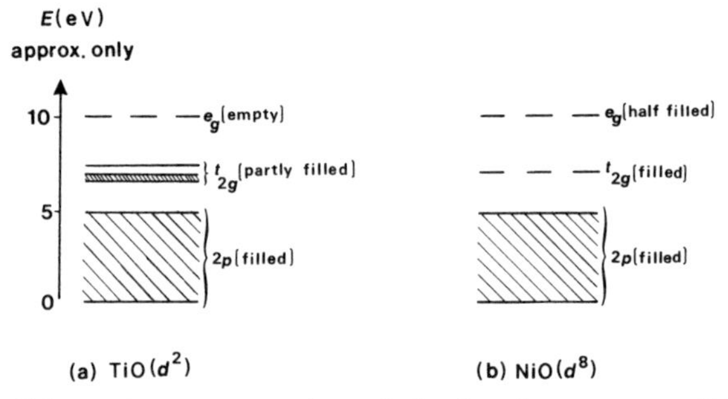

Figure 5-42 Schematic representation of the band structures of (a) TiO, (b) NiO. (After A. T. Howe and P. J. Fenshaw, *Quart. Rev.*, **21**, 507 (1967))

the t_{2g} conduction band (i.e. as a tendency towards discrete levels). At MnO, the orbital contraction is such that the electrical conduction mechanism switches from metallic to that of NiO (hopping via discrete levels): there seems to be a critical orbital contraction below which there is a sharp change of conduction mechanism.

Changes in magnetic behaviour accompany the above properties. Broadly, two types of behaviour are exhibited: in the metallic conductors there is no paramagnetic† moment (in the absence of an applied field) because of the absence of magnetic interactions between delocalized electrons, but from MnO to NiO cooperative paramagnetism is present (ferro-, ferri-, or antiferromagnetism). At fairly high temperatures the paramagnetic moments are those expected on the basis of the number of unpaired electrons, as computed by the 'spin only' formula, $\mu_{s.o.} = 2\sqrt{S(S+1)}$ where S = total spin. Below

† Paramagnetism is associated with unpaired electrons. It is strong in comparison with diamagnetism which is a property of closed electron shells and is possessed by all matter.

a critical temperature (known as the Néel point) the paramagnetic moments drop with decreasing temperature as shown schematically in Figure 5-43. In order to understand this phenomenon it is helpful to consider the simpler case of a two-centre interaction.

Several transition metals (Cr, Cu, Mo, Ru, Rh) form dimeric acetates $M_2(OCOCH_3)_4$ which have the structure of Figure 5-44. The M–M distances

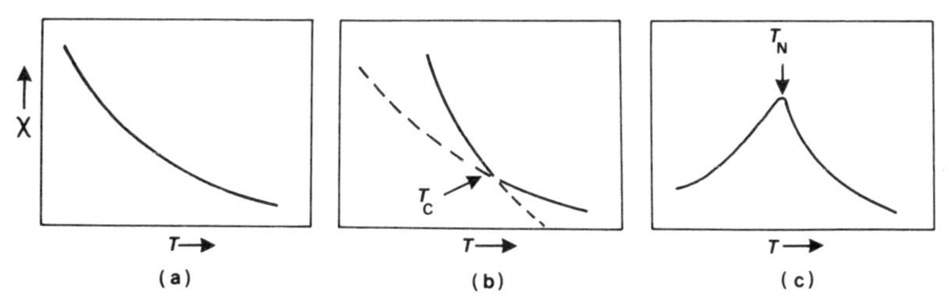

Figure 5-43 Schematic representation of the variation of paramagnetic moment, χ, with temperature for: (a) simple paramagnetism, (b) ferromagnetism, (c) antiferromagnetism. (Cotton and Wilkinson, 1972)

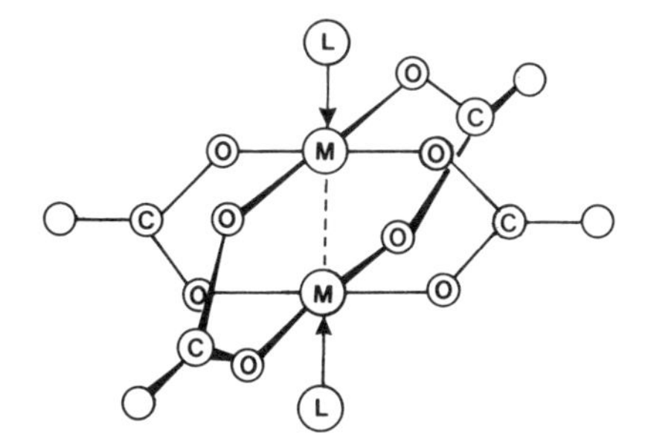

Figure 5-44 The structure of dimeric acetates $M_2(OCOCH_3)_4$, (M = Cr, Cu, Mo, Ru, Rh). L, when present, is H_2O, pyridine, or other donor ligand

in them vary in an interesting way but we concern ourselves with the Cr and Cu complexes only. The metal atoms are sufficiently close for their unpaired *d*-electrons to interact. When they do so in such a way as to *lower* the magnetic moment below the spin-only value the process is referred to as antiferromagnetism. It is a cooperative effect—i.e. involves two (or more) magnetic centres. The process is considered in terms of an equilibrium between populations of singlet and triplet levels. Thus, for Cu^{II}, d^9 which can be considered as a one-electron-hole system:

$\uparrow\ \uparrow$ Triplet

ΔE

$\uparrow\downarrow$ Singlet

If ΔE is $\leq kT$ the upper level will be well populated and the observed magnetic moment per metal atom will be near that expected on the basis of the spin-only formula: this is what happens in $Cu_2(OAc)_4$. Upon cooling, kT is lowered, the population of the lower level increases at the expense of the upper one, the paramagnetic moment drops. The Néel temperature marks the point at which this decrease in μ begins. For $Cr_2(OAc)_4$ $\Delta E \gg kT$ at room temperature; thus, although Cr^{II} is d^4 and would exhibit paramagnetism in a mononuclear complex, in this compound it is paramagnetic.

Returning to the oxides MnO to NiO, it is found that all exhibit some form of cooperative paramagnetism (Ferromagnetism results from parallel coupling of spins, ferrimagnetism from non-equivalent antiparallel coupling; Morrish, 1965.) The transition temperature from normal paramagnetic behaviour (at high temperature) to cooperative paramagnetism increases

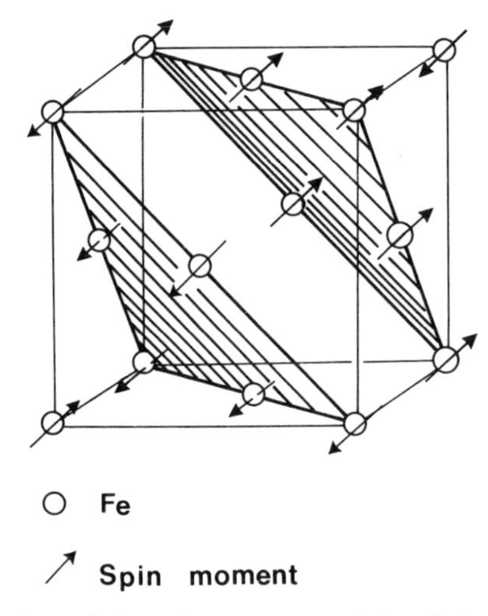

Figure 5-45 Orientation of the spin moments of the *d*-electrons on iron in FeO (oxygen atoms not shown). (From *Fundamentals of Inorganic Crystal Chemistry* by H. Krebs. © 1968 McGraw-Hill Book Company (UK) Ltd. Used with permission)

from Mn to Ni, the order of increasing *d*-electron localization. We consider in a little more detail the mechanism of this magnetic ordering for $Fe_{1-x}O$, which shows antiferromagnetism.

Magnetic ordering in FeO (which is quicker to write than $Fe_{1-x}O$) does not take place via direct interaction of t_{2g} *d*-electrons on nearest-iron neighbours as these functions are too localized; it takes place via oxygen. For this to happen the e_g *d*-electron orbitals must be used; these have lobes pointing directly at the six nearest-neighbour oxygens. There is sufficient overlap between these e_g and the p-orbitals on oxygen to cause spin alignment by virtue of spin polarization of the lobes of the oxygen orbitals, thereby coupling neighbouring *cation* spins via pairs of oxygen p-electrons. Since, of necessity, the members of an electron pair in a p-orbital are antiparallel, this ensures alternating antiparallel alignment of d-electron spins throughout the lattice (Figure 5-45). The process is termed 'super exchange'. As described, this is equivalent to saying that there are small contributions to the total wave function of the system from e_q–p overlap; this small contribution is, nevertheless, adequate to cause alignment. It is as if a p-electron unpairs from one oxygen and transfers to the neighbouring cation thereby aligning its spin antiparallel to its own. The unpaired electron left on oxygen then forms a bond with its opposite cation neighbour, resulting in alignment antiparallel to that of the other cation. The process continues throughout the lattice. Schematically it can be represented as:

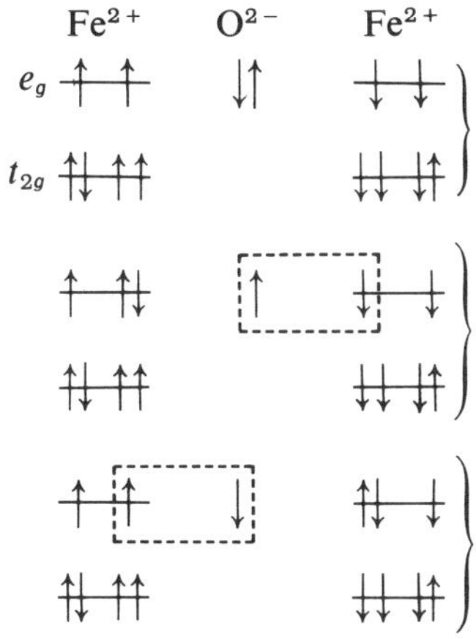

Fe^{2+} O^{2-} Fe^{2+}

e_g

t_{2g}

(*a*) Random spin orientation

(*b*) Transient O(p)-Fe(e_g) bond formation

(*c*) As (*b*)

5.5.5 MA Carbides and Nitrides

The rock salt structure is also adopted by many carbides, MC, and nitrides, MN, which show rather similar properties, such as very high metallic conductivity and lustre. They have melting points among the highest recorded (e.g. TiC 3410°C) and are very hard (typically 8–9 on a scale 0–10 where 10 = diamond), qualities which account for their industrial uses. Clearly, the bonding in these solids is highly covalent; they adopt the rock salt lattice because its high symmetry is perfect for the bonding scheme described below. The small size of the non-metal atoms (which are formally described as occupying octahedral holes in a close-packed metal lattice) makes for efficient use of space and contributes towards mechanical hardness. We outline the bonding in TiC following Krebs (1968).

Carbon uses its three $2p$-orbitals which are aligned with the Ti—C bond directions. As in FeO, there is σ bonding via the $e_g(d_{x^2-y^2}, d_{z^2})$ orbitals on Ti which are given further directional character by hybridization with the metal s-orbital. π Bonding is also possible by overlap of the t_{2g} d-orbitals with carbon p-orbitals as shown in the scheme of Figure 5-46. Both σ and π schemes are mesomeric.

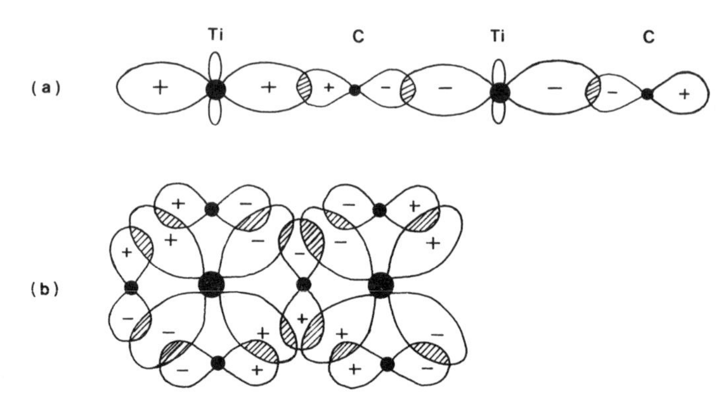

Figure 5-46 (a) σ-Bonding scheme in TiC with d^2s hybrids on Ti and p-orbitals on C. (b) π-Bonding scheme in TiC

5.5.6 Dioxides: Rutile and Related Structures

The rutile structure is adopted by dioxides of a number of B-Group (non-transition) metals and by several transition metals with incomplete d shells, see p. 71. The properties of these materials are far from simple but many can be understood in outline. Table 15 shows the occurrence of the rutile structure in transition-metal dioxides, and of closely related variants.

In the distorted variants indicated by a in Table 15 metal–metal distances alternate along the c-axis so that pairs of metal atoms form and the structure contracts along c due to creation of these bonds. There is a compensating

Table 15. Crystal structures of transition-metal dioxides, $M^{IV}O_2$

d^0	d^1	d^2	d^3	d^4	d^5	d^6
Ti	V[a]	Cr	Mn			
Zr[b]	Nb[c]	Mo[a]	Tc[a]	Ru	Rh	
Hf[b]	Ta	W[a]	Re[a]	Os	Ir	Pt[d]

These have the normal tetragonal rutile structure except for:
[a] An axially distorted monoclinic form—see text.
[b] A unique seven-coordinated structure (baddeleyite).
[c] Superstructure of rutile.
[d] An hexagonal layer lattice.

expansion along a so that although the axial ratio, c/a, varies considerably the volume of the unit cell remains remarkably constant throughout the series.

Note particularly that the formation of metal–metal pairs results in displacement of the metal atoms from the octahedral centres. Since the octahedral inversion centre is lost in this process this distortion is clearly distinguished from those of the Jahn–Teller type (in which the centre of inversion must be retained).

There is clearly a need to explain why a group of five dioxides have the monoclinic form. Figure 5-47 shows that a plot of axial ratio against d-electron number has pronounced minima at d^1 (3d series), d^2 and d^3 (4d and

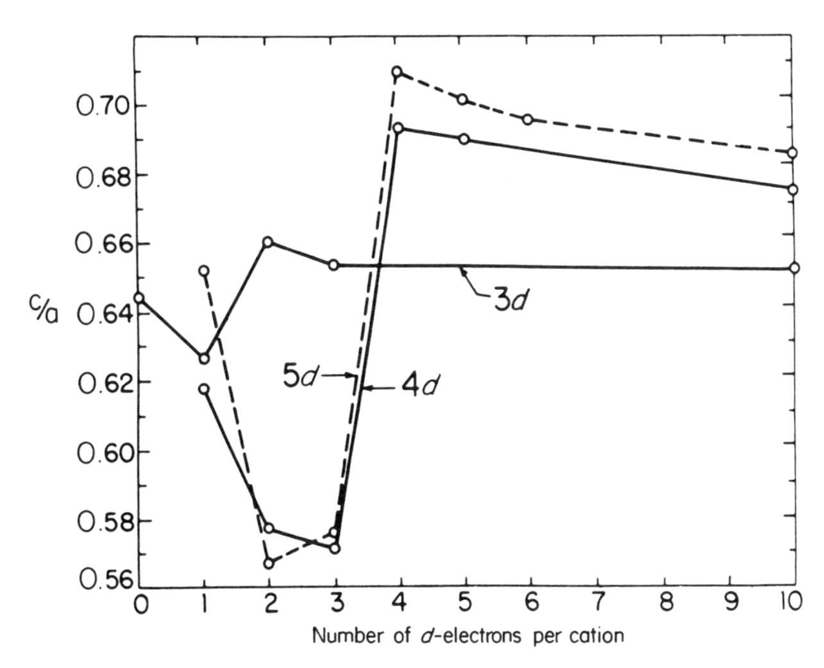

Figure 5-47 Variation of axial ratio, c/a, with d-electron number, for rutile-like dioxides. (From D. B. Rogers and others (1969))

5d series), correlating exactly with formation of the monoclinic variant of the rutile structure. d-Electron configuration is therefore the key to this structural change. Further, conduction properties vary widely and also need elucidation. Thus, TiO_2 is effectively an insulator (cf. TiO, which is a metallic conductor), VO_2 has metallic conductivity in its normal tetragonal rutile form but below 340 K it transforms to the monoclinic type with loss of metallic conduction. The Cr, Mo, W, Re, Ru, (Rh), Ir and Os dioxides are metallic conductors.

These features have been accommodated by a model due to Goodenough and elaborated by Rogers, Shannon, Sleight and Gillson (1969). Consider the schematic one-electron energy diagrams in Figure 5-48. A doubled formula unit M_2O_4 is taken so that M—M doublet formation in the monoclinic forms can be discussed on the same model. Considered as Ti^{4+}, $2 \times O^{2-}$ the metal ion has no valence electrons, but oxygen has a complete octet. (This is a formalism only but is convenient for counting electrons.) We consider rutile in terms of a σ-bonded framework made up by the overlap of sp^2-hybrid orbitals on oxygen (since it is planar three-coordinate, p. 72) and $d^2(e_g{}^2)sp^3$ hybrids on the octahedrally coordinated metal; in addition a π-bonded scheme is possible using the remaining p function on oxygen with two t_{2g}-metal orbitals of appropriate orientation, see Figure 5-46b. The remaining t_{2g}-orbital (with lobes directed along c) is suitably oriented for M—M bond formation, and this will occur *if* the M—M separation is sufficiently small. Since we are dealing, not with an isolated molecule but with a crystal, there is an infinite number of hybrid levels which are therefore broadened into *bands* of finite width. The number of electrons that can be accommodated in each band per M_2O_4 unit is given in square brackets.

For TiO_2 (no d-electrons) the Ti—O σ and π bands are just filled: metallic conductivity is therefore not expected or observed. VO_2 has a single t_{2g} d-electron per metal atom which goes into the overlapping V—V σ bond and V—O π^*-conduction bands thus resulting in metallic conduction. Below 340 K the structure changes to the monoclinic form with alternating short (2·62 Å) and long (3·17 Å) V—V distances along c due to formation of strong V—V bonds. The band scheme of Figure 5-48a must now be modified as shown in (b). The V—V σ-bond band is split into two *levels* localized between pairs of vanadium atoms, M—M (σ bonding) and M—M (σ^* antibonding). Two electrons from a pair of metal atoms exactly fill this level; metallic conduction is no longer possible. The high temperature (tetragonal) form of VO_2 has a paramagnetic moment corresponding approximately to one unpaired electron per metal atom, while the monoclinic form is diamagnetic, both properties being in accord with the above band scheme. CrO_2 with a d^2 configuration has the largest axial ratio for the 3d series indicating minimal metal–metal interaction. This is not unexpected as there is a steady contraction of 3d-orbitals across the series so that the t_{2g}-electron functions are probably quite localized on chromium.

MoO_2 is rather splendid; it has the monoclinic form with Mo–Mo pairs but *also* shows metallic conductivity, cf. VO_2. Since it has four d-electrons per metal pair the M—M σ-bonding level is filled leaving two electrons available for metallic conduction in the M—O π^* band which is then partly filled.

Looking again at Figure 5-47 we see that c/a values fall into three sets. (a) Those which we may regard as 'normal', Ti, Cr, Mn in 3d; d^4 and onwards for the second and third transition series. (b) The d^1 oxides VO_2 and NbO_2

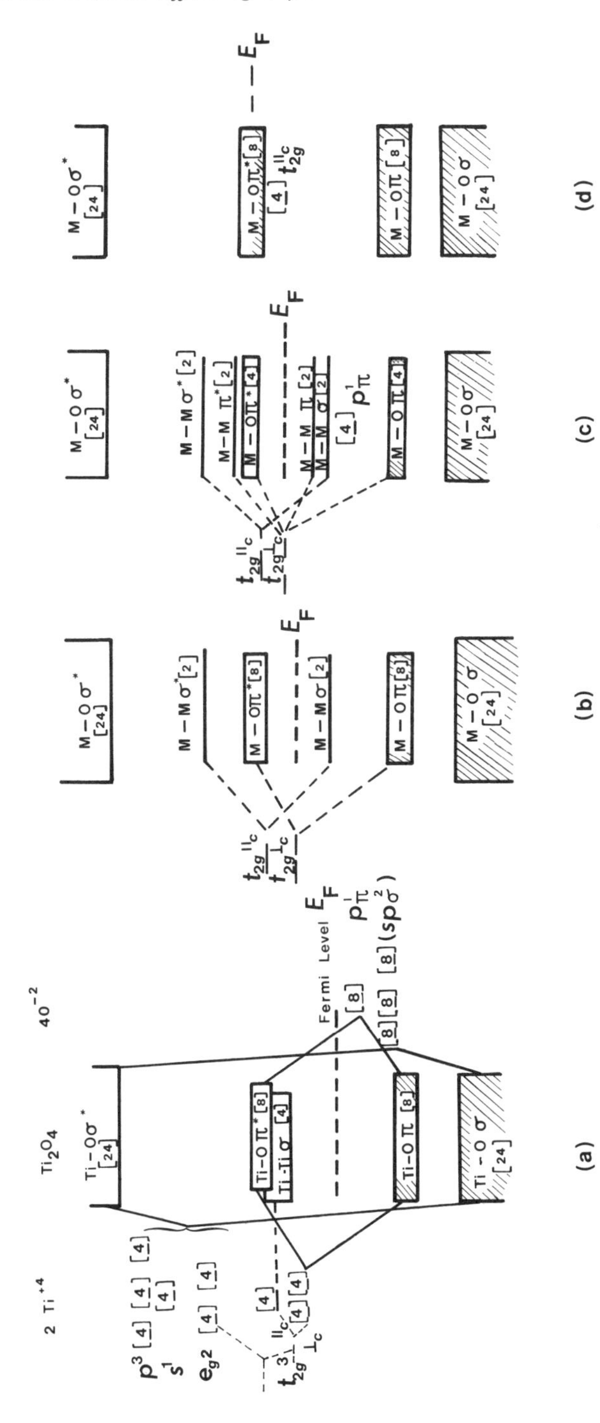

Figure 5-48 Schematic one-electron energy diagrams for rutile-like dioxides. (a) Undistorted rutile structure with strong metallic interactions parallel to the *c*-axis. (b) A distorted structure with single M–M σ-bonds. (c) A distorted structure with both M–M σ- and π-bonds. (d) An undistorted structure with negligible direct M–M interactions; t_{2g} states are non-bonding. (From D. B. Rogers and others (1969))

in which M—M pairs have separations of 2·62 and 2·80 Å respectively. (*c*) The d^2 and d^3 oxides of Mo, W, Tc and Re, which are notable for their very short M—M distances which are all ca 2·50 Å. It is believed that the metal–metal bonds in VO_2 and NbO_2 are approximately of order one, but that in (*c*) the situation corresponds to double bonding. A revised band scheme for this situation is shown in Figure 5-48c, which now includes M—M π bonding *levels*. This represents an extreme situation in which no ligand character is allowed in the M—M π scheme and would clearly not admit the occurrence of metallic conduction. We must therefore conclude that the true situation in MoO_2 (and in the W, Tc, Re compounds) is represented by a scheme intermediate between those of (b) and (c), thereby accommodating both metallic conduction and high M—M bond order.

The second- and third-row metal cations with d^4 to d^6 configurations show little, if any, metal–metal interaction as indicated by Figure 5-47. The appropriate band scheme is given in Figure 5-48d. For these electron configurations the t_{2g}-orbitals parallel to the *c*-axis are filled (4 electrons per metal pair) and hence non-bonding. The remaining *d*-electrons consequently partly fill the M—O π^* band resulting in metallic conduction. In PtO_2 (which has a slightly different structure) the M—O π^* band would also be filled so that metallic conduction is not expected, but this has not yet been fully confirmed. In itself the change of structure from IrO_2 (normal rutile) to PtO_2 (hexagonal layer structure) shows up a further subtle effect of *d*-electron configuration upon the adoption of structure type. The change cannot be a cation-size effect as both larger and smaller cations are found in the rutile structure. It *could* be that at platinum the bond directionality has fallen just below the critical value for adoption of a rutile-type structure, due to the progressive increase in stability of *d*-electron levels with increase of atomic number, making the less ionic layer structure necessary. This would be consistent with trends in the general chemistry of the platinum metals and with the not inconsiderable covalent character revealed in the Pt–halogen bonds by nuclear quadrupole spectroscopy. Alternatively, we could be seeing here the abrupt disappearance of a stabilizing effect due to change in band structure.

BIBLIOGRAPHY

Cotton, F. A., and G. Wilkinson, *Advanced Inorganic Chemistry*, 3rd ed., Interscience, New York, 1972.

Dehlinger, U., *Theoretische Metallkunde*, Springer, Berlin, 1955.

Dickens, B., and W. E. Brown, *Inorganic Chem.*, **9**, 480 (1970).

Dickens, B., *J. Inorg. Nuclear Chem.*, **27**, 1495 (1965).

Dunitz, J. D., and L. E. Orgel, *Advances Inorganic Radiochem.*, **2**, 1 (1960).

Grdenič, D., *Quart. Rev.*, **19**, 303 (1965).

Greenwood, N. N., *Ionic Crystals, Lattice Defects and Nonstoichiometry*, Butterworths, London, 1968.

Johnson, D. A., *Some Thermodynamic Aspects of Inorganic Chemistry*, Cambridge University Press, 1968.
Kaufman, L., and H. Bernstein, *Computer Calculation of Phase Diagrams*, Academic Press, New York, 1970.
Kettle, S. F. A., *Coordination Compounds*, Nelson, London, 1971.
Morrish, A. H., *The Physical Principles of Magnetism*, Wiley, New York, 1965.
Mooser, E., and W. B. Pearson, *Acta Cryst.*, **12**, 1015 (1959).
Orgel, L. E., *J. Chem. Soc.*, **1958**, 4186.
Parthé, E., *Zeit. Krist.*, **115**, 52 (1961).
Penn, D. R., *Phys. Rev.*, **128**, 2093 (1962).
Phillips, J. C., *Covalent Bonding in Crystals, Molecules and Polymers*, University of Chicago Press, 1969.
Phillips, J. C., *Rev. Mod. Phys.*, **42**, 317 (1970).
Rogers, D. B., R. D. Shannon, A. W. Sleight and J. L. Gillson, *Inorg. Chem.*, **8**, 841 (1969).
Sharpe, A. G., *Endeavour*, **27**, 120 (1968).
Tateno, J., *J. Phys. Chem. Solids*, **31**, 1641 (1970).
Truter, M. R., *Chem. Brit.* **1970**, 203.
van Vechten, J. A., *Phys. Rev.*, **187**, 1007 (1969).
van Vechten, J. A. and J. C. Phillips, *Phys. Rev. B*, **2**, 2147; 2160 (1970).
Waddington, T. C., *Advances Inorganic Radiochem.* **1**, 157, (1959).
Wells, A. F. *Structural Inorganic Chemistry*, 3rd Ed., The Clarendon Press, Oxford, 1962.

CHAPTER 6

Molecular and Hydrogen-bonded Crystals

We begin this chapter by taking up the question of packing in molecular crystals from the point at which we left off in Chapter 3. In that chapter the packing of identical spheres, non-directionally bonded, was seen to lead to cubic and hexagonal closest-packed structures. We now relax first the shape restriction, and then admit specific additional bonding interactions due to polar groups, giving especial attention to hydrogen bonding.

6.1 MOLECULAR CRYSTALS

The term 'molecular crystal' is applied to a solid composed of molecules which are stable in their own right. They are soft, have low melting points and dissolve in organic solvants. Many of the molecular properties change little from gas to liquid to crystal, emphasizing the weakness of intermolecular bonding relative to that within each unit. For example, the principal features of the infrared and Raman spectra of molecular crystals can be accounted for on the basis of a model which treats them as an 'oriented gas'. On the other hand electronic spectra show features characteristic of the crystalline state, reminding us of the cooperative nature of bonding even in the very simplest cases. Figure 6-1 compares the electronic spectrum of solid xenon with that of the gas. Explanation of the new features is couched in terms of 'excitons', transitions localized over several molecules or atoms rather than one (Dexter and Knox, 1965).

Bonding between non-polar molecules is non-directional to a good approximation, so that for crystals in which this type of bonding predominates the structure will be determined by the way in which molecules can be packed. Molecular crystals are always close-packed: the total volume occupied varies from 60 to 80% (taking van der Waals radii). Volume occupation below 60% represents inability to find a periodic mode of packing of sufficiently high density for which the energy of interaction is more favourable than that of the amorphous state. Thus, very awkwardly shaped molecules tend to form glasses rather than crystals: e.g. 2,6-di-n-octylnaphthalene only crystallizes with difficulty, and when it does so, occupies only 59·5% of space, the lowest on record (Kitaigorodskii, 1961). The highest volume occupation is found for graphite (p. 221), 88·7%. Occasionally a structure is formed that would have been open (less than 60% space used) but for inclusion of a 'foreign' molecule. For example, the clathrates β-quinol:argon or $Ni(CN)_2(NH_3)_2:C_6H_6$ consist of very open frameworks with an atom (argon) or molecule (C_6H_6) trapped in each cage. No primary bond is formed between host cage and trapped entity.

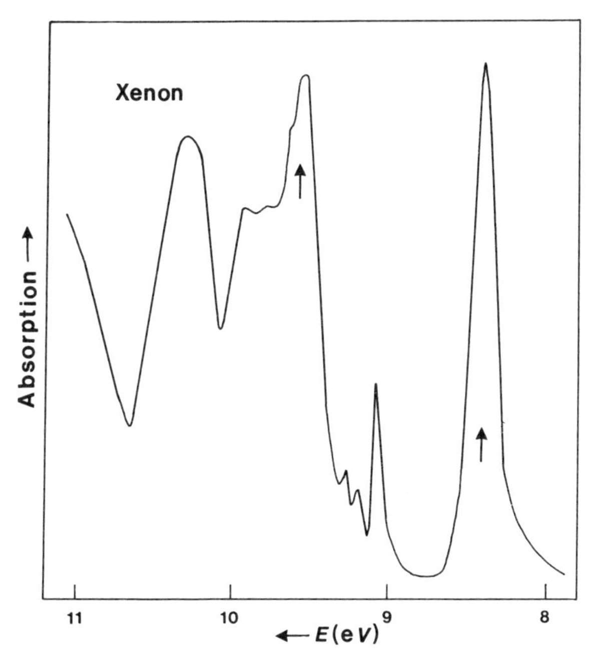

Figure 6-1 Electronic absorption spectrum of solid xenon. Arrows indicate the lines present in the spectrum of gaseous xenon. (Dexter and Knox (1965))

As with every other type of bond we have considered, there are many cases which are intermediate between molecular and covalent or metallic or ionic; we shall meet examples of all of these. For crystals composed of molecules having some polar character the additional interactions possible (dipole–dipole; dipole–induced dipole, etc.) may have an important modifying or even determinative effect upon the structure adopted.

6.1.1 Packing of Molecules: Kitaigorodskii's Theory

Although molecular crystals have been considered for many years as more or less distorted variants of close-packed structures, it was not until 1955 that the noted Russian crystallographer A. I. Kitaigorodskii laid down the basic principles clearly and followed through their implications. His work dealt specifically with organic crystals but there is no reason to limit its generality. Just why this took so long to come about is hard to say but a contributory factor may be associated with the conventional way of drawing crystal structures. As in most of the diagrams in this book, atoms in structures are normally represented by points or small circles; this is done for clarity. But if instead we draw in each atom complete with its electron-cloud boundary surface at the van der Waals radius, we get a much better representation of the molecular packing. Of course, this also makes it difficult to draw more than a small portion of the structure clearly. Furthermore the inability to determine H-atom positions accurately by X-ray crystallography† meant that very often it was not clear that the molecules were packed closely.

Molecular crystals do not contain voids; rather, the individual molecules are so packed as to make maximum use of space. In general, molecules are asymmetric—they have portions which protrude and, correspondingly, regions of low electron density. The simple principle enunciated by Kitaigorodskii states that: 'the protrusions on one molecule will pack into the hollows of the next'. Simple! But this commonsense idea has far-reaching consequences.

The number of *point* contacts between any two molecules can vary widely. However, if instead we ask 'how many neighbouring molecules are there which make *one or more* contacts with a reference molecule' then we can define a molecular coordination number (m.c.n.). In the great majority of molecular crystals the m.c.n. is found to be twelve. Further, these structures can be regarded as composed of layers in which each molecule has an m.c.n. of six (cf. closest-packed layers).

The importance of Kitaigorodskii's idea is that it places severe restrictions upon the number of space-group symmetries which molecular crystals may adopt. Of the 230 space groups, only 13 are 'most probable' for crystals composed of molecules of arbitrary shape. The proof of this remarkable statement is straightforward but lengthy and therefore beyond the scope of this book. However, we can at least indicate the type of argument em-

† This comment applies mainly to earlier work; nowadays increased precision associated with automatic data collection, and the ease of carrying out many cycles of refinement by computer, makes detecting hydrogen atoms easier. Although we are not concerned with experimental methods in this book, mention should be made of one very special technique. The scattering of neutrons by hydrogen and deuterium takes place with different efficiencies which are described quantitatively by quoting scattering amplitudes, f. For hydrogen and deuterium f_H and f_D differ in *sign* as well as magnitude ($f_H = -0{\cdot}38$; $f_D = +0{\cdot}65$). Thus, if a sample is deuterated in the mole ratio $H/D = 0{\cdot}650/0{\cdot}378 = 1{\cdot}72$ (i.e. the *inverse* of the ratio of the factors f_H, f_D) the deuterium scattering exactly cancels that of hydrogen by destructive interference. Hence, comparison of the diffraction pattern of a 'normal' compound with that of the same compound having $H/D = 1{\cdot}72$ reveals, by difference, the H-atom positions since these only affect the pattern from the normal compound (Pimentel and McClellan, 1960).

ployed. For simplicity we consider only the two-dimensional problem. That is, we wish to find which periodic arrangements of an *arbitrarily* shaped molecule are possible in two dimensions such that close packing (i.e. an m.c.n. = 6 in the plane) is possible. The general theory of plane lattices shows that such arrays can be built only from a square, a rectangle, a parallelogram or a rhombus of angle 60° (Section 2.1.1). Of these, only the rectangle and the parallelogram are allowed lattice types for molecular crystals composed of *arbitrarily* shaped molecules. This is because with both the square and the rhombus the repeat distance in one direction (see below) automatically determines the other repeat distance of the lattice. The implications for a square lattice of *arbitrarily* shaped molecules are illustrated by Figure 6-2: because the molecules are of arbitrary shape, any linear periodic array (or chain) of them will have arbitrary dimensions *both* along and normal to the chain axis. The rectangle and the parallelogram can accommodate this situation.

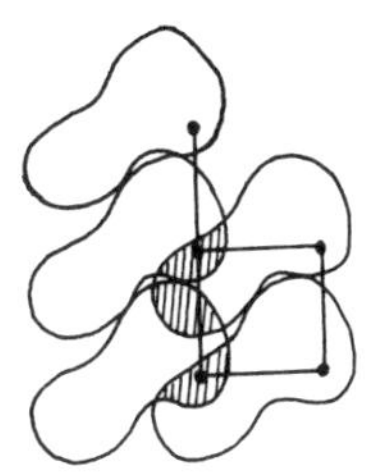

Figure 6-2 Illustration of the unsuitability of a rectangular lattice for arranging arbitrarily shaped molecules. (Kitaigorodskii (1961))

We shall treat the plane figures as composed of variously combined chains (or one-dimensional figures). Such chains may be formed by simple translations of an aribtrarily shaped object (which has one side white, one black):

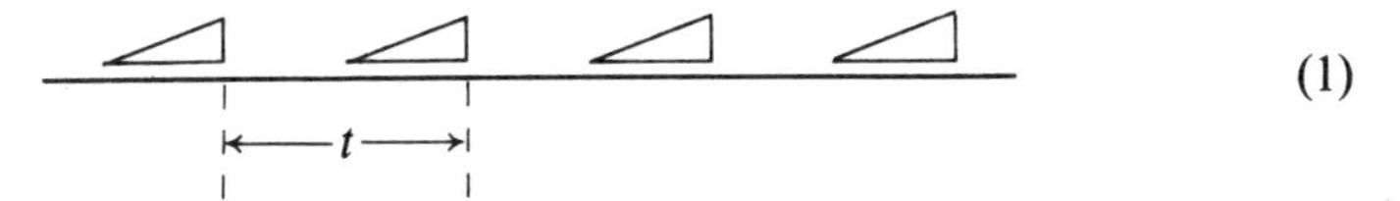

(1)

or a screw axis (2_1):

t

(2)

Further chains can be constructed by introducing other elements of symmetry normal to the chain axis. For example, introducing mirror planes:

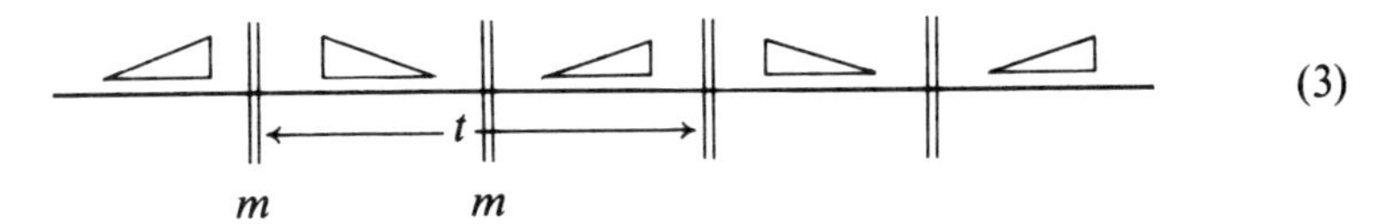

(3)

Two-dimensional chains can be constructed from one-dimensional chains by adding other symmetry elements. Thus, introduction of a mirror plane parallel to the chain axis of (1) or (3) gives:

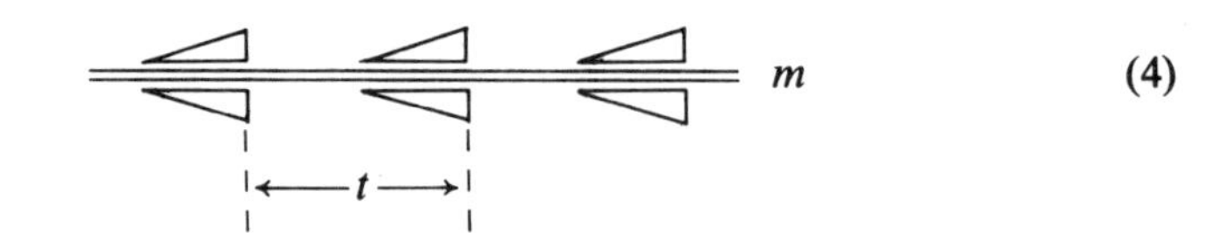

(4)

and

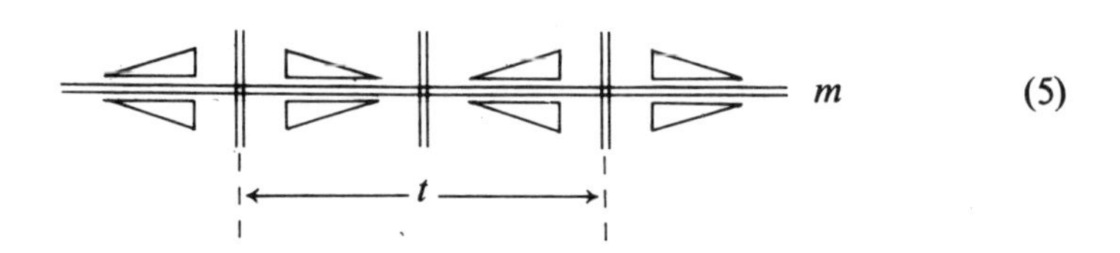

(5)

Let us now combine some of these chains into two-dimensional arrays with the aid of other elements of symmetry. Using the very simple chains (1), we can align them as shown in Figure 6-3. In this arrangement we see that (*a*) the protrusions of one molecule fit into the hollows of the next throughout the lattice; (*b*) the m.c.n. = 6; (*c*) the plane lattice is of the parallelogram

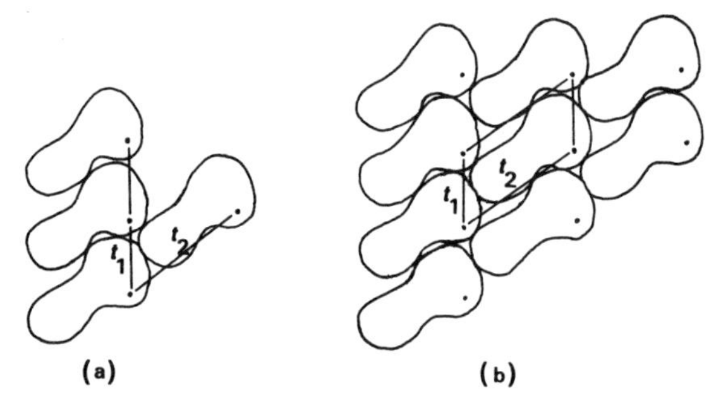

Figure 6-3 A close-packed planar layer of non-directionally bonded molecules of arbitrary shape. (a) A chain of type 1 (see text) to which one additional molecule has been added, determining the second translational repeat, t_2. (b) More of the layer. (Kitaigorodskii (1961))

type with two arbitrary repeat directions. Such an arrangement is therefore to be expected in molecular crystals. Another allowed arrangement can be generated from the same simple chain by the use of a twofold glide line normal to the chain axis as shown in Figure 6-4. If a plane array is built on the basis of the two-dimensional chain (5), the result, Figure 6-5, shows two highly undesirable features: (*a*) large voids within the structure, (*b*) m.c.n. = 4 only.

Following such arguments through for all possible combinations of symmetry elements leads to a list of allowed planar structures. These sheets are then stacked according to the same packing principle (i.e. the hollows of one accept protrusions of the other—just as in the stacking of closest-packed layers of atoms) to give the allowed three-dimensional arrays, which can be described in terms of their space-group symbols. Thus, Kitaigorodskii was

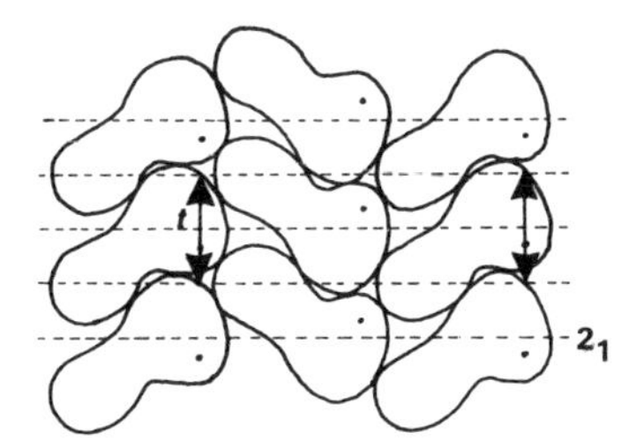

Figure 6-4 Another close-packed layer of molecules formed from chains of type 1. (Kitaigorodskii (1961))

led to the remarkable (and little known) result that only thirteen space groups are 'most probable' for molecular crystals containing molecules of *arbitrary* shape: of these, eight to ten are commonly found.

We now have real insight into the problem which we set ourselves at the beginning of this section. The systematic application of simple geometrical principles has shown that, for molecular crystals in which intermolecular bonding is of the dispersive type, packing is of the closest-packed type with

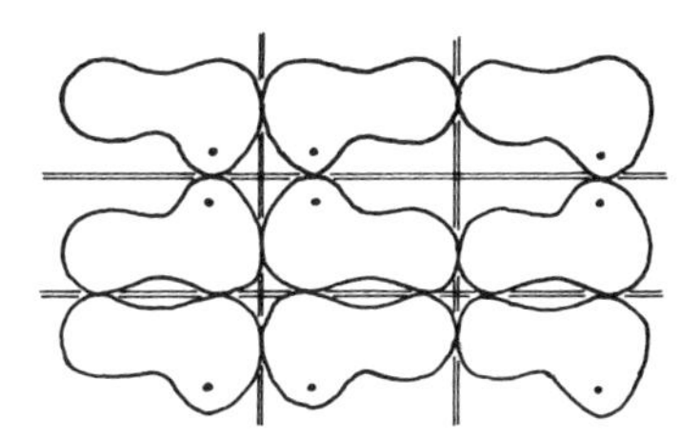

Figure 6-5 An undesirable form of molecular packing derived from chains of type 5. (Kitaigorodskii (1961))

m.c.n. values of 12. Further, only a small number of different spatial arrangements are probable. These arrangements were determined on the premise that the structure adopted will be that which is closest packed, i.e. of minimum volume. Experimental results have confirmed this premise.

Of course, many molecules have inherent symmetry (i.e. are not arbitrarily shaped). In such cases some of the above restrictions are lifted and other

Table 1. Most-probable space groups for molecular crystals (A. I. Kitaigorodskii, *Organic Chemical Crystallography*, Consultants Bureau, New York, 1961)

Inherent molecular symmetry	1, 2, *m*		$\bar{1}$, 2, *m* *mmm*		*mmm*				222	
Molecular symmetry in crystal	1		$\bar{1}$		2		*m*		2	
	Space group	*Z*	Space group	*Z*	Space group	*Z*	Space group	*Z*	Space group	*Z*
Space group and multiplicity (*Z*) of position occupied by the molecule	$P\bar{1}$	2,4	$P\bar{1}$	1,2	$C2/c$	4	Pmc	4	$C2/c$	4
	$P2_1$	2,4	$P2_1/c$	2,4	$P2_12_12$	2,4	Cmc	4	$P2_12_12$	2,4
	$P2_1/c$	4	$C2/c$	4	$Pbcn$	4	$Pnma$	4	$Pbcn$	4
	Pca	4	$Pbca$	4						
	Pna	4								
	$P2_12_12_1$	4								

P represents a primitive unit cell, i.e., atoms at corners only. A *C*-centred cell has twice as many atoms as its equivalent cell but the axes are more conveniently arranged.

space groups are available. The situation has been examined rigorously, along the lines described for arbitrarily shaped molecules and a list of allowed space groups drawn up, Table 1.

6.1.2 Energetics of Molecular Crystals

The success of Kitaigorodskii's theory in predicting a short-list of allowed space groups for crystals composed of molecules with a given symmetry encourages one to ask for even more: is it possible to predict the *actual* space group adopted, and the correct molecular orientations within the unit cell? Attempts to answer these questions have led to a more detailed understanding of molecular crystals, although to date nearly all of this work is on organic materials. An important application of such work is in refining crystal structures to obtain most-probable positions for hydrogen atoms, given the heavy-atom positions.

Equilibrium positions of molecules in crystals are the result of a balance between attractive and short-range repulsive forces. The pairwise interaction energy is most simply represented by an expression of the form

$$U(r) = A/r^n - B/r^m$$

although determination of the exact form of $U(r)$ is a thorny problem. For non-polar molecules the attractive potential is due effectively to London or dispersion forces. It has been approximated in various ways. Thus considering interacting molecules as spherically polarizable systems the method of atom–atom interactions (used successfully by Kitaigorodskii and by D. E. Williams) uses an expression of the form

$$V = -\frac{3\alpha_1\alpha_2}{2r^6} \cdot \frac{E_1E_2}{E_1 + E_2}$$

where α_1 and α_2 are polarizabilities of the two systems separated by distance r between their centres of gravity; E_1 and E_2 are their mean transition energies. Taking anisotropic interactions into account, for systems of cylindrical symmetry Rae and Mason (1968) employed

$$V = -c(\theta, \phi, \theta', \phi')/r^6$$

where c is a complex expression involving angular coordinates of the two interacting molecules, and their parallel and perpendicular polarizabilities. The latter expression lends itself to treatment of crystal potential energy in terms of bond–bond rather than atom–atom interactions. For example, when applied to benzene, three axially symmetric electron clouds were considered: the C—H and C—C σ bonds both consisting of localized electrons, and the ring π-electrons taken as a single-charge cloud symmetric about the sixfold axis of the molecule. The bond–bond type of calculation is rather easier to carry out than the atom–atom variety and seems likely to be important in the future.

The repulsive part of the potential is of overriding importance in determining the detail and, often, the nature of the packing in molecular crystals; physically, this reflects its higher-order dependence upon r (i.e. $n > 6$). In calculations of equilibrium crystal structures the position adopted corresponds to that most favourable to the repulsive forces, whilst the attractive forces (which are effectively non-directional) are in fact *not* minimized at that configuration. Expressions used to describe the repulsive potential vary and need not concern us in detail, although we note that in order to assess this function, particularly accurate molecular wave functions are necessary. In order to avoid this difficulty, a semi-empirical approach suggests a potential of the form $A \exp(-Cr)$ where A and C are constants which must be estimated. Although the repulsive forces determine molecular orientation, it appears that the precise form accorded them does not greatly affect the calculated orientation.

Using expressions for attractive and repulsive potentials such as those above, summation of all the pairwise molecular interactions (divided by two—otherwise each one is counted twice) yields the lattice energy of the structure. If this is minimized with respect to (*a*) molecular orientation, (*b*) unit cell dimensions taken at 0 K, the lattice energy is that of the equilibrium structure at 0 K and is equivalent to the heat of sublimation, ΔH_s (0 K). In practice a value of ΔH_s (0 K) for comparison with that calculated is estimated from the observed value by

$$\Delta H_s (T\,\mathrm{K}) = \Delta H_s (0\,\mathrm{K}) + 2RT$$

We should note the important qualification, that lattice energy can only be equated with heat of sublimation if the molecule has the same configuration in both vapour and solid states: degrees of intramolecular rotational freedom will cause major differences otherwise (Davies, 1971). These calculations all relate to the enthalpy, H, whereas at any temperature T above absolute zero, the preferred crystal structure is that of lowest free energy $G = H - TS$. Therefore, the calculations only apply strictly to structures at 0 K. However, the effect of the entropy term is usually not too important and nowadays there are increasing numbers of X-ray structure determinations carried out at very low temperature, which is an acceptable approximation to 0 K. For calculations at constant entropy and volume, the above method of minimization of energy is therefore a reasonable approximation.

We now consider the results of three recent calculations illustrating the importance of this work.

Kitaigorodskii (1971) has calculated the equilibrium conformations of the molecules EPh_4, (E = C, Si, Sn, Pb), all of which crystallize in the tetragonal system with *two* molecules per cell. The calculation begins with knowledge of the E atom positions and with the restriction that the orientation of the four tetrahedrally disposed phenyl groups be such as to give the molecule the highest possible symmetry: in this case the requirement is for a fourfold rotation–reflection axis. The entire calculation can then proceed in terms of two angles: θ, which measures the angle of rotation of the phenyl groups about the E–C bonds, and ϕ, which measures the orientation of the molecule about its fourfold axis. Figure 6-6 shows how the interaction energy $U(r)$ varies with θ, while Figure 6-7 shows potential-energy contours at 4·18 kJ mol^{-1} intervals for $SnPh_4$. The minimum lies at $\theta = 61°$, $\phi = 7°$, compared with the experimental values $\theta = 59°$, $\phi = 7°$. Thus, the calculation confirms that the structure is the one of lowest energy, and incidentally supports the method of calculation.

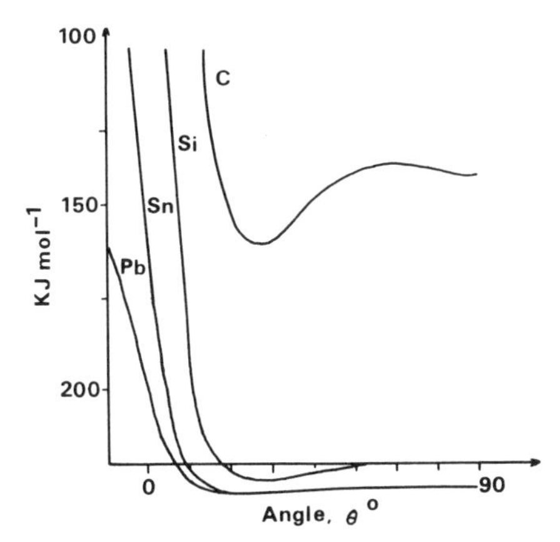

Figure 6-6 Variation of the interaction energy $U(r)$ for $SnPh_4$ with θ, the angle of rotation of the phenyl groups around the Sn—C bonds. (Ahmed, Kitaigorodskii and Mirskaya (1971))

Two particularly important calculations have recently been made by Mason and Rae (1968) using their bond–bond interaction method described above. In the first, the structure of crystalline benzene was studied over a range of orientations about the known molecular lattice positions. At the energy minimum, the computed unit-cell constants were within 3% of the observed values. The computed lattice energy at 270 K at the same minimum was 52·3 kJ mol^{-1} compared with the experimental latent heat of sublimation of 44·4 kJ mol^{-1}. This was made up of 73·6 kJ mol^{-1} from the attractive dispersion forces and 21·3 kJ mol^{-1} due to repulsive forces. In view of the approximations involved at various stages of the calculation, the agreement with experiment is very gratifying.

In their second calculation, Mason and Rae sorted out a long-standing problem. Given the task of packing benzene molecules to form a crystal, surely the most obvious way to try is to stack them in columns, which can then be packed to form the crystal. This is an obvious method to try since plane space can be filled by hexagons. But benzene crystallizes, as shown in Figure 6-8, in a 'herringbone' packing mode which is almost always found for aromatic hydrocarbons and for many other molecules, such as ethylene.

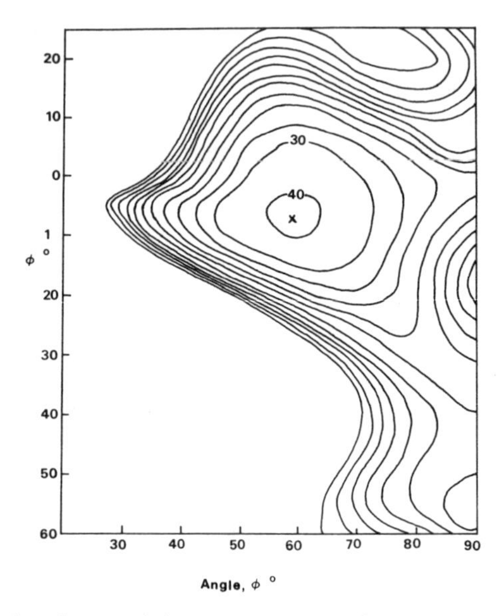

Figure 6-7 Calculated potential energy contours for crystalline $SnPh_4$. (Ahmed, Kitaigorodskii and Mirskaya (1971))

In contrast, heterocyclic molecules crystallize in a variety of packing modes; *sym*-triazine in particular is composed of vertical stacks offset relative to each other, as shown in Figure 6-9, to make most efficient use of space. Since the principal difference between benzene and *sym*-triazine molecules

sym-triazine

is replacement of three CH groups by nitrogen, in accounting for their crystal structures suspicion at first falls upon their different charge distributions caused by the greater electronegativity of nitrogen which might therefore stabilize the observed *sym*-triazine structure by multipole–multipole interaction. Further, the lone pairs on nitrogen might also be expected to make a

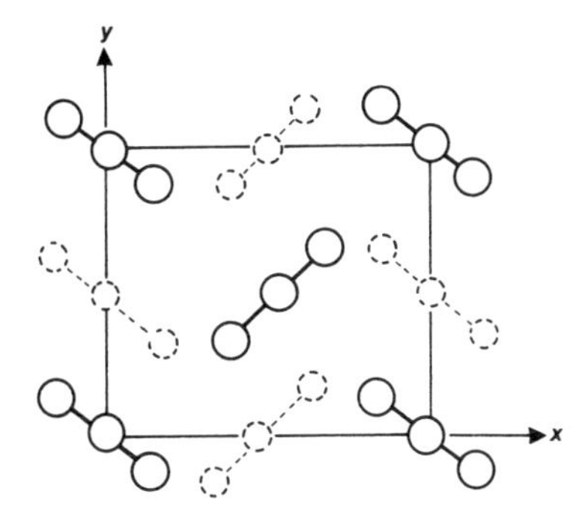

Figure 6-8 The crystal structure of benzene. Two superimposed layers are shown

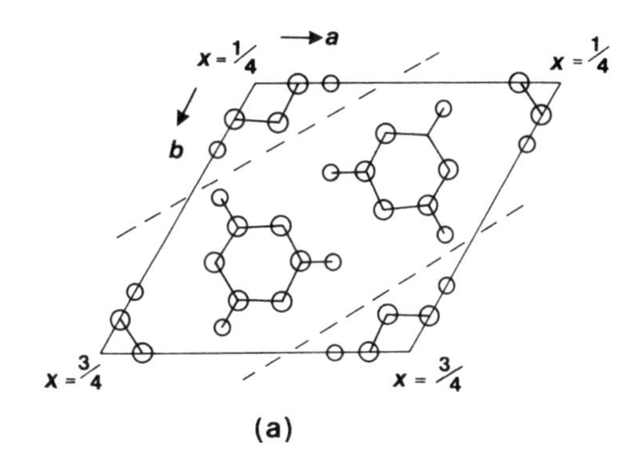

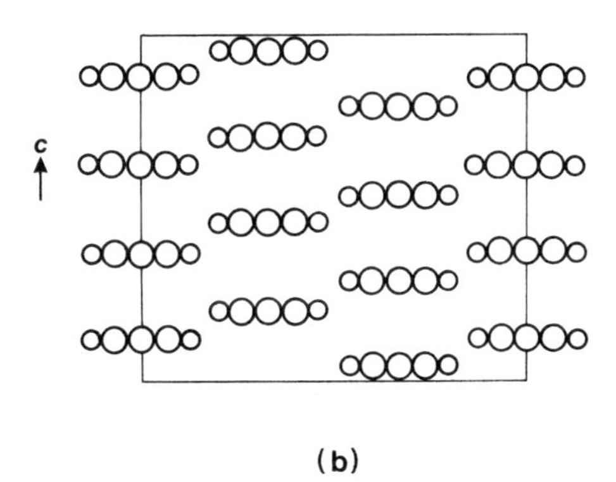

Figure 6-9 The crystal structure of *sym*-triazine. (a) Plan view. (b) Showing the columnar stacking. (Mason and Rae (1968))

large electrostatic contribution to the total lattice energy. Calculation showed that neither of these effects is large and that they do not have a determinative effect on the crystal structure: together they contribute a small net attractive term to the total calculated lattice energy of 39·3 kJ mol^{-1}, which is close to the experimental value.

The reason for the difference between these two crystals arises from the need to find a structure for benzene in which H . . . H repulsions are minimized. This is accomplished more effectively in the observed structure than in the columnar stacking mode of *sym*-triazine which would bring the H atoms of adjacent benzene columns closer to each other. This was elegantly demonstrated by calculating the lattice energy of benzene in the *sym*-triazine lattice: it was found that the potential energy was not minimized until the *a*-axis of the unit cell was 1·6 Å larger than the observed *sym*-triazine value (i.e. it was necessary to separate the molecular columns). If benzene had such a structure, the large *a* spacing would remove much of the barrier to rotation of the rings, leading to a disordered structure. On the other hand, reduction of *a* towards the *sym*-triazine value raises the repulsive potential to unacceptable values.

The importance of this result cannot be overemphasized. One awaits with interest many more such calculations, as they will enable us to refine our understanding of the packing modes in molecular crystals.

In **summary**, our knowledge of the principles underlying structures of non-polar molecular crystals is most encouraging. Kitaigorodskii's work has laid out the symmetry requirements for packing, whilst fuller appreciation of factors affecting particular structures is beginning to come from calculations of lattice energy. The crystallography of molecular solids is seen to be approaching the status of a branch of quantitative solid-state science.

6.1.3 Inorganic Molecular Crystals

The theoretical framework outlined in the two preceding sections covers molecular crystals composed of any kind of molecule: organic, organometallic, or inorganic. In this section, we review a few inorganic examples chosen partly for their importance and especially to allow discussion of bonding of intermediate types.

Diatomic molecules usually form a closest-packed lattice at temperatures just below the melting point: in these structures the molecules rotate freely and behave as spherical atoms. Thus,

h.c.p.	H_2, β-N_2 CO
c.c.p.	γ-O_2, β-F_2

Upon cooling (typically, in the range 20–70 K) molecular rotation is restricted and other, lower symmetry, lattices are commonly adopted. **Chlorine, bromine** and **iodine** all have the same orthorhombic structure shown in Figure

6-10. This structure is a classic illustration of Kitaigorodskii's packing theory: thus, chains of molecules related by a twofold screw axis (see (2) on p. 171) are packed alongside each other to give layers as shown in Figure 6-9a. These layers are stacked (b) so that the hollows in the first layer accept the protrusions of the next: i.e. they are offset relative to each other. Interatomic distances show that bonding within each close-packed layer is stronger

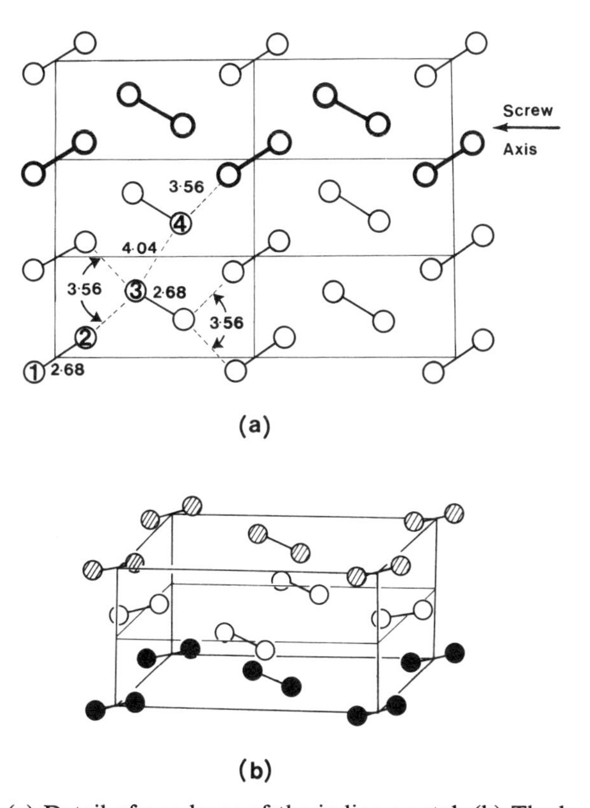

Figure 6-10 (a) Detail of one layer of the iodine crystal. (b) The layer stacking found in crystalline chlorine, bromine and iodine. Note the formal analogy with c.c.p. (From *Fundamentals of Inorganic Crystal Chemistry* by H. Krebs. © 1968 McGraw-Hill Book Company (UK) Ltd. Used with permission)

than between layers. Thus, within each layer, 3·56, 4·04 Å; between layers 4·35, 4·40 Å. This difference explains the ease with which crystals of iodine are cleaved. Comparison with van der Waals radii (r_W) shows that distances within the layer are shorter than $2r_W$ indicating a covalent-bond contribution, although bonding between layers is of the van der Waals type. Krebs has explained these features along the following lines, although he prefers the term mesomerism.

Consider the p-orbitals on iodine. Two are filled but the third is singly occupied and is used principally to form the strong covalent bond to the other atom of the diatomic molecule (Ψ_1). However, other contributions to the total wave function come from overlap with a p-orbital of the next-nearest iodine neighbour in the crystal:

$$\begin{array}{llll} \mathrm{I} & \mathrm{I—I} & \mathrm{I} & \Psi_1 \\ \mathrm{I—I} & & \mathrm{I—I} & \Psi_2 \end{array}$$

with Ψ_1 having more weight than Ψ_2. Each iodine molecule is then associated by two shortish bonds (3·56 Å) to approximately co-linear neighbours forming a scheme in which the bond lengths run:

$$\mathrm{I} \overset{3{\cdot}56}{\text{-------}} \mathrm{I} \overset{2{\cdot}68}{\text{———}} \mathrm{I} \overset{3{\cdot}56}{\text{-------}} \mathrm{I} \qquad \rightarrow z, \ \downarrow x$$

In a direction perpendicular to this, but within the layer, overlap of p_x-orbitals takes place sideways resulting in the formation of longer bonds, 4·04 Å. Although both p-orbitals are formally filled, it is considered that a wave function, Ψ_3, makes a contribution in which the electron hole spends some of its time in the p_x-orbital, thereby making possible the formation of longer (since this wave function only makes a small contribution) bonds, 4·04 Å. Thus:

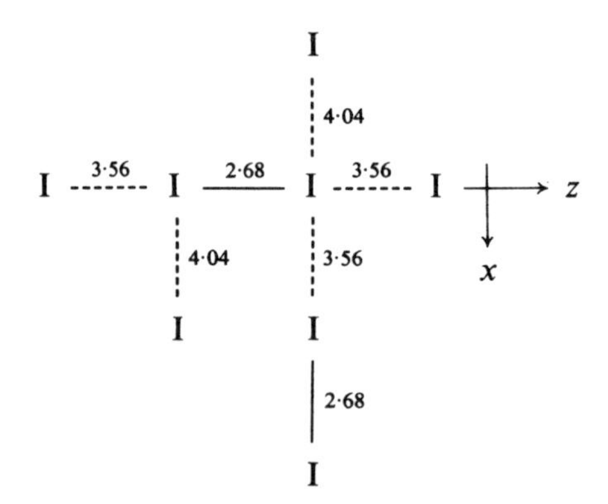

Due to the packing arrangement, there is an alternation of the three bond lengths as shown above, and in Figure 6-10. Although we have considered p-orbitals only, there is almost certainly some contribution from $d(t_{2g})$–p-orbital overlap. In bromine and chlorine crystals the bonding is less pronounced within the layers, so that the intramolecular distances are more nearly equal, the effect falling off in the order of decreasing atomic weight.

According to the 8–N rule, **sulphur** can form two covalent bonds to neighbours: joining them can therefore lead to formation of either rings or chains, and both happen in polymorphs of sulphur and selenium. In neither case is it necessary to consider the topological requirements since a molecular crystal is formed whichever way the atoms are linked. Sulphur has quite a complex allotropy: the thermodynamically stable form consists of S_8 crown-shaped rings in close-packed layers, m.c.n. = 6, within *each layer*; see Figure 6-11.

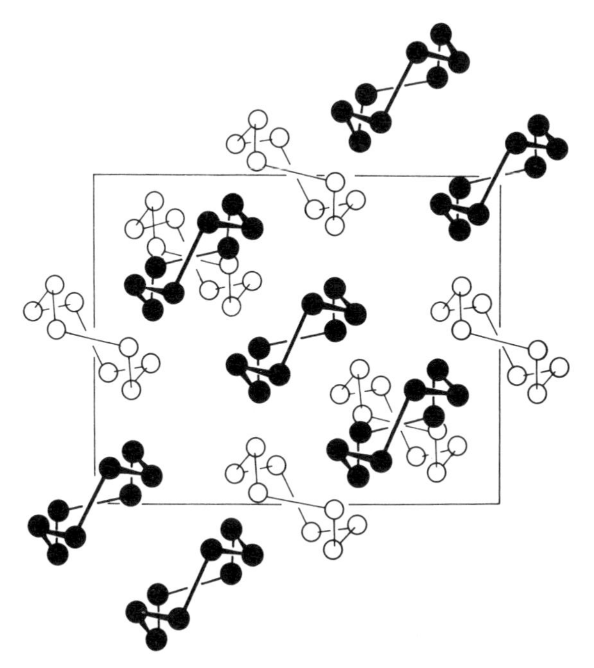

Figure 6-11 The packing of S_8 rings in crystalline orthorhombic sulphur. (Modified from *Fundamentals of Inorganic Crystal Chemistry* by H. Krebs. © 1968 McGraw-Hill Book Company (UK) Ltd. Used with permission)

An unstable form of selenium has the same structure. Metastable allotropes of sulphur are known in which S_6 or S_{12} rings are present, whilst 'fibrous' sulphur consists of spiral chains interleaved with S_8 rings (Figure 6-12).

Crystalline CO_2 and N_2O have a cubic lattice in which the carbon atoms (of CO_2) define a face-centred cubic array. The molecules are close-packed into layers stacked very tightly, m.c.n. = 12, each molecule making one C···O contact with each neighbour.

In contrast $\mathbf{HgCl_2}$, although its molecules are also linear, packs in a different way. Its space group, *Pnma*, is adopted by some twenty molecular compounds. A projection of the structure is shown in Figure 6-13, from which it can be seen that the individual layers are of the same type as in iodine (each molecule has six nearest neighbours in the plane; total m.c.n. = 12 taking in contacts with the neighbouring layers), but that their stacking is different.

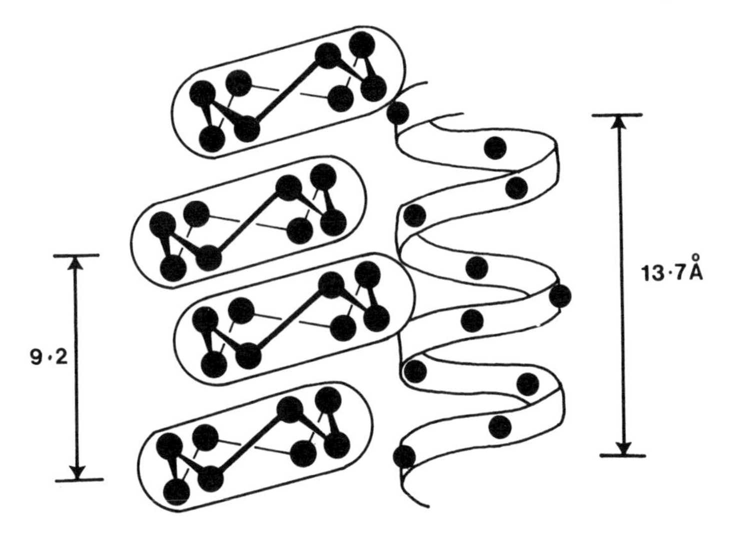

Figure 6-12 The structure of 'fibrous' sulphur. (After W. E. Addison, *The Allotropy of the Elements*, Oldbourne Press, London, 1964)

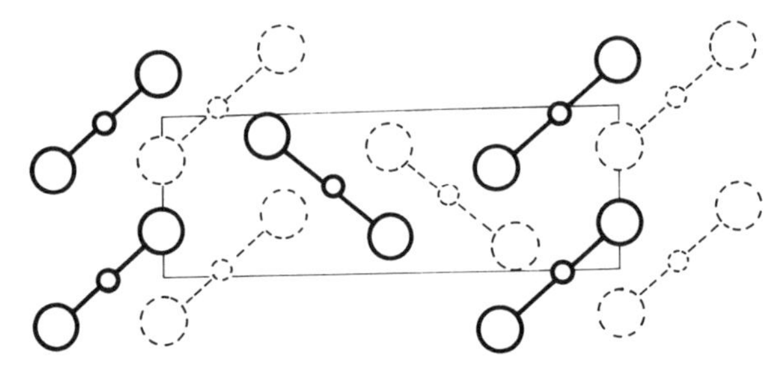

Figure 6-13 The crystal structure of $HgCl_2$. (From A. F. Wells *Structural Inorganic Chemistry*, The Clarendon Press, Oxford, 1962)

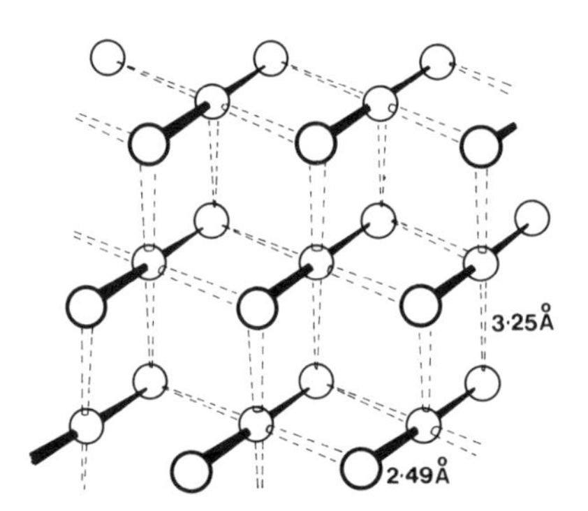

Figure 6-14 The crystal structure of $HgBr_2$; compare with the $CdCl_2$ layer structure, Figure 3-18b

Although $\mathbf{HgBr_2}$ might have been expected to adopt the same lattice as $HgCl_2$, its packing is completely different and denser, leading to non-bonded inert-atomic distances *shorter* than in the chloride. The reason for this is that bromine can indulge in bonding by use of filled *p*-orbitals donating into vacant 6*p*-orbitals on mercury in a way that is much less important for chlorine: cf. the Cl_2, Br_2, and I_2 structure. Indeed, it is probably better not to regard $HgBr_2$ as a molecular crystal but to relate it to the layer structure of $CdCl_2$ as shown in Figure 6-14; it is genuinely intermediate between covalent, ionic and molecular, in bond type.

Replacement of one halogen by methyl leads to adoption of yet another structure type, Figure 6-15. Rotation of the methyl groups gives the molecules axial symmetry: in such cases the molecules are always aligned with a rotation

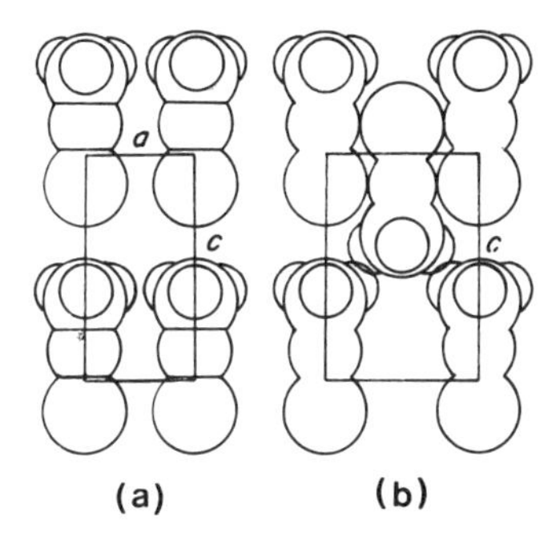

Figure 6-15 The structure of CH_3HgCl, showing two different projections. (Kitaigorodskii (1961))

axis of the unit cell (Kitaigorodskii, 1961). It seems that beyond a certain length it becomes easier to pack molecules parallel to each other rather than in any of the other forms we have encountered. This occurs in the mercurous halides Hg_2X_2 (X = F, Cl, Br, I); each mercury atom has four next-nearest neighbour contacts with other molecules completing highly distorted octahedral coordination.

Cl Cl

Cl—Hg $\overset{2{\cdot}53}{\text{—}}$ Hg $\overset{2{\cdot}52}{\text{—}}$ Cl

3·17 Å

Cl Cl

A large number of **other molecular halides** crystallize in structures which are close packed, or nearly so. Since fluorine is the least likely of all the halogens to indulge in extra-molecular bonding of the type described above for crystals of the other halogens, we might reasonably expect to find simple close-packed structures for crystalline molecular fluorides. The hexafluorides

MF_6, (M = Mo to Rh and W to Pt), crystallize in a b.c.c. lattice near room temperature; although this is not strictly close packed, it is probably stabilized by the additional interactions possible with the next-nearest neighbours. The considerably fewer hexachlorides seem to have similar structures: α-WCl_6 and UCl_6 can be considered as h.c.p. lattices of chlorine with metal atoms in one-sixth of the octahedral holes. The tetrameric pentafluorides $(MF_5)_4$ can be similarly described: for M = Ru, Os, Rh, Ir, the structures consist of approximately h.c.p. fluorine with M in one-fifth of the octahedral sites, while for M = Nb, Ta, Mo, W, the fluorine lattice is approximately cubic close packed. Although description of all these molecular halides in terms of their relations to c.c.p. and h.c.p. lattices is helpful in terms of visualization, it should be realized that the equivalent description of molecular packing in terms of Kitaigorodskii's principles is the more realistic physically. The pentafluorides have the structure

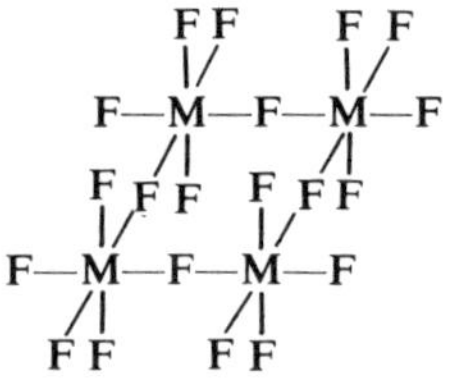

```
          F F    F F
           |/     |/
       F—M—F—M—F
          /|     /|
       F F F   F F F
        |/     |/
   F—M—F—M—F
      /|     /|
     F F    F F
```

with the bridge M—F bonds longer than the terminal ones (2·03 compared with 1·84 Å, for OsF_5).

The SnI_4 structure, also adopted by $TiBr_4$ and the tetraiodides of Si, Ge and Ti, can be described as c.c.p. iodine with metal atoms in one-eighth of the tetrahedral holes. Again, this is an equivalent description to one in terms of close packing of tetrahedral molecules. Wells (1962) has drawn attention to the formal relation between SnI_4 and CuI (wurtzite structure): in both, the c.c.p. lattice of iodine is the same, but the number and arrangement of occupied tetrahedral holes is different.

Early in the second and third transition series the *d*-orbitals are large and diffuse, resulting in formation of many cluster compounds such as $[Ta_6Cl_{12}]^{2+}2Cl^-$ in which M—M bonding is partly responsible for stability of the structure. Rather surprisingly, platinum(II) also forms a cluster halide, Pt_6Cl_{12}, but in this there is no metal–metal bonding: the structure is shown in Figure 6-16. Each platinum atom uses four dsp^2 hybrids (directed towards the corners of a square) to bond to four chlorines; it is the bridging chlorines which hold the cluster together.

The bonding in $ReCl_3$ is mixed, making it intermediate between molecular and covalent halides, rather like $HgBr_2$. It consists of trimeric units Re_3Cl_9 which retain their identity up to 600°C in the gas phase. In the solid these units are held together by shared chlorine atoms as shown in Figure 6-17. The structure of the bromide is similar.

As a final example, we consider the packing of two **phthalocyanine complexes**. Phthalocyanine, and its Ni^{II} and Pt^{II} salts all crystallize in the same monoclinic lattice. However, the differences are significant. Due to the shape

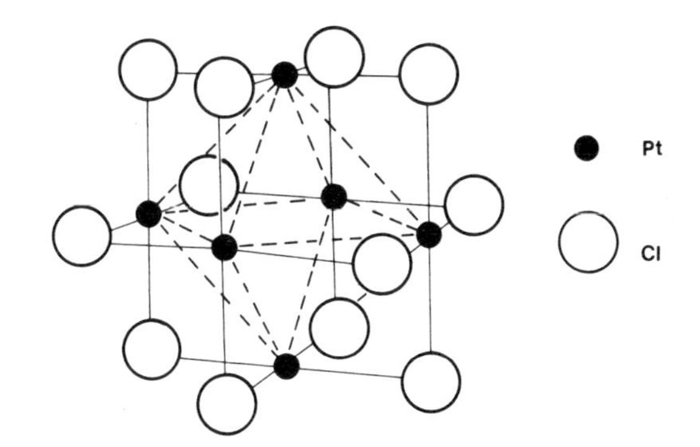

Figure 6-16 The structure of Pt_6Cl_{12}

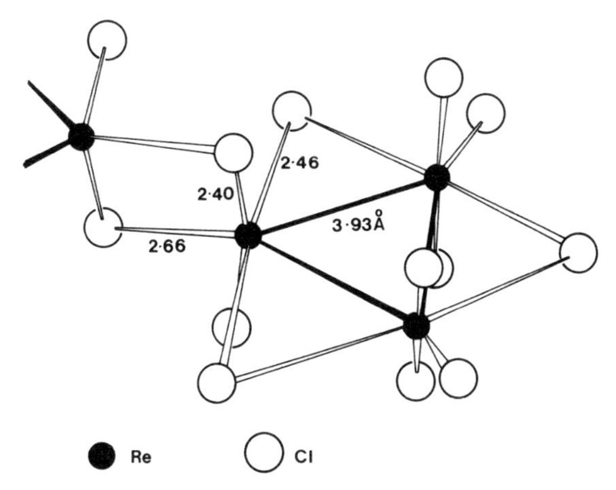

Figure 6-17 The structure of solid rhenium trichloride, showing trinuclear units held together by longer bonds (2.66 Å). (Cotton and Wilkinson (1972))

of the individual molecules the *b*-axis of the cell is short in comparison with the others:

	a(Å)	*b*(Å)	*c*(Å)
phthalocyanine (PC)	19·85	4·72	14·8
Ni(PC)	19·90	4·71	14·9
Pt(PC)	23·9	3·81	16·9

This brings the molecules rather close to each other. Looking along the *b*-axis there is a difference in stacking of the Ni and Pt salts (Figure 6-18). For Pt the arrangement is staggered so that the metal just has four square-planar disposed bonds (1·97 Å) to the four PC nitrogen atoms, but for Ni the stacking brings one nitrogen from each of the two next-neighbour molecules along *b* directly over the metal giving it a distorted octahedral arrangement (four Ni—N at 1·83; two Ni---N at 3·38 Å). It seems possible that there is additional electrostatic energy stabilizing the nickel structure.

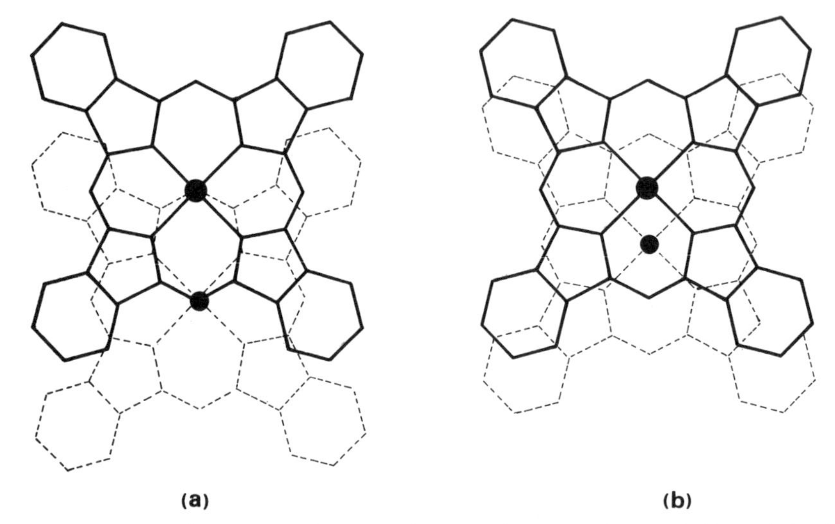

Figure 6-18 The stacking of the phthalocyanine complexes of (a) nickel, (b) platinum. (Kitaigorodskii (1961))

6.1.4 Molecular Crystals Containing Chains

Chain molecules always pack with their axes parallel. In the case of polymers which have degrees of flexibility (e.g. rotation about certain bonds) a variety of conformations may be adopted in solution, but in the crystal the chain orientation is such that it has the maximum possible length consistent with valence theory and the need to prevent steric hinderance between substituent groups. For molecules with end groups (such as long-chain alcohols) the end groups form a plane layer which is inclined to the chain axes by an angle θ which may be either 90° or oblique. Packing of chains of arbitrary cross-section according to Kitaigorodskii's principles is most probable in the simple lattices of Figures 6-3 and 6-4, although if chains are roughly spherical in cross-section (e.g. a helix viewed along its axis) the packing will be like that of a closest-packed layer, Figure 3-1b. Viewed normal to their axes, chains will tend to pack as shown in Figure 6-19a and 6-19b. Further considerations of layer stacking enter in the case of chains of finite length but we shall only be concerned with chains that are, in principle, of infinite extent.

Helical Chains

Tellurium and the stable form of selenium both form structures composed of helical chains packed together to give a high-symmetry hexagonal lattice. Along their lengths the chains are aligned as in Figure 6-20a with the 'elbow' of one chain fitting into the recess of the next. In accord with the 8–*N* rule each

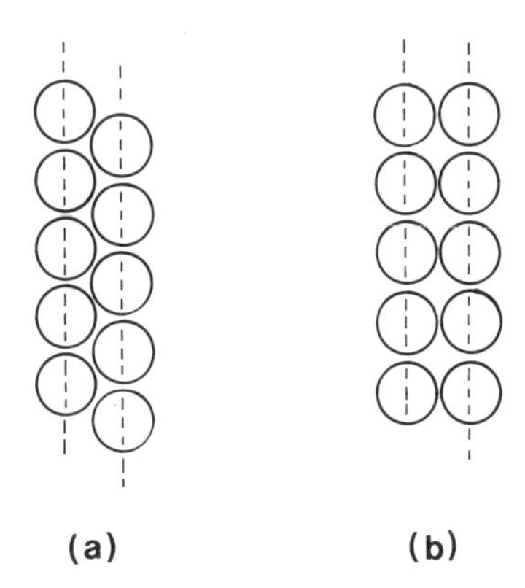

Figure 6-19 Two possible ways of packing linear chains in crystal.

atom has two nearest neighbours covalently bonded to it. A distorted octahedron is completed about each atom by neighbours from *three* nearest chains (Figure 6-20a). With the tendency towards dehybridization down a group, the ratio of nearest: next-nearest neighbour distances decreases resulting, for the heaviest members, in formation of a metallic structure. This trend is well shown by Se, Te and Po, the latter having (in its α-phase) the simple cubic structure (i.e. rock salt with *all* positions occupied by Po); see also Table 2. Since the observed interchain distances are a little shorter than the sums of van der Waals radii, bonding between chains of Se and Te clearly has a slight covalent contribution in addition to dispersive forces. Comparison with the simple cubic lattice of α-Po is a profitable one: the *trend* towards it is clearly seen through the series of three structures.

If pure *p* functions were used, the chain angles would be 90° (Figure 6-20b)

Table 2. Distances (Å) and angles for crystals of Group VIb elements (H. Krebs, *Fundamentals of Inorganic Crystal Chemistry*, McGraw-Hill, London, 1968)

	Nearest, d_1	Next nearest, d_2	d_2/d_1	Bond angle
Se	2·374	3·426	1·443	102° 50′
Te	2·878	3·451	1·199	101° 46′
α-Po	3·359	3·359	1·000	90°

but the need to accommodate the stereochemical effect of two non-bonding pairs opens up the angle to ca 102° (Se and Te have configurations [core]s^2p^4).

Down the triad S, Se, Te, dehybridization becomes progressively more important, and not all of the s^2p^4 electrons are used in s/p combinations. For Se, $4d$-orbitals are only a little higher in energy than $4p$ and can readily enter sp^3d^2 hybrids. The chain structure and packing of Se is therefore to be understood in terms of bonding in the chains, which is almost s^2p^4, and of weak

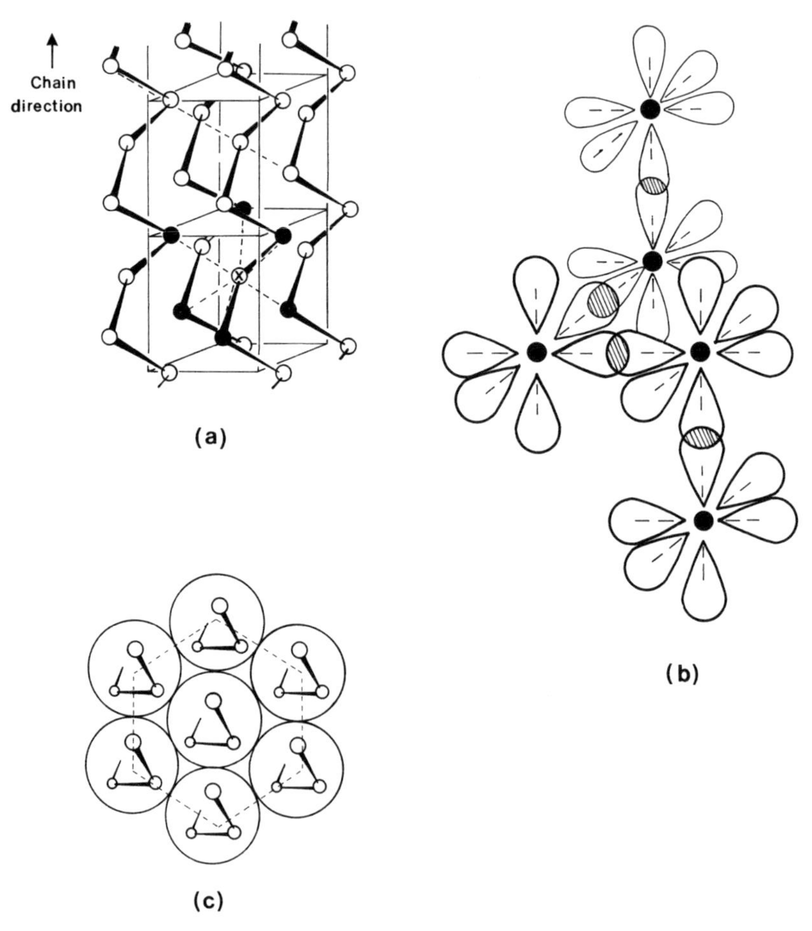

Figure 6-20 (a) The crystal structure of hexagonal selenium and tellurium. Note the distorted octahedron (black atoms) formed about each atom (x is taken as the example). The dotted lines indicate the directions in which the sp^3d^2 hybrids are overlapped. From *Fundamentals of Inorganic Crystal Chemistry* by H. Krebs. © 1968 McGraw-Hill Book Company (UK) Ltd. Used with permission). (b) The structure each chain would have if only p-orbitals were used in bonding. (c) Illustration of the packing of the chains (cf. Figure 3-1b)

interchain bonding via a very low population of electrons in sp^3d^2 hybrids, overlapped to give a band.

The helical chain structure of cinnabar, α-HgS, is shown in Figure 5-33. Bonding between the chains can be described by means analogous to that for selenium.

Tetrahedral Chains

The structure of $BeCl_2$ consists of chains of tetrahedrally coordinate Be atoms bridged by chlorines. SiS_2 has the same structure, as does $BeMe_2$ if the hydrogen atoms are omitted. Methyl is held in a bridging position by three-centre bond formation using sp^3 hybrids on both Be and C.

Since these chains are approximately cylindrical in section, the packing is very simple with the chains offset as in Figure 6-19a. The structure is shown in Figure 6-21. The same type of chain is found in $K^+[FeS_2]^-$ and in TlS which is correctly formulated $Tl^I[Tl^{III}S_2]$. In these complex ionic chain crystals the cations are between chains, in eight-coordination.

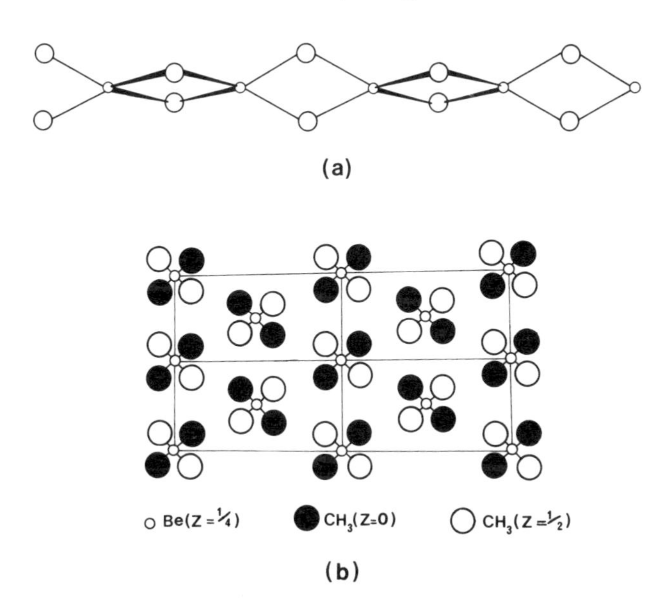

Figure 6-21 (a) Schematic diagram of a chain in which tetrahedrally coordinated atoms are doubly bridged. (b) The packing of chains in $\{Be(CH_3)_2\}_n$. (Kitaigorodskii (1961))

We shall consider examples of other anionic chains based upon tetrahedra in Chapter 7 (silicates). A molecular chain is present in α-SO_3 which can be considered as a series of (SO_4) tetrahedra linked by sharing vertices; CrO_3 has a similar structure. The γ-form of SO_3 has the ring structure shown in Figure 6-22a.

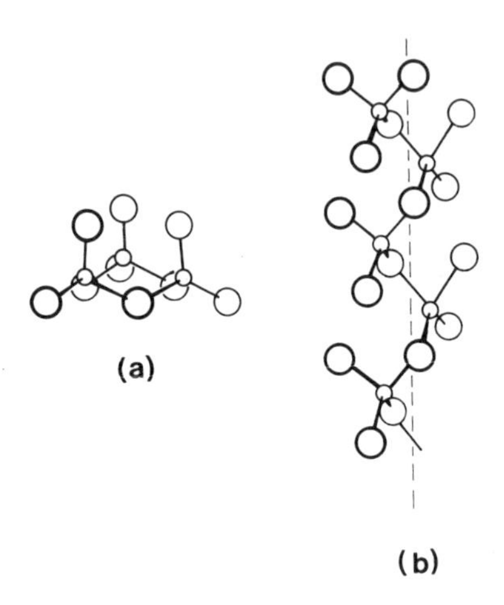

Figure 6-22 The structure of (a) γ-SO_3, (b) α-SO_3. (From A. F. Wells, *Structural Inorganic Chemistry*, The Clarendon Press, Oxford, 1962)

Planar Chains are formed by $PdCl_2$ and one polymorph of $PtCl_2$, and by their bromides. The chains are planar to accommodate the bonding requirements of d^8 Pt^{II} which uses dsp^2 hybrids to bond to four chlorine neighbours. The chain packing is in accord with Kitaigorodskii's principles, Figure 6-23, with the chains offset as in Figure 6-19a.

Pleated Chains are present in HgO, PbO (yellow) and Ag(SCN), all of which were discussed in Chapter 5. A rather special kind is exhibited by SbSBr and by nine other sulpho- and seleno-halides of antimony and bismuth. Pleated $(SbS^+)_n$ chains lie between Br-ions. Each antimony atom is bonded to three sulphur nearest neighbours and vice versa in accordance with the requirements of valence theory. Chains of a related type are present in Sb_2O_3. (N.B. As_2O_3 has a different structure, p. 228.) They are as shown in Figure 6-24 and represent, topologically, one of the simplest three-connected systems.

Octahedral Chains

There are many molecular chains which can be considered as formed by sharing of vertices, edges or faces of octahedra. We list examples without much further discussion as no new packing principles are raised: Figure 6-25 summarizes their structures. Joining octahedra by a common *vertex* has no significant restrictions placed upon it in terms of bonding, other than the

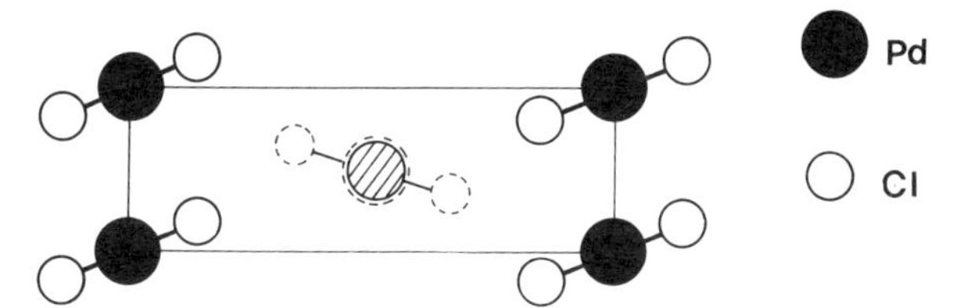

Figure 6-23 The chain packing in $(PdCl_2)_n$

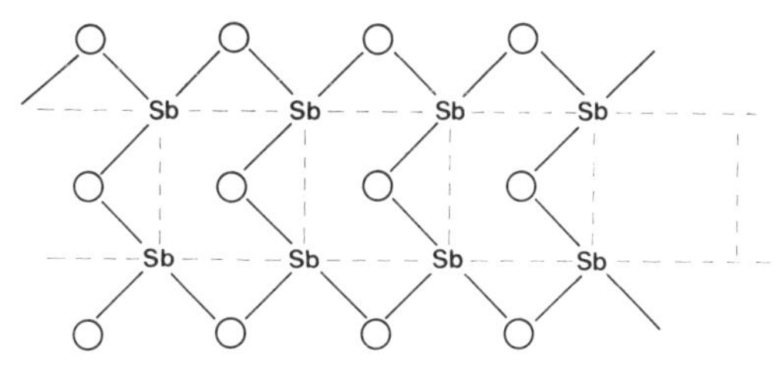

Figure 6-24 The structure of the chains in valentinite, Sb_2O_3. The simple three-connected net upon which it is based is also shown (-----)

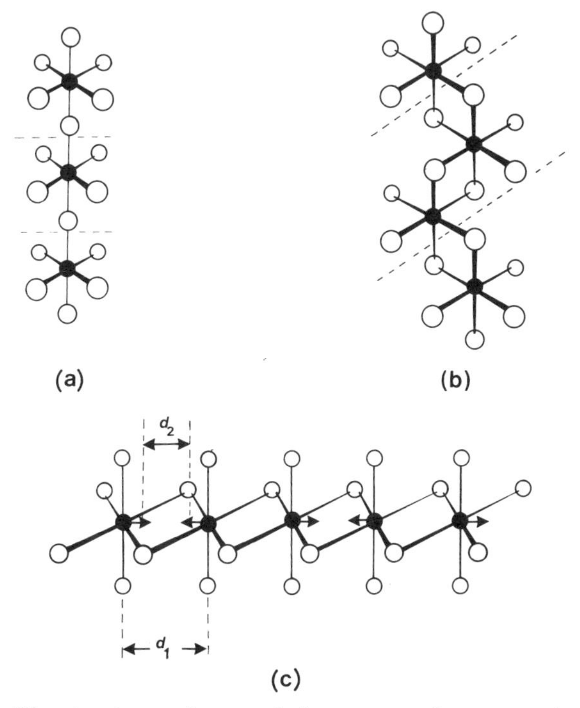

Figure 6-25 The structures of some chain compounds composed of octahedra. (a) An MX_5 chain sharing opposite vertices. (b) A chain formed by sharing edges. (c) A chain formed by sharing opposite octahedral edges. In NbI_4 the metal atoms occur in pairs and are displaced from the octahedron centres as shown

obvious one that the bridging atom or group must be able to bond to two metal atoms. Thus, both transition and non-transition metal atoms are found in such chains, e.g. BiF_5 and α-UF_5; the anion chain in Na_2AlF_5 is similar. A closely related structure is that of $WOCl_4$ in which short and long W—O bonds alternate along the chain: this is best regarded as a molecular

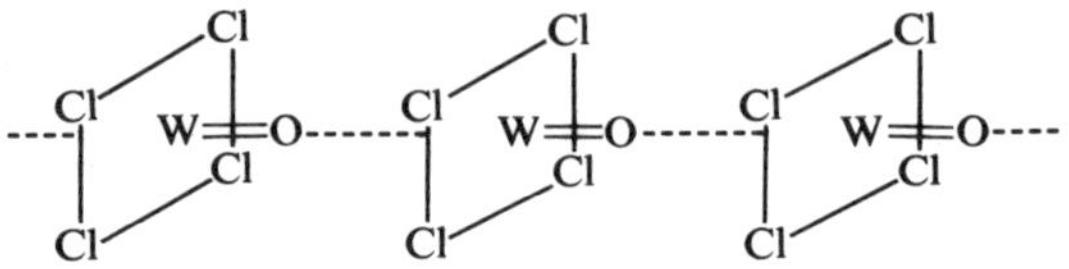

crystal in which strongly dipolar molecules $WOCl_4$ associate in chain form to give the strongest possible electrostatic interaction. In contrast WOF_4 molecules condense in the solid state via W---F interactions, not W---O, because fluorine is more electronegative than oxygen. The resulting structure is a fluorine-bridged tetramer; cf. MF_5's, p. 186.

Sharing *opposite edges* brings the central atoms closer together ($l\sqrt{2}$ compared with $2l$, where l = length of a bond from the central atom to one of its six ligands, all bonds being assumed equal) with the result that transition metals can, in principle, form additional bonds along the chain via t_{2g}-orbital overlap (since t_{2g}-orbitals point at the edges of an octahedron). Sharing opposite edges is found in NbI_4, see Figure 6-25c, which should be paramagnetic since it is $d^1(t_{2g}^1)$; it is diamagnetic. Since Nb is near the beginning of its transition series, the *d*-orbitals are quite large, allowing some overlap with neighbouring atom orbitals. However, the most energetically favourable situation is formation of metal–metal bonds between pairs of atoms, as shown in the figure. This results in a shift of metal atoms from the octahedral centres, the strength of the metal–metal bond more than compensating for the spin-pairing energy.

Many complexes of the type MX_2L_2, where X = Cl, Br or I, and L = neutral donor ligand, are composed of octahedra sharing opposite edges via halogen bridges. Typical examples are α-$CoCl_2Py_2$, $CoX_2.2H_2O$, $MnX_2.2H_2O$, $CdCl_2(NH_3)_2$. There is no metal–metal bond in these compounds.

Octahedral *edges* that are neither opposite nor adjacent are shared in the structure of $TcCl_4$ and of $ZrCl_4$, Figure 6-25b.

Sharing *faces* is a highly unfavourable process electrostatically as this brings the metal atoms very close together ($1{\cdot}15l$, where l is defined as above). When it does occur the chains contain quite strong covalent bonds, although it is possible to form binuclear anions *without* metal–metal bonds by sharing octahedral faces, as in $Tl_2Cl_9^{3-}$ or $Cr_2Cl_9^{3-}$, by slight displacement of the metal atoms *away* from octahedral centres. For the chromium complex the absence of a metal–metal bond is confirmed by observation of a paramagnetic moment corresponding to three unpaired electrons per atom (Cr^{III} is t_{2g}^3). As the size of the atom is increased while keeping chlorine as the bridging group, metal–metal interaction becomes inevitable. Thus, in $W_2Cl_9^{3-}$ not only is a bond formed but the metal atoms are displaced *towards* each other to increase its strength; the bond is formed by overlap of all three t_{2g}-orbitals (each is singly occupied since W^{III} is d^3) as shown in Figure 6-26. The same

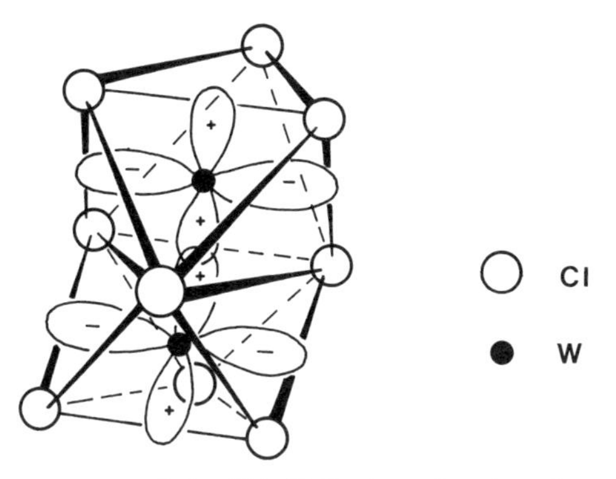

Figure 6-26 The structure of the $[W_2Cl_9]^{3-}$ ion, showing the orientation of the t_{2g} orbitals which form the metal–metal bond. Only one of the three sets of t_{2g} orbitals is shown for clarity

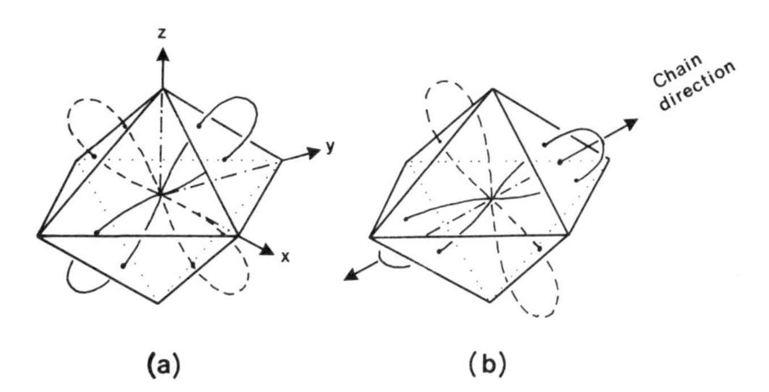

Figure 6-27 Linear combinations of t_{2g} orbitals used in the β-$ZrCl_3$ structure (a) $d_{(1/\sqrt{2})z(x+y)}$, (b) $d_{(1/\sqrt{2})\{\cos 30°\,(1/\sqrt{2})z(x+y)+(\sin 30°)xy\}}$. (From *Fundamentals of Inorganic Crystal Chemistry* by H. Krebs © 1968 McGraw-Hill Book Company (UK) Ltd. Used with permission)

kind of bonding can, and does, occur in the β-$ZrCl_3$ structure (p. 56). Thus, the t_{2g}^3 Mo^{III} compounds MoI_3 and $MoBr_3$ form strong metal–metal bonds in just this way. For d^1(TiX_3, ZrX_3, HfX_3) and d^5(t_{2g}^5)($RuCl_3$) compounds the bonding is a little more complex. Consider the d^1 case. Since only one of the three orbitals is half-occupied and suitable for bonding, the metal–metal bond can be considered as involving contributions from all the possible mesomeric structures. However, Krebs (1968) considers that better overlap is obtained if a linear combination of the t_{2g}-orbitals is taken which points directly at the shared octahedral faces, Figure 6-27.

6.2 TWO-DIMENSIONAL MOLECULAR CRYSTALS: MONOLAYERS AND ADSORBED GASES

There are two further important aspects of structural chemistry which we should mention here, since the basic principles of packing are also operative in them.

Monolayers on Liquid Surfaces

When a small drop of a molecular material with a hydrophilic group (e.g. cetyl alcohol) is placed on the surface of a water-filled Adam trough, the hydroxyl group is attracted into the surface of the water leaving the hydrophobic hydrocarbon chain sticking up in the air. The amount of cetyl alcohol is chosen such that there will not be quite enough to form a monolayer. The surface film may then be compressed by moving a boom along the trough so that a complete monolayer is formed, thereby creating a two-dimensional film in which the long chains are packed efficiently. By suitable modifications of technique it is possible to form layers of (for example) lead salts of long-chain fatty acids and pick them off onto a glass substrate. Repetition of this process leads to build-up of crystals formed by superimposition of these monolayers: the crystals consequently have regular sheets of lead atoms at large spacings (the length of the hydrocarbon chains) and are of value in calibrating soft X-ray spectrometers.

Gases Adsorbed on Crystal Surfaces

Although the physical and chemical interactions of gases with solid surfaces has been the object of intensive study for a very long time, due in large measure to the technological importance of heterogeneous catalysis, it is only within the last decade or so that substantial progress has been made in observing directly the initial interaction processes. Such work is done under ultra-high vacuum conditions so that uncontaminated metal and other surfaces can be prepared. Further, a whole battery of elaborate (and expensive) techniques have reached the stage of commercial development (e.g. l.e.e.d.—low energy electron diffraction) so that there has recently been a great expansion of our knowledge of the processes that occur when volatile materials are deposited on solid surfaces. The structures of the surface 'complexes'—which may be considered as two-dimensional crystals—can be studied in much the same way as normal crystals, though as yet with somewhat lower precision. It is found that both chemisorption (a process in which a strong bond is formed between adsorbent and substrate, characterized by a large heat of adsorption) and physical adsorption of gases on solids yields predominantly ordered structures. One fascinating feature of the results is that the surface layers have structures that are determined by the substrate and differ from those which the adsorbent would have if allowed to form crystals in the absence of surface.

Some basic rules for the formation of monolayer structures have recently been formulated by Somorjai and Szalkowski (1971). They may well be modified and extended in time, but the principles upon which they are based are quite fundamental and give the proposed rules a ring of authority.

It is assumed that the surface structures formed by gases on solids will exhibit maximum interactions, both adsorbate–substrate and adsorbate–adsorbate; in other words, the closest-packing principle which formed the basis of our study of molecular crystals. Now clearly, closest packing of gas atoms will not be maintained unless the surface mobility (or diffusion energy E_D) is sufficiently low; otherwise long-range order will be destroyed,† and the substrate-adsorbate bonds must also be of such a strength that the heat of desorption is greater than E_D. Given these conditions, two rules appear to be observed.

(*a*) Adsorbed atoms or molecules prefer close-packed arrangements and tend to form surface structures characterized by the smallest unit cells permitted by molecular dimensions, adsorbate–adsorbate and adsorbate–adsorbent interactions.

(*b*) Adsorbed atoms or molecules in monolayer thickness tend to form ordered structures characterized by unit cells closely related to those of the substrate. In other words, the surface structures bear a greater similarity to the geometry of substrate than to that of the crystal structure which the adsorbate would form by itself. When multilayer adsorption occurs, as the influence of the solid surface is screened by successive monolayers, the structure of the pure adsorbate crystal will eventually assert itself.

These rules can be given greater precision through application of lattice theory.

Although the bulk of this chapter deals with three-dimensional molecular crystals, we have seen that the principles of space-group theory by which their constituent layers are constructed are also applicable to the special (and

† There appears to be a rather tricky problem here! Consider a film of gas atoms (or molecules) adsorbed on a surface at substantially less than monolayer concentration and assume that they arrived at their adsorption sites by a random process. If they have sufficient thermal energy to diffuse they may then form surface complexes with other atoms. But then we must assume that the adsorbent–adsorbent interaction is sufficiently large that E_D for the pair is greater than the available thermal energy if the nucleus of a stable surface complex is to be formed. This argument is oversimplified but one can see that some amusing problems of energetics are involved.

enormously important) case of adsorbed gas monolayers on solid surfaces. Indeed, two-dimensional space-group theory was applied by Somorjai and Szalkowski in their paper in which the above rules were developed. The interested reader may care to speculate upon the process of the transition from surface-determined to normally (Kitaigorodskii) packed molecular crystal structures consequent upon multilayer deposition; a wealth of curious physical properties should exist.

6.3 CRYSTALS COMPOSED OF POLAR AND HYDROGEN-BONDED MOLECULES

When crystals are formed from polar molecules, interactions in addition to those due to van der Waals attractive and repulsive forces are possible. Alignment of permanent dipoles relative to each other to minimize electrostatic energy will commonly result in molecular orientation different from that which would have been assumed by a hypothetical molecule of the same shape but without polarity. Such interaction may altogether determine the mode of packing of molecules in the sense of forming a restriction in *addition* to the requirements of molecular packing theory. Thus, the relative orientation of the molecules may be mainly determined by the need to minimize electrostatic energy, but their packing will still make efficient use of space in accord with packing principles, quite possibly using a different space group. In this way the additional (polar) interactions *add* to the lattice energy; in so doing, most of the attractive energy from dispersion forces will be retained (since close packing is maintained), although some price *may* be paid in increased repulsive energy but never more than can be compensated by the gain due to minimization of electrostatic interactions. An example is provided by comparison of the heats of sublimation of H_2O (46·7 kJ mol^{-1}) and (H_2S 16·9 kJ mol^{-1}); strong hydrogen bonds are formed in ice but not in solid hydrogen sulphide. While H_2S has a simple molecular lattice structure, ice (which has many polymorphs) crystallizes normally in a diamond-like structure. It would be of great interest to have results of calculations of lattice energy for crystals in which related molecules adopt different modes of packing. By hypothetical replacement of one in the lattice of the other (cf. Mason and Rae's work on benzene/*sym*-triazine) it would be possible to follow the balance of energy terms quantitatively.

The most important form of additional (to van der Waals) bonding interaction is undoubtedly the H bond. 'A hydrogen bond exists when a hydrogen atom is bonded to two or more other atoms.' Although exhaustive attention has been given to the study of H bonding (for authoritative reviews see Pimentel and McClellan, 1960; and Hamilton and Ibers, 1968) remarkably little is known of the detailed energetics determining structures in which it plays a role. Our understanding of the relative importance of other types of interaction between polar molecules in crystals (Section 6.3.3) also borders on the primitive. A shining exception is the recent work of Rae (1969) on HCN, which we consider in some detail since all the important types of interaction were explicitly included.

Molecular Interactions in HCN Crystals

This dipolar molecule ($\mu = 1{\cdot}174$) has two crystalline modifications, both consisting of parallel linear H-bonded chains ---H—C≡N---H—C≡N--- H—C≡N--- with a transition between the two forms at 170 K resulting in slightly different chain packing (Figure 6-28). The crystal internal energy

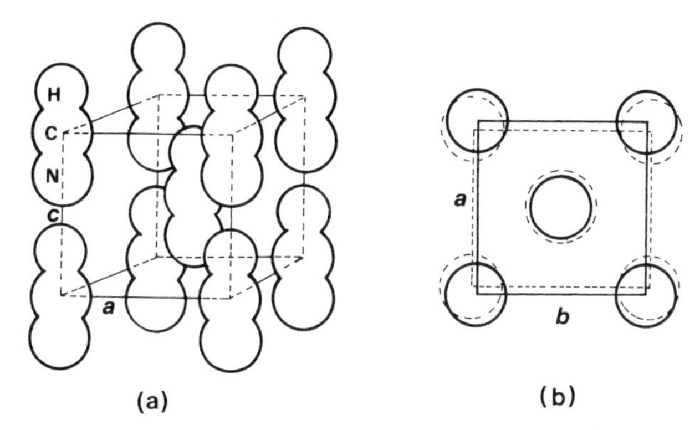

Figure 6-28 The structures of crystalline HCN. (a) The chains of H-bonded molecules. (b) The slightly different modes of packing in the high (—) and low (---) temperature phases. (Rae (1969))

was computed in terms of the following contributions. The assumption is made that interactions within the crystal can be divided into long-range and short-range respectively. Short-range energy is due to overlap of wave functions of neighbouring molecules; long-range energy is not affected by this and can be subdivided into (*a*) Coulomb interaction between dipolar molecular charge clouds, (*b*) dipole-induced dipole energy due to polarization of one molecule by the field of the rest of the lattice, (*c*) dispersion energy (due to mutual polarization of pairs of molecules). Of these contributions the short-range energy is the most difficult to assess directly and a semi-empirical method was used. Note that the effect of H bonding is taken care of by the interactions listed, the eventual success of the calculation confirming its mainly ionic nature in this case.

The really subtle move comes in estimating the dipole-dipole interaction energy. Because intermolecular distances are comparable with those within the molecule, it is probable that a sum of dipole–dipole interactions will not accurately represent the electrostatic energy, since quadrupole–quadrupole and other higher moments will not have fallen off to negligible values when separation between interacting poles is small. In principle, a correction can be calculated from knowledge of molecular wave functions, but the mathematics is awkward. Instead, Rae used the following approach.

The object of the exercise is to make calculation of dipole–dipole interaction energy a good approximation to electrostatic energy by taking dipoles that are small in comparison with intermolecular distances. These small dipoles were estimated by subtracting electron densities appropriate to H, C and N, *atoms* from the known electron-density distribution of the HCN molecule. This difference electron density, shown in Figure 6-29, was found to divide into alternating positive and negative regions which were approximated by replacing them pairwise with small dipole moments as shown.

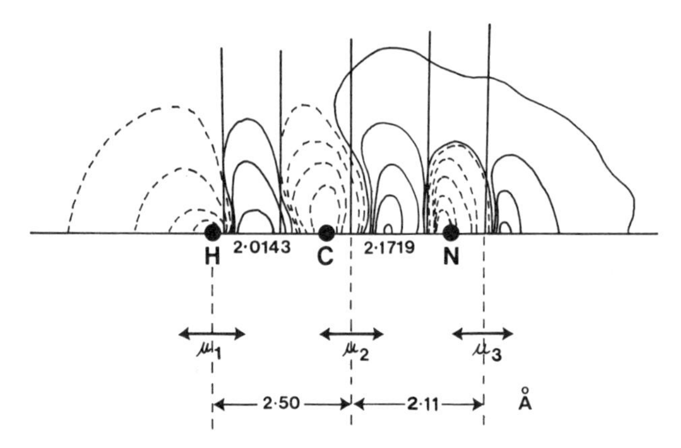

Figure 6-29 Electron-density difference map for solid HCN. (Modified from Rae (1969))

Summing all the above energy terms yielded the lattice energy which was computed as a function of unit-cell size for both high and low temperature modifications. Figure 6-30 shows the well-defined minimum obtained when the *c*-axis was varied (i.e. affecting HCN---HCN intermolecular contacts). The calculated value of *c* was within experimental error of the observed value 8·20 ± 0·04 Å for both phases. Agreement with experiment was a little less impressive when *c* was held constant at the experimental value but $a = b$ (tetragonal high-temperature phase) or *a* and *b* (orthorhombic low-temperature phase) varied (a = 8·5 compared with 8·75 Å) and the minimum was much shallower. The pronounced differences between the two curves are due primarily to differences in dipole–dipole and dipole-induced dipole energies. Computed lattice energies were in good agreement with those determined.

The success of this calculation implies that all of the important interactions have been considered and that the model should be widely applicable. Further investigations of this nature would be most valuable.

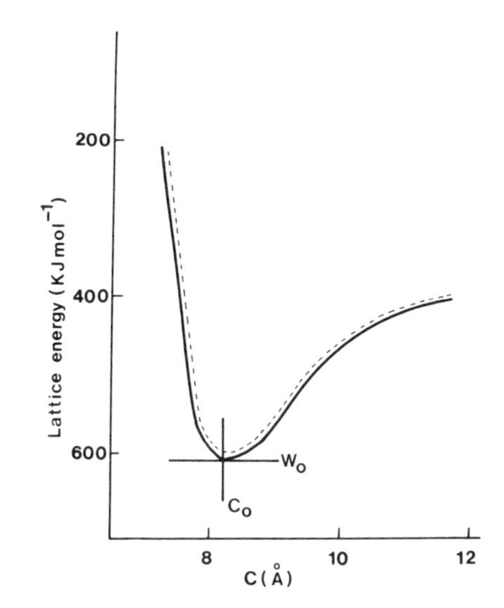

Figure 6-30 Calculated potential energy minima for HCN. The two lines refer to the two different phases. (From Rae (1969))

6.3.1 Hydrogen-bonding in Crystals

We have already indicated the importance of the H bond in structural chemistry, and the attention it has received. H bonds can be either asymmetric, with the hydrogen atom closer to one of the two atoms with which it is associated, A—H···B (as in the HCN crystal) or symmetric as in HF_2^-, $[F—H—F]^-$. The presence of H bonding is most reliably detected by observation of 'at least two heavy atom–hydrogen atom distances less than the sum of van der Waals radii' (Hamilton and Ibers, 1968), as borne out by the data of Table 3. The differences, Δ, between calculated and observed H···B distances vary from ca 1 Å for the strongest asymmetric H bonds, to 0·3 for the very weak C—H···O situations. A hydrogen-bonded system A—H···B need not be linear and is commonly found to be bent.

To this criterion for detection of an H bond may be added a crystallographic one. Kitaigorodskii has noted that where molecules associate in pairs (or in larger multiples) due to H bonding, the pair is to be regarded as the basic 'molecule' upon which packing considerations are then operative. In

Table 3. Van der Waals contact distances and observed heavy-atom separations (Å) for some common H bonds (Hamilton and Ibers, 1968)

A—H---B	A—B (calc.)	A—B (obs.)	H---B (calc.)	H---B (obs.)	Δ^a
F—H---F	2·7	2·4	2·6	1·2	1·4
O—H---O	2·8	2·7	2·6	1·7	0·9
O—H---F	2·8	2·7	2·6	1·7	0·9
O—H---N	2·9	2·8	2·7	1·9	0·8
O—H---Cl	3·2	3·1	3·0	2·2	0·8
N—H---O	2·9	2·9	2·6	2·0	0·6
N—H---F	2·9	2·8	2·6	1·9	0·7
N—H---Cl	3·3	3·3	3·0	2·0	1·0
N—H---N	3·0	3·1	2·7	2·2	0·5
N—H---S	3·4	3·4	3·1	2·4	0·7
C—H---O	3·0	3·2	2·6	2·3	0·3

[a] The difference between H---B (calc.) and H---B (obs.)

consequence, although the structures formed are those summarized in Table 1, the number of molecules in the cell (Z) is greater. Thus in α-naphthylamine, molecules associate in sixes, the space group, *Pbca*, being one of those allowed (Table 1); instead of the normal $Z = 4$, it is 24.

This criterion is not restricted to H bonded association. Mason (1965) has contrasted the crystal structures of anthracene and acridine. Although anthracene packs in the 'herringbone' fashion characteristic of aromatic hydrocarbons, acridine (which differs only in having one nitrogen atom in place of a *meso* CH) adopts entirely different forms. One modification of crystalline acridine is composed of dimers, as shown below, in which electrostatic attraction is maximized. These dimers form the packing unit from which the crystal is then constructed.

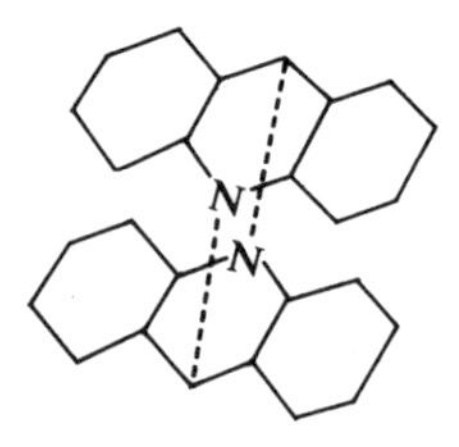

In chemical terms, H bonds are associated with the presence of an acidic hydrogen atom and a basic acceptor: in other words, the hydrogen is attached to an atom or group with electron-withdrawing properties, and therefore tends to associate with regions of high electron density, typically with non-bonding electron pairs. Although it is generally accepted that an ionic model of H bonding cannot accommodate all the effects associated with the phenomenon, it seems likely that in understanding the major structural consequences of H bonding and crystal energetics, it is at least a good starting point and often a very good approximation, as in Rae's calculations on HCN (above). The value of the ionic approximation is further discussed below. Nevertheless, in general it is best to think of the H bond in terms of a molecular orbital (MO) scheme since this is of greater generality and also makes a formal link with the now familiar three-centre bonding in boranes.

As an example, consider Pimentel's description of the linear HF_2^-. Defining the molecular axis as z, overlap schemes between the p_z-orbital on each fluorine and the $1s$-orbital on hydrogen are constructed. First, we prepare linear combinations of the p_z-orbitals suitable for bonding to hydrogen. Thus:

$$\Psi_1 = p(\text{A}) + p(\text{B})$$

$$\Psi_2 = p(\text{A}) - p(\text{B})$$

Clearly, only Ψ_2 is of the right symmetry to overlap with an s-orbital. The combinations $\Psi_2 \pm s$ are therefore constructed leaving Ψ_1 as a non-bonding MO. Thus:

$$\Psi_3 = \Psi_2 + ks$$

$$\Psi_4 = \Psi_2 - ks$$

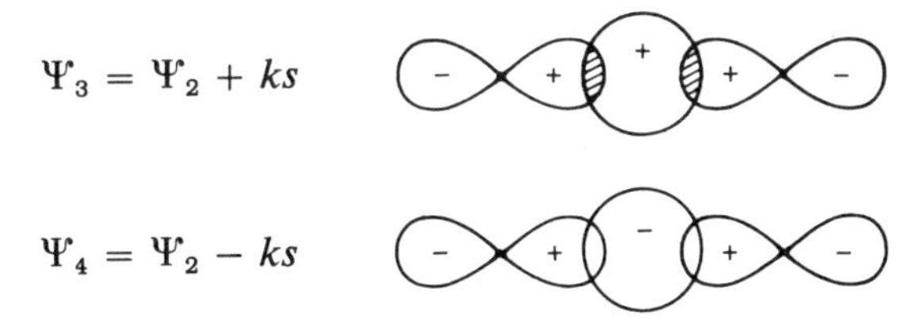

With each fluorine 'prepared' in the configuration $(1s)^2(2s)^2(2p_x)^2(2p_y)^2(2p_z)^1$, two $2p_z$ electrons, the hydrogen $1s$ and a fourth electron contributed from the cation, must all be fed into the energy manifold as shown below. From this

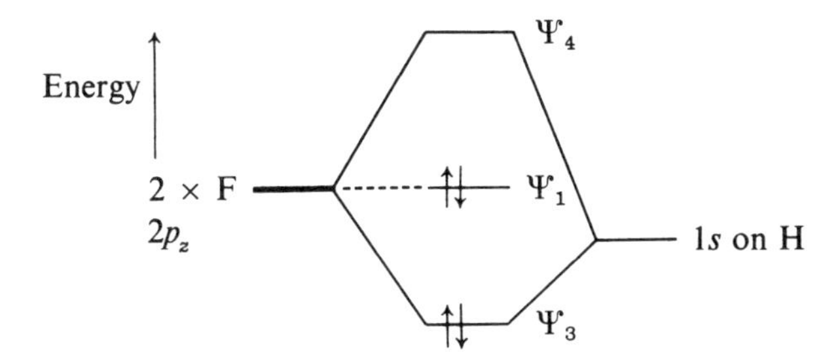

it is evident that one electron pair is non-bonding (and located on the fluorines) and that the other (in Ψ_3) provides the bonding by spreading itself over three atoms. The scheme is simply modified by addition of coefficients to describe asymmetrical H bonds. Since it is necessary to accommodate a lone pair on the terminal atoms, these must be quite highly electronegative, otherwise no H bond will be formed, or it will be weak.

The success of models based upon an ionic description of the H bond can be understood on the basis of the MO description. Two bonding electrons are spread over three atoms: for the most usual case of an asymmetric bond A—H···B the bonding orbital Ψ_3 will be concentrated at one end of the H bond (A in this case). Hence, its replacement by either a localized dipole (as in Rae's theory) or a point charge (as in Baur's theory, see below) is a good approximation to the electron-density distribution.

If both acidic hydrogen and basic acceptors exist in a molecule it almost invariably follows that the crystal structure adopted is one in which H bonding is possible. However, we emphasize that close-packing principles are still operative and the efficiency of use of space is not lower than 60%, i.e. comparable with packing densities of non-polar molecular crystals. Being *predominantly* ionic, and hence non-directional in character, the H bond is formed subject to the constraints of molecular packing. It usually turns out that molecular conformation can be arranged and a suitable space group found such that a H bond can be formed. Nevertheless, the frequent departures from linearity and the variation in A to B distances in A—H···B bonds for a given A–B pair witness to the prior satisfaction of molecular packing requirements, see Table 4. In inorganic hydrates, an especially important class from our viewpoint, the water molecule exhibits substantially constant geometry: H bonding in this class of compound is therefore adapted to this geometry and to the packing requirements of the ions and other constituents.

Classification of Hydrogen-bonded Structures

Such is the number and variety of H-bonded structures that attempts have naturally been made to classify them. Because, as we have discussed above, H-bond formation constitutes a constraint *in addition to* other packing requirements (a term which is here meant to include requirements appropriate to the type of bond of the constituents, such as the need to form coordination polyhedra in ionic situations), H-bond networks will commonly reflect the

Table 4. Variation in O—H···O distances in some inorganic crystals (After Pimentel and McClellan, 1961)

	A—H···B	*d*(O···O) Å
Ni dimethylglyoxime	N—O—H···O	2·44 ± 0·02
$HNO_3 \cdot H_2O$	N—O—H···OH_2	2·68
$NH_4H_2PO_4$	O—H···O=P	2·49
KH_2AsO_4	As—O—H···O	2·53
$NaHCO_3$	C—O—H···O	2·55
H_2O_2	O—H···O	2·78

high symmetry and method of construction appropriate to lattices of equivalent constituents without acidic and basic groups. We outline one approach to H-bond classification. The subject is discussed at greater length by Wells (1962) and by Gallagher and Ubbelohde (1955).

Consider the structures *possible* given the number of H atoms per molecule available for H-bond formation.

(*a*) *One* H-*atom between two molecules or ions*. For example HF_2^-, or the compounds M^IAH_2, where M^I = alkali-metal cation, A = carboxylate. Many structures of the latter type have been investigated by Speakman: they contain short symmetric H bonds.

$$Na^+ \quad CH_3{-}C(=O){-}O\cdots H\cdots O{-}C(=O){-}CH_3^- \quad \text{(schematic)}, \quad O\cdots O = 2{\cdot}443\ \text{Å}$$

Formally related is the CrO_2H structure. This has layers related to the CdI_2 type; Cr is in octahedral sites in a close-packed oxygen array. The difference is that the H atoms form a layer *between* the oxygen layers which themselves are stacked directly over one another; neutron diffraction has been used to show that 50% of the bonds in the deuterate are O—D···O, the others being O···D—O.

(*b*) Given *one* H *atom per molecule* it is possible to form: (*i*) dimers, (*ii*) chains, (*iii*) rings. In the two cases (*a*) and (*b*, *i*), the dimeric units pack according to normal principles having already satisfied their H-bonding requirement.

There are many examples of H-bonded chains: e.g. HCN. HF (at –125°C) forms planar zigzag chains as shown:

$$F\cdots F\cdots F\cdots F\cdots F \quad (\text{angle } 120°,\ F\cdots F = 2{\cdot}49\ \text{Å})$$

H-atom positions have not been determined. The bond angle at fluorine clearly implies use of sp^2 hybrids: the bonding in this H-bonded chain is therefore of considerable directionality. Note that the molecular positions closely resemble those of the chains from which the iodine structure is composed, showing that close packing is retained.

In $NaHCO_3$ the anions are linked to form chains between which Na^+ ions are accommodated (Figure 6-31a), whilst in $KHSO_4$ both chains and dimer rings are formed in equal proportions. However, $KHCO_3$ is constructed of dimers analogous to those of $KHSO_4$.

$$O{-}C\begin{matrix} O\cdots H\cdots O \\ O\cdots H\cdots O \end{matrix}C{-}O, \quad O\cdots O = 2{\cdot}61\ \text{Å}$$

An equivalent way of satisfying H-bond requirements leads to formation of chains in β-oxalic acid and several other dibasic acids:

$$>C-R-C\begin{matrix}O\cdots O\\ \\ O\cdots O\end{matrix}C-R-C\begin{matrix}O\cdots O\\ \\ O\cdots O\end{matrix}C-R-C<$$

(*c*) Given *two or more* H *atoms per molecule*, a very large number of structural possibilities arise. One simple solution was shown in the chain structure of β-oxalic acid (above). More commonly, two- and three-dimensional structures are formed. Wells (1962) has classified many of these in terms of nets.

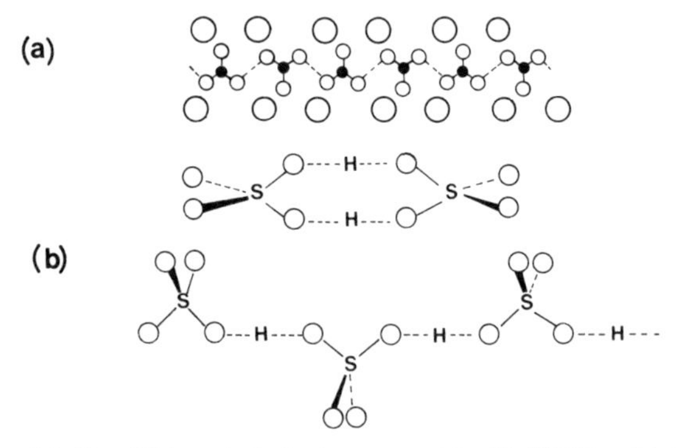

Figure 6-31 (a) The H-bonded chain present in $NaHCO_3$. (b) Schematic representation of the dimers and chains in $KHSO_4$

Hydrogen-bonded sheets are commonly formed by polybasic inorganic acids. H_2SO_4 and H_2SeO_4 consist of puckered layers of EO_4^{2-} groups joined to four others by H bonds, Figure 6-32a, and a related four-connected net is shown by selenious acid, H_2SeO_3, Figure 6-32b.

Inorganic acids often exist in hydrated forms, which have structures different from those of the anhydrous materials. $H_2SO_4.H_2O$ has a more complex layer structure in which each SO_4^{2-} now has five H bonds while H_3O^+ forms three, Figure 6-32c. Nitric acid monohydrate is best regarded as containing the pyramidal hydroxonium ion, i.e. $H_3O^+NO_3^-$, as three equivalent protons are indicated by nuclear magnetic resonance: see Figure 6-32d. Sodium sesquicarbonate, $Na_2CO_3.NaHCO_3.2H_2O$ has a complex structure in which H-bonded anion pairs are joined into layers by water molecules each forming only two H bonds.

$$O-C\begin{matrix}O\cdots(OH_2)\cdots O\\ \\ O\end{matrix}\;\;\;\;C\begin{matrix}O\\ \\ -O\end{matrix}$$

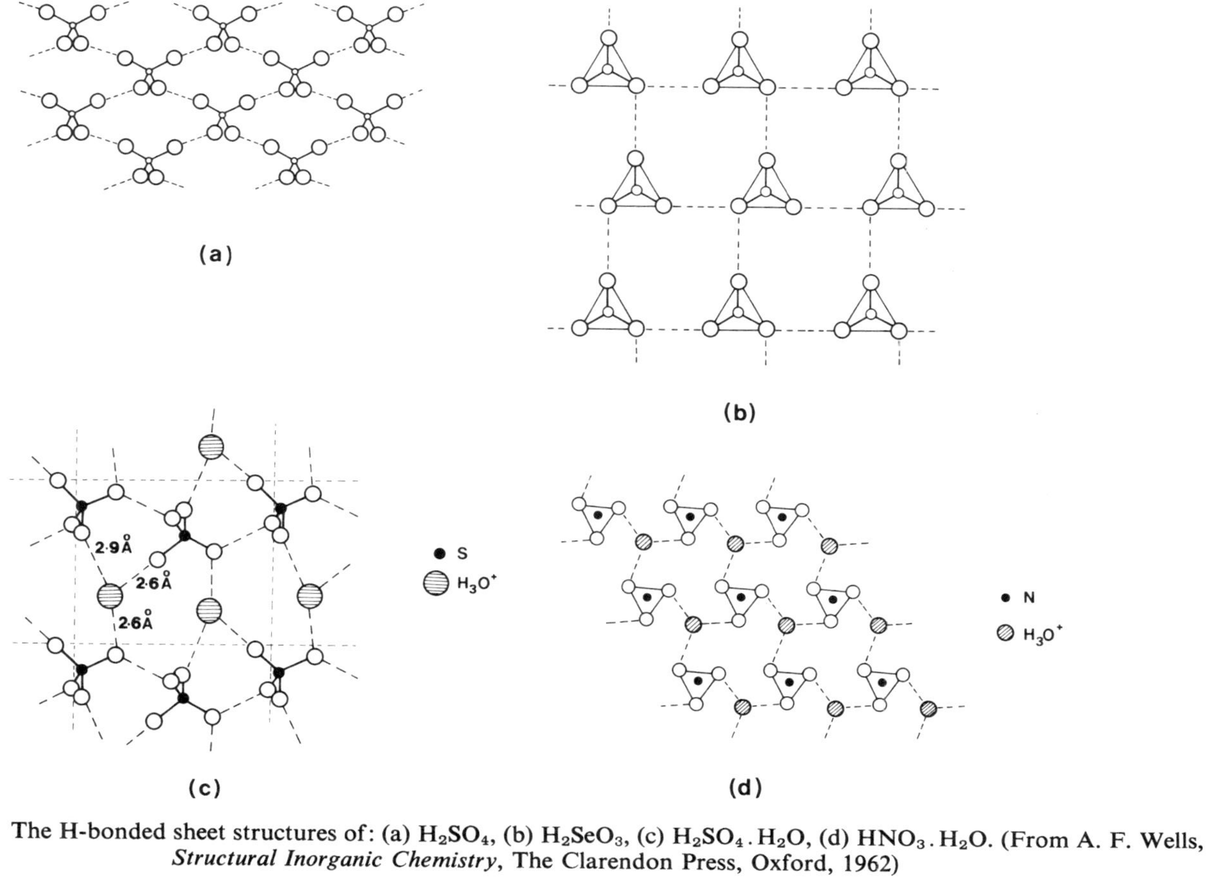

Figure 6-32 The H-bonded sheet structures of: (a) H_2SO_4, (b) H_2SeO_3, (c) $H_2SO_4.H_2O$, (d) $HNO_3.H_2O$. (From A. F. Wells, *Structural Inorganic Chemistry*, The Clarendon Press, Oxford, 1962)

We leave H-bonded sheets with a brief look at the beautiful hexagonal net structures of boric ($B(OH)_3$) and metaboric (HBO_2) acids, Figure 6-33. Boric acid consists of *molecules* H-bonded to give nearly planar layers. Some difficulty has been experienced in locating the H atoms, but the best evidence seems to indicate that they are co-linear, O—H---O. Trimers $B_3O_3(OH)_3$ are present in metaboric acid.

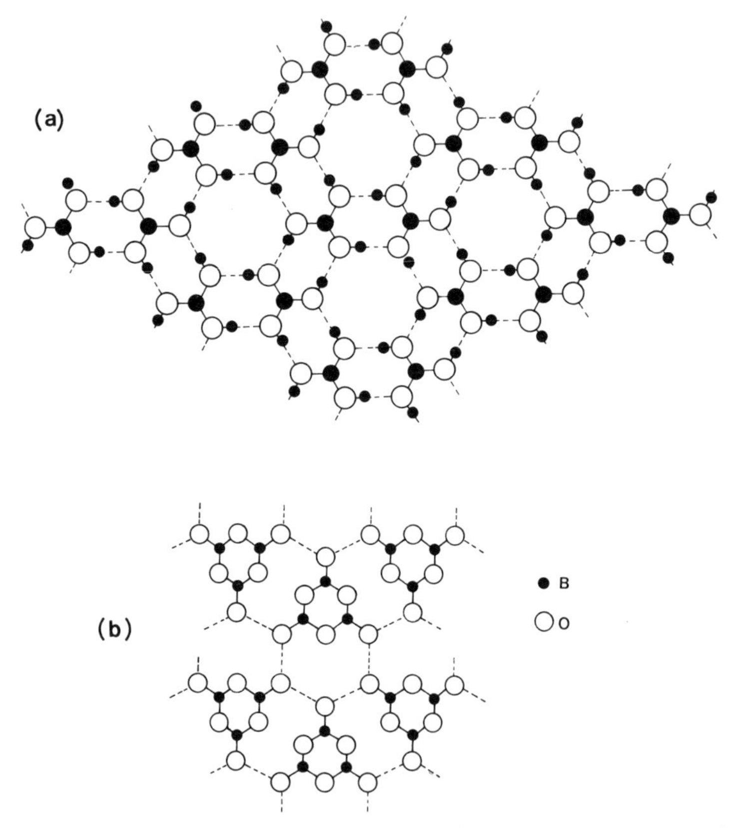

Figure 6-33 The H-bonded sheet structures of: (a) H_3BO_3, (b) $B_3O_3(OH)_3$. (From A. F. Wells, *Structural Inorganic Chemistry*, The Clarendon Press, Oxford, 1962)

Three-dimensional Hydrogen-bonded Networks

Out of the enormous number of three-dimensional H-bonded structures, we select a few examples of inorganic interest.

Ice (I*h*). The classic three-dimensional network is that of ice, which shows a variety of structures (nine in all) when prepared at various temperatures and pressures. Ordinary ice (called ice I*h*, where *h* stands for hexagonal) has a hexagonal unit cell in which each oxygen has four equidistant oxygen neighbours at 2·75 Å. The oxygens occupy the intersections of the hexagonal net shown in Figure 4-1c, i.e. the positions of Si in the tridymite structure. The hydrogens are located along (and possibly slightly off) the O---O vectors but, owing to the stoichiometry, on average, there can only be two hydrogens associated with any one oxygen at a given instant. The many possible ways of permuting the hydrogen positions gives rise to disorder which is reflected in the entropy of the structure. The difference between the experimentally determined entropy and that estimated from the third law of thermodynamics is $R \log \frac{3}{2}$, corresponding exactly with the above account.

Ammonium Salts

Next to water, the ammonium ion is the commonest H-bonded group found in inorganic compounds. The structures of the salts we consider below rather beautifully illustrate the principles discussed in the early part of Sections 6·3 and 6.3.1. Due to its shape the most favourable H-bonding schemes for NH_4^+ will clearly be those in which the other terminal atom of each H bond is along one of the four tetrahedral vectors. But according to the principles outlined earlier the formation of H bonds will not result in a structure that is unfavourable so far as packing of the constituents is concerned.

$(NH_4)_2SiF_6$ has a cubic phase with the K_2PtCl_6, Li_2O or anti-fluorite structure, if the H atoms are neglected. In the anti-fluorite structure the cation is tetrahedrally coordinated, but in hexahalo complexes each tetrahedral vector from the cations meets *three* atoms from the face of an MX_6^{2-} octahedron, not a single atom as in Li_2O. Hence, NH_4^+ in this salt is associated with twelve crystallographically equivalent fluorines at the vertices of a partially truncated cube; see Figure 6-34. This may be more readily appreciated by recalling that the K_2PtCl_6 structure ($\equiv KPt_{1/2}Cl_3$) is alternatively described as a c.c.p. lattice of potassium and chlorine; hence each potassium (equivalent to NH_4^+ in $(NH_4)_2SiF_6$) must be twelve coordinate. Ammonium can be thought of as forming a multicentre overlap scheme with three fluorines,

```
 F
  \
F---H—N
  /
 F
```

at each of four cube corners. It is more likely, however, that there is more direct association with each fluorine in turn since the fluorine *p*-orbitals are not ideally oriented for simultaneous overlap. Hamilton and Ibers suggest that this might be accomplished via bending vibrations of NH_4^+, which would allow the hydrogens to form four definite N—H---F bonds at the corners of a distorted tetrahedron, during part of each vibration cycle.

We may assume that the reasons for adoption of the anti-fluorite type of

Table 5. Phase transformations and structure types in ammonium halides (after Hamilton and Ibers, 1968)

	Phase I	T (°C)	Phase II	T (°C)	Phase III	T (°C)	Phase IV
NH_4Cl	R	184	CsCl(a)	−31	CsCl(b)		
NH_4Br	R	138	CsCl(a)	−38	CsCl(c)		
ND_4Br	R	125	CsCl(a)	−58	CsCl(c)	−104	CsCl(d)
NH_4I	R	−14	CsCl(a)	−42	CsCl(c)		

R = rock salt. CsCl (a, b, c, d) = variants of CsCl structure, see text.
T = transition temperature.

structure are no different from those applying to alkali-metal salts $M^I_2MF_6$ or $M^I_2MCl_6$. The structure of $(NH_4)_2SiF_6$ illustrates one way in which H-bonding requirements are accommodated within the structure required by packing considerations.

The halides $\mathbf{NH_4X}$ (X = Cl, Br, I) have quite different radius ratios† and would therefore be expected to adopt either the rock salt or CsCl structures.

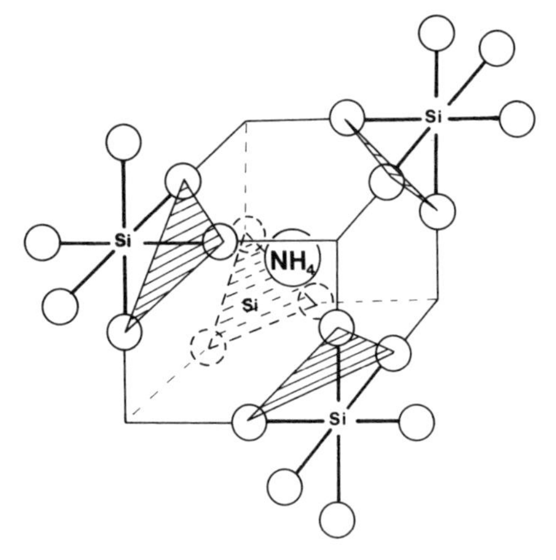

Figure 6-34 The structure of $(NH_4)_2SiF_6$ showing the ammonium ion in twelve-coordination

This is what happens, although the complex polymorphism of these compounds (see Table 5) reflects various ways in which the two basic structures (adopted by virtue of the packing requirements of the ions concerned) are modified to allow additional stabilization by H-bond formation. Our discussion of these three salts follows that of Hamilton and Ibers (1968) closely.

† This statement begs the question to some extent as we have not defined the size of NH_4^+. Nevertheless it is obvious qualitatively that SiF_6^{2-} is much larger than, for example, Cl^-.

The highest temperature phase in each case has the rock-salt structure, but the lower temperature phases are all based upon CsCl. Transition temperatures at which the change is made decrease from Cl to Br to I, following the order of H-bond strength. The most detailed study has been made of ND_4Br. Its phase IV has an ordered CsCl structure: each NH_4^+ is in cubic eight-coordination with bromine and makes *four* N—H···Br bonds tetrahedrally (Figure 6-35). Each Br^- accepts *four* H bonds, also tetrahedrally. Phase II is very similar to phase IV but there is now twofold disorder caused by differing

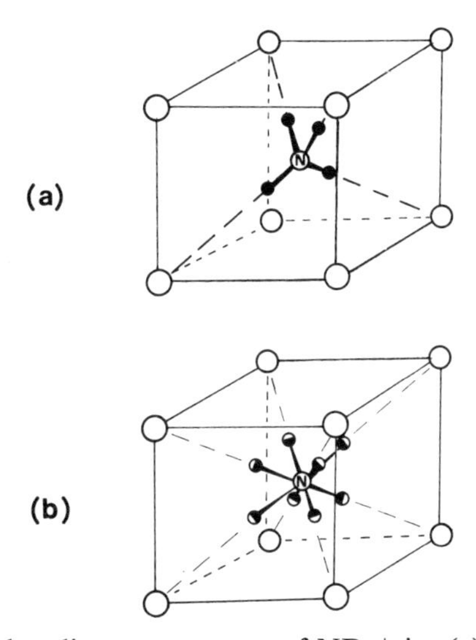

Figure 6-35 The H-bonding arrangement of ND_4^+ in: (a) phase IV of ND_4Br, (b) phase II of ND_4Br (half-shaded atoms indicate that ND_4^+ is disordered, choosing one or other tetrahedral set)

choice of the tetrahedral set of anions to which H bonds are formed. Phase III is another ordered CsCl structure but the unit cell is twice the size of that of CsCl in order to accommodate NH_4^+ ions with two different orientations. Each Br^- is the acceptor of four H-bonds at the corners of a square, and the structure is distorted so that four Br^- ions are closer to each NH_4^+ than the other four of a very slightly distorted cube. The reader may care to speculate upon the possibility of a phase with the PtO structure.

The rock salt (Phase I) structure of the ammonium halides is not as simple as might appear. Several lines of evidence show that the cations are not freely rotating but that each forms *one* strong H bond about which it can rotate. Since this N—H···X bond direction can be chosen from one of six octahedrally, the structure is highly disordered. The data also fit a model involving static disorder around the strong N—H···X bond rather than rotation, see Figure 6-36.

In all of these ammonium salts we see various ways of accommodating H bonding within the basic lattice types that would be adopted by these compounds in the absence of H bonding. However, the N—H···F bond is much stronger than that of N—H···Cl, so much so that it exerts a determinative effect upon the structure of NH_4F, which is that of wurtzite (the more ionic of the two 4:4 lattices) with hydrogen atoms along all N···F vectors. It appears that in this compound the covalency of the H bond, i.e. its *directionality*, is too high for one of the more ionic lattices (cf. the directed H bonding in HF).

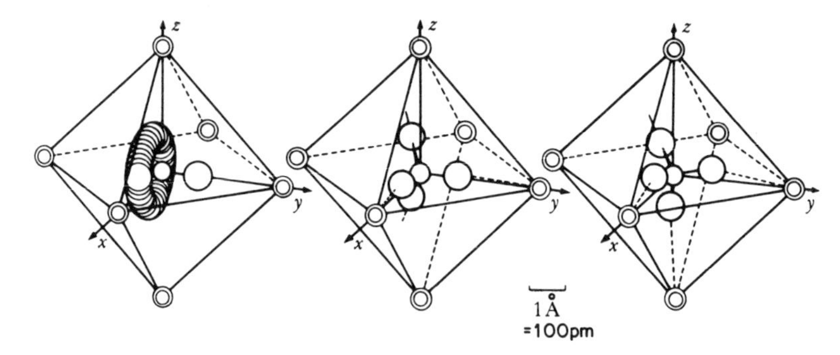

Figure 6-36 Various models for the arrangement of NH_4^+ in phase I of the ammonium halides. The NH_4^+ groups may be in almost free rotation about individual threefold axes, or they may exhibit multiple disorder. (Levy and Peterson (1953))

6.3.2 Water in Inorganic Crystals: Baur's Electrostatic Theory

Vast numbers of inorganic materials occur, or are crystallized from aqueous solution, as hydrates. In the majority of cases the presence of water is essential for stability of the structure in one of two senses. (*a*) The compound may not be formed in the absence of water: e.g. $[Fe(H_2O)_6][SiF_6]$ is known but Fe^{2+} $[SiF_6]^{2-}$ is not known, being unstable with respect to SiF_4 and FeF_2. (*b*) Removal of water from a hydrate nearly always results in collapse of the structure and formation of a new (anhydrous or partially hydrated) one. The term 'hydrate' should, strictly speaking be reserved for compounds in which a definite number of water molecules can be distinguished rather than, for example, clays and other aluminosilicates in which various amounts of water may be sorbed.

Water is most commonly found associated with cations: coordination via lone pairs on oxygen is a means of reducing effective cationic charge. Sidgwick noted that the tendency of an ion to hydrate is greater, (*a*) the greater its charge, (*b*) the smaller its size if a cation *or* the greater its size if an anion. A survey of the extent of hydration of various ions bears out the general truth of this statement. Thus, salts of K^+, Rb^+, Cs^+, Tl^+ and NH_4^+, are almost invariably anhydrous but those of Na^+ and Li^+ are usually hydrated. The effect of ionic charge is well illustrated by the series CsCl (anhydrous), $BaCl_2$ (known as mono- and di-hydrates), $LaCl_3$ (6 and 7 H_2O).

Due to its electronic structure, water commonly has a tetrahedral environment in crystalline hydrates (the same applies to H_3O^+). The two hydrogens H bond to anionic groups or to the lone pairs of neighbouring water molecules, whilst the lone pairs on oxygen are associated with cationic species, either metal ions or the hydrogens of other water molecules. Both lone pairs may be used to bond to the same cation. These situations are summarized in Figure 6-37.

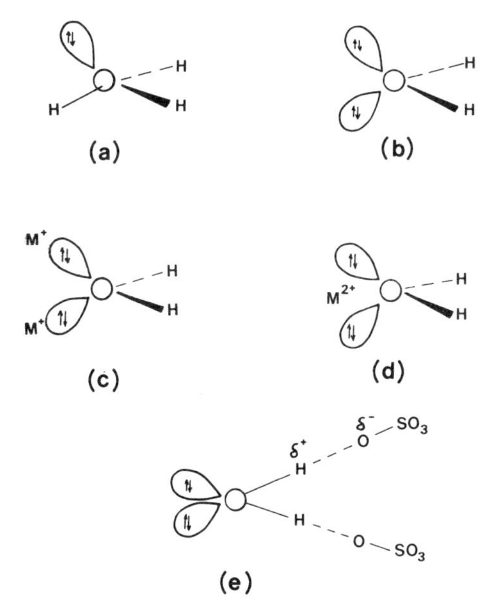

Figure 6-37 The approximately tetrahedral arrangement of electron pairs in (a) H_3O^+, (b) H_2O, (c) to (e) Various means of coordination to water in solids

Hydrates have been classified on the basis of the environment of water. This can be difficult in cases where more than one type of water environment is found. We follow a classification due to Wells (1962) based upon the number of water molecules per formula unit compared with the c.n. of the metal ion: i.e. $x:m$ in $M_mA_n.xH_2O$.

Hydrates Containing Oxyanions

C.n. $= x:m$. $BeSO_4.4H_2O$ contains the tetrahedral $Be(OH_2)_4^{2+}$ ion. The structure approximates to that of CsCl in its packing. Each water molecule is bonded to oxygens of two sulphate groups as in Figure 6-37e. $AlCl_3.6H_2O$ contains octahedral $Al(OH_2)_6^{3+}$ ions. $CaO_2.8H_2O$ and $Sr(OH)_2.8H_2O$ have square-antiprismatic $[M(OH_2)_8]^{2+}$ groups, while $[Nd(OH_2)_9]^{3+}(BrO_3^-)_3$ has a nona-coordinate cation of the type shown in Figure 6-38.

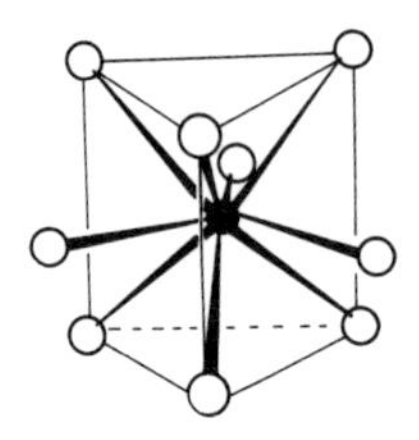

Figure 6-38 The form of nona-coordination found in $[Nd(H_2O)_9]^{3+}$

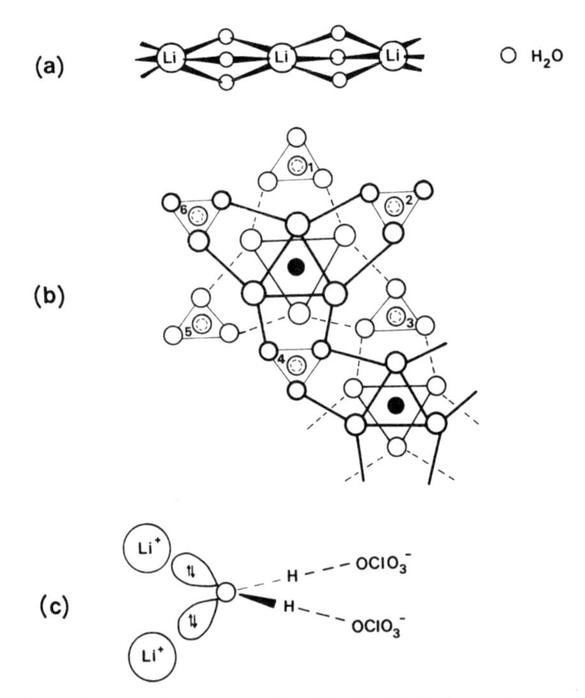

Figure 6-39 The columnar hydrates, $LiX.3H_2O$. (a) Schematic representation of the $\{Li^+(H_2O)_3\}_n$ columns. (b) Plan view showing packing of the columns in $LiClO_4.3H_2O$, linked via sets of ClO_4^- groups in two planes (1,3,5), (2,4,6). ● = Li^+. (c) The environment of water in $LiClO_4.3H_2O$

C.n. $< x{:}m$ In $NiSO_4.7H_2O$ there is one mole of water held in the lattice between SO_4^{2-} and $[Ni(OH_2)_6]^{2+}$ ions. The detail of the structure is complex.

C.n. $> x{:}m$. This class of hydrate is of particular fascination. From the formulae and a knowledge of preferred coordination arrangements of the cations involved, it is clear that either mixed coordination spheres must be involved or that water acts as a bidentate ligand.

There are many salts $LiA.3H_2O$ (A = Cl, Br, I, ClO_3, ClO_4, NO_3, BF_4, etc.). They consist of columns $\{Li(OH_2)_3^+\}_n$ in which water is bidentate (Figure 6-37c and 6-39) giving Li^+ a c.n. of six octahedrally. The columns are linked together via H bonds to anionic groups. Consider the perchloric salt. Each water molecule has the environment shown in Figure 6-39c, in which all O—H···O bonds lie in planes normal to the column axes.

In $LiOH.H_2O$, lithium achieves four-coordination to oxygens of two OH^- groups and two water molecules (cf. Li_2O) in which water has an environment similar to that of $LiClO_4.3H_2O$ (Figure 6-40).

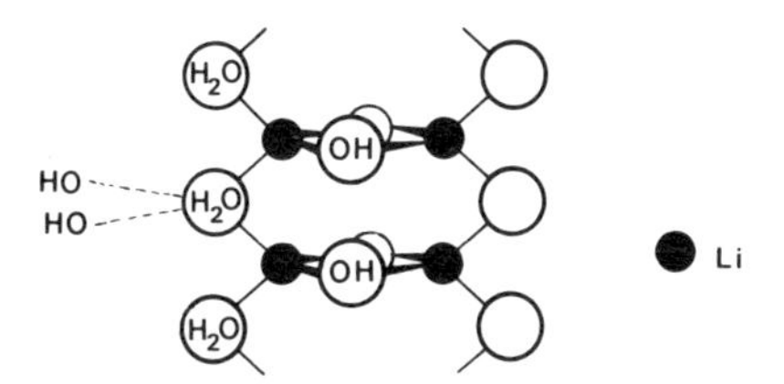

Figure 6-40 The structure of $LiOH.H_2O$, showing tetrahedral coordination at Li^+, H_2O, and OH^- (including lone pair)

The structure of $MgSO_4.4H_2O$ is of some interest in that it contains a 'bifurcated' H bond. Two questions need to be answered concerning the H bonding in this structure. (*a*) Why the bifurcated H bond?, and (*b*) why not adopt an alternative structure in which normal length H bonds can be formed? There is another orientation which the second type of water molecule could adopt. Baur (1965) has answered these questions by some calculations based upon an electrostatic model of the H bond (see discussion in Section 6.3.1).

The heavy-atom (i.e. all atoms other than hydrogen) positions were taken as known and charges assigned: Mg^{2+} + 2*e*, sulphate oxygens − 0·5*e* each. It was found best to place charges 0·5*e* on each hydrogen and −*e* on the oxygen of water. Using this point-charge distribution the electrostatic energy of the system was computed as a function of hydrogen position. The minimum-of-energy positions agreed extraordinarily well with those determined by neutron diffraction. Of especial interest is the fact that the bifurcated H bond structure is actually of lower energy than the alternative.

Baur used this model with great success on a variety of hydrates, including $BaCl_2.2H_2O$, discussed below. Organic acids were treated as point-charge systems by assigning each atom a charge estimated from simple MO calculations.

Hydrated metal halides show a great variety of structures. Wells points out that structures of fluorides almost always differ from those of the corresponding chlorides, bromides and iodides. For example:

$$\begin{cases} CaF_2 \text{ (fluorite)} & FeF_2 \text{ (rutile)} \\ CaCl_2 \text{ (distorted rutile)} & FeCl_2 \text{ (}CdCl_2\text{)} \end{cases}$$

They also differ in the degree of hydration, which may go either way. Thus,

$$\begin{cases} CaF_2 & KF.2H_2O,\ KF.4H_2O & AgF.2H_2O,\ AgF.4H_2O \\ CaCl_2.6H_2O & KCl & AgCl \end{cases}$$

In $KF.2H_2O$ the coordination spheres are:

$$K^+:\ 4H_2O,\ 2F^- \qquad F^-:\ 4H_2O,\ 2K^+$$

in each case forming a distorted octahedron. Water has its preferred tetrahedral coordination ($2K^+$, $2F^-$), see Figure 6-41.

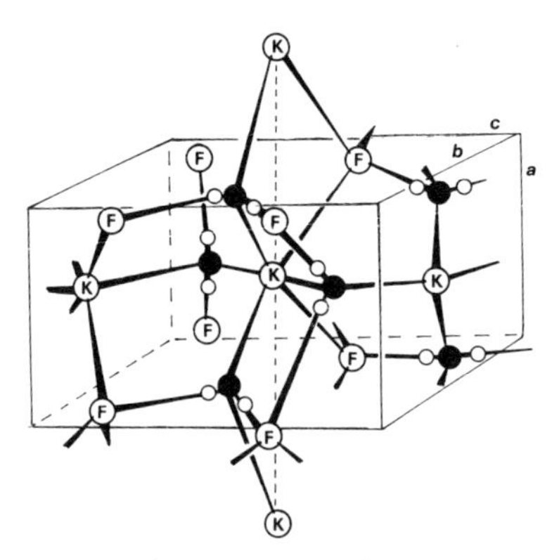

6-41 The structure of $KF.2H_2O$

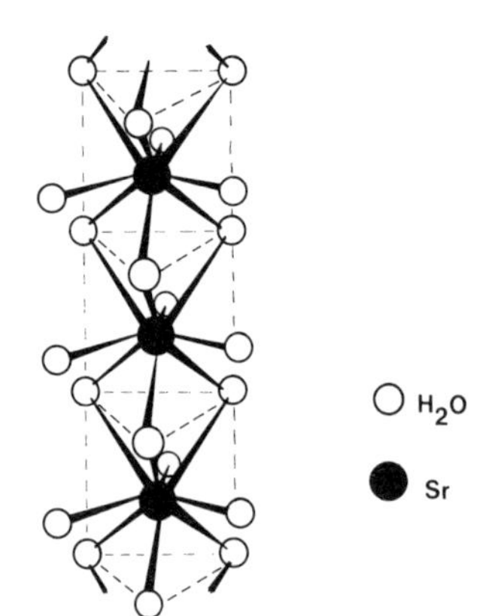

Figure 6-42 Structure of the columns $\{Sr(H_2O)_6{}^{3+}\}_n$ in $SrCl_2.6H_2O$

Hydrated metal-chloride structures are known in which discrete ions, chains or sheets are present. We consider one or two examples of each. $[Mg(OH_2)_6]Cl_2$ is really an example from the class $x{:}m$ = c.n. of M, but it is unwise to assume that formula is a reliable guide to structure. Thus, $CoCl_2.6H_2O$ contains *trans*-$[CoCl_2(OH_2)_4]$ molecules plus two moles of lattice water and the same *trans* grouping is present in $FeF_2.4H_2O$ and $FeCl_2.4H_2O$ (one of the few examples of the same structure for fluoride and chloride). *Cis* arrangement of coordinated water is not common but is found in $CsMnCl_3.2H_2O$ which has chains of distorted octahedra sharing opposite vertices:

```
       Cl    Cl
         \  /
—Cl——Mn——Cl—
        ↗  ↖
     OH2    OH2
```

Size considerations alone would require $[Mg(OH_2)_6]Cl_2$ to have the fluorite structure, but more favourable H bonding is possible in a monoclinic variant as found.

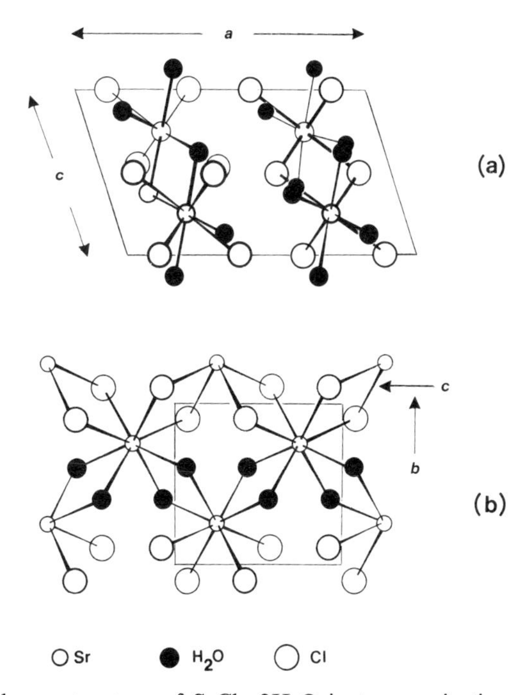

Figure 6-43 The layer structure of $SrCl_2.2H_2O$ in two projections. (a) Shows the layers edge-on. (b) Gives a plan view of one layer

Chains are found in another hexahydrate, $SrCl_2.6H_2O$ (cf. the $M(OH_2)_6^{2+}$ and $MX_2(OH_2)_4$ structures above). In this, Sr^{2+} is nona-coordinate (cf. $Nd(OH_2)_9^{3+}$), Figure 6-42. Finally, we note the layer structure of $SrCl_2.2H_2O$, Figure 6-43: Sr^{2+} is square-antiprismatically eight-coordinate ($4Cl^-$, $4H_2O$), the complexes being linked via chlorine bridges to give chains which are bonded together by shared water molecules.

6.3.3 Non-hydrogen-bonded Semipolar Interactions: Alcock's Secondary Bonds

The semipolar interactions discussed in preceding sections for hydrogen-bonded systems Y—H···X are really only a special case of the more generalized interaction Y—A···X. By direct analogy with the hydrogen-bonded situation, we recognize an A···X interaction as significant if the internuclear distance is shorter than the expected van der Waals radii sum. Such interactions may occur between the constituents of a crystal whether they are of the same type or different. Remarkably little attention has been given to the codification and understanding of these interactions, despite (or, perhaps, because of) the very wide range of compounds in which they are found. In a very recent review, Alcock (1973) has made an important start on the problem. Although it is desirable that eventually this work be integrated into both the specific theories of hydrogen-bonded crystals (as outlined above), and the general packing theory of Kitaigorodskii, the clarification achieved by Alcock is extremely valuable. We now outline his main observations, although a quick look at a couple of examples may make the nature of the problem clearer.

(*a*) Reaction of ClF_3 with SbF_5 yields a compound very approximately described as $ClF_2^+SbF_6^-$. However, the environment of chlorine is rather curious: it is roughly *square-planar* coordinated to two fluorines at normal bonding distances (1·52 Å) and two more rather further away (2·38 Å).

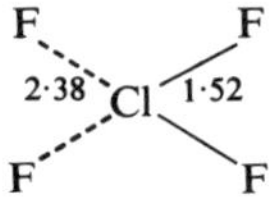

F F
2·38 Cl 1·52
F F

The expected Cl···F contact distance is 3·22 Å (i.e. van der Waals radii sum). Evidently the Cl···F bond is intermediate in type. $BrF_2^+SbF_6^-$ is similarly constructed.

(*b*) In crystalline Me_2SnCl_2 (the molecules are tetrahedral in the vapour phase), the environment of the tin atom is as shown in Figure 6-44. It remains basically tetrahedral but the C—Sn—C angle has opened to 123° and two longer bonds (3·5 Å) to neighbouring molecules help to complete a distorted octahedron. The van der Waals radii sum for a non-bonded Sn···Cl interaction is 3·85 Å, again much longer than the observed separation.

How are we to understand these intermediate bond lengths, and is it possible to classify such interactions and rationalize their geometry and occurrence? Alcock refers to them as 'secondary bonds' and discusses the bonding using a model closely related to that of Pimentel and McClellan for HF_2^- (p. 203). The most likely form of interaction is a dative bond formed by donation of a lone pair on X into an empty σ orbital of the primary Y—A bond:

such interactions are capable of more exact description on the basis of MO calculations. What is quite clear is that an electrostatic or 'ionic' model is totally inadequate to account for the observations.

The following rules appear to form a good basis for description and codification of structures exhibiting 'secondary bonds.'

1. The geometry of the primary bonds to the central atom is determined by the numbers of lone and bonding electron pairs (i.e. the familiar Gillespie–Nyholm theory).

2. Secondary bonds may form in any direction *in line* with primary bonds, **but**

3. not in the same direction as a lone pair on the central atom. Although (2) states that 'in-line' interaction is preferred, deviations of up to 15° from 180° are known. It must also be admitted that packing considerations may

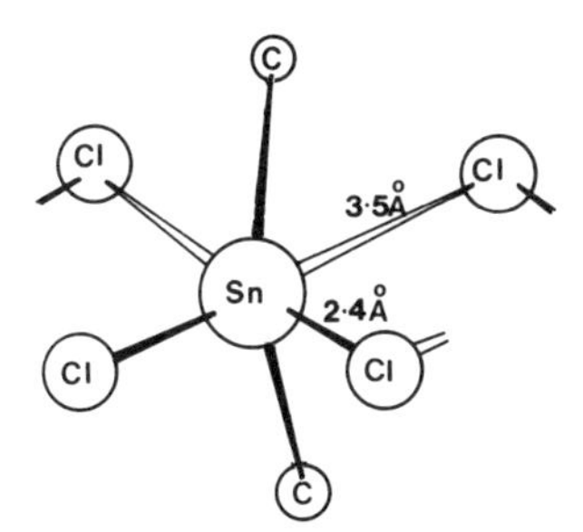

Figure 6-44 The environment of tin in Me_2SnCl_2

limit the *number* of secondary bonds to any atom which may otherwise be formed. In practice, secondary bonds to P, S and Cl, are rare, which points to an important additional factor, viz. the electronegativities of A and X. The strongest secondary bonds are formed when A is most electropositive, e.g. F—Br···F in BrF_2^+ compounds; a positive charge on A is also beneficial. However, making X more electronegative does not necessarily strengthen the secondary bonds, indeed in general the reverse seems to be true.

The evidence upon which these rules are based is laid out in Alcock's review (1973), to which the reader is referred for further details. It is to be hoped that this work will open up more systematic investigation of this kind of structure.

BIBLIOGRAPHY

Alcock, N. W., *Advances Inorganic Radiochemistry*, **15**, 2 (1972).

Ahmed, N. A., A. I. Kitaigorodskii and K. V. Mirskaya, *Acta Cryst. B*, **27**, 867 (1971).

Baur, W., *Acta Cryst.*, **19**, 909 (1965)

Cotton, F. A., and G. Wilkinson, *Advanced Inorganic Chemistry*, 3rd ed., Interscience, New York, 1972.

Craig, D. P., R. Mason, P. Pauling and D. P. Santry, *Proc. Roy. Soc. A*, **286**, 98 (1965)

Davies, M., *J. Chem. Educ.*, **48**, 591 (1971)

Dexter, D. L., and R. S. Knox, *Excitons*, Interscience, New York, 1965

Hamilton, W. C., and J. A. Ibers, *Hydrogen Bonding in Solids*, Benjamin, New York, 1968

Kitaigorodskii, A. I., *Organic Chemical Crystallography*, Consultants Bureau, New York, 1961

Kitaigorodskii, A. I., *J. Chim. Phys.*, **63**, 9 (1966)

Krebs, H., *Fundamentals of Inorganic Crystal Chemistry*, McGraw Hill, London 1968

Levy, H. A., and S. W. Peterson, *J. Amer. Chem. Soc.*, **75**, 1536 (1953)

Mason, R., *Proc. Roy. Soc. A*, **258**, 302 (1960)

Mason, R., and A. I. M. Rae, *Proc. Roy. Soc. A*, **304**, 501 (1968)

Pimentel, G. C., and A. L. McClellan, *The Hydrogen Bond*, Freeman, San Francisco, 1960

Rae, A. I. M., and R. Mason, *Proc. Roy. Soc. A*, **304**, 487 (1968)

Rae, A. I. M., *Molecular Physics*, **16**, 257 (1969)

Somorjai, G. A., and F. J. Szalkowski, *J. Chem. Phys.*, **54**, 389 (1971)

Ubbelohde, A. R., and K. J. Gallagher, *Acta Cryst.*, **8**, 71 (1955)

Wells, A. F., *Structural Inorganic Chemistry*, 3rd ed., The Clarendon Press, Oxford, 1962

Williams, D. E., *Acta Cryst. A*, **25**, 464 (1969)

CHAPTER 7

More about Mainly Covalent Crystals

The common feature of the crystals considered in this Chapter is that the directed bonds in them determine much of the structure. This does not exclude concurrent presence of ionic or any other kind of bonding. The division of material between Chapters 6 and 7 is to some extent arbitrary; in this chapter we are concerned with layer structures, to which some of the comments made about chains are also applicable. Discussion of silicates is brought under this heading mainly to emphasize the importance of directed bonding in (SiO_4) groups, and the determinative effect this has on the resulting structures, in the sense of dictating the broad lines along which the rings, chains or sheets, must be packed.

7.1 LAYER STRUCTURES

In layer structures, the number and disposition of bonds to *nearest* neighbours is determined by the available electrons and described by valence theory. The more distant neighbours are arranged in conformity with the requirements of topology (see Section 4.2.2). The stacking of layers is simple, there being few possibilities, but bonding between them raises some interesting questions: many of these have been elucidated by Krebs (1968), whose treatment we follow in most cases.

7.1.1 Structures Based upon Planar Three-connected Nets

The net of Figure 4-2a is suitable for layer structures formed from atoms which require three covalent bonds to nearest neighbours. The classic example of this is carbon in graphite, in which intralayer bonds are formed via sp^2 hybrids exactly as in benzene. p-Orbitals normal to the plane overlap (as in benzene) to form π-MOs covering the entire layer: i.e. they form a partly filled conduction band. This is the reason for both the colour (black) and the metallic electrical conductivity within the layers of graphite. The layers are widely separated (3·354 Å) and bonded by van der Waals forces only; electrical conductivity normal to them is therefore extremely small. Due to the weakness of interlayer bonding, graphite is readily cleaved; this same property is also associated with its use as a heavy-duty lubricant, although the mechanism of its lubrication is far from simple.

Any *two* graphite layers can be stacked in only one way, viz. with three alternate atoms of one ring above ring centres in the lower layer, Figure 7-1. The third layer can either be above the first ($\overline{AB}$, AB,. . ., giving hexagonal graphite) or displaced with respect to it ($\overline{ABC}$, ABC,. . ., rhombohedral graphite). However, the ABC stacking sequence is never found to extend throughout any one crystal; a complex form of disorder is therefore possible.

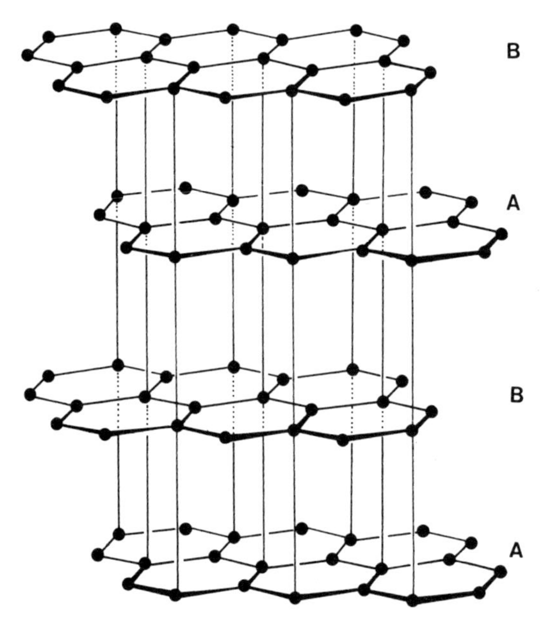

Figure 7-1 The structure of hexagonal graphite. (From *Fundamentals of Inorganic Crystal Chemistry* by H. Krebs. © 1968 McGraw-Hill Book Company (UK) Ltd. Used with permission)

Planar layers of the same kind are also present in white BN, which exhibits diamorphism analogous to that of carbon, which is not entirely unexpected since they are isoelectronic, having an average of four valence electrons per atom. Alternate boron and nitrogen atoms occur in graphite-like rings in BN but the layer stacking is not the same (Figure 7-2) and must result from weak directed bonding via *p*-orbitals normal to the layers. It is stable up to high temperatures in air and is therefore used for making refractory objects, and as a high-temperature lubricant. The higher electronegativity of nitrogen (3·00 compared with 2·00 for boron on Phillip's scale) results in greater localization of π electrons than in graphite: consequently BN is an insulator.

At high temperature and pressure (40 kbar), white BN is transformed to a cubic form with the diamond structure and greater hardness.

The same three-connected planar net is found in other situations. Thus, the structure of AlB_2 is based upon this net with vertical layer stacking (as in BN) required by the need to form bonds to metal atoms between the sheets,

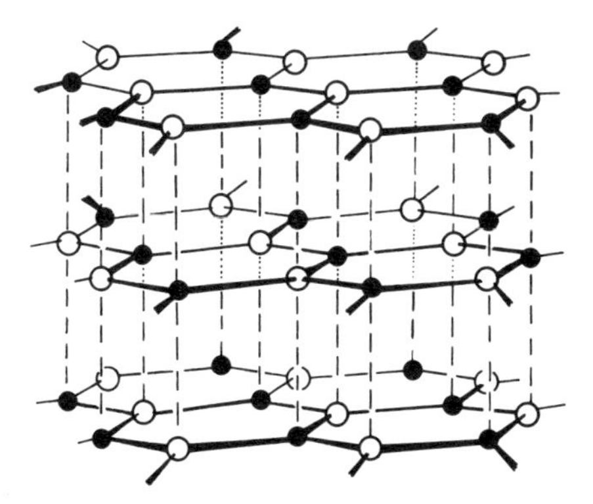

Figure 7-2 The structure of hexagonal BN. (From *Fundamentals of Inorganic Crystal Chemistry* by H. Krebs. © 1968 McGraw-Hill Book Company (UK) Ltd. Used with permission)

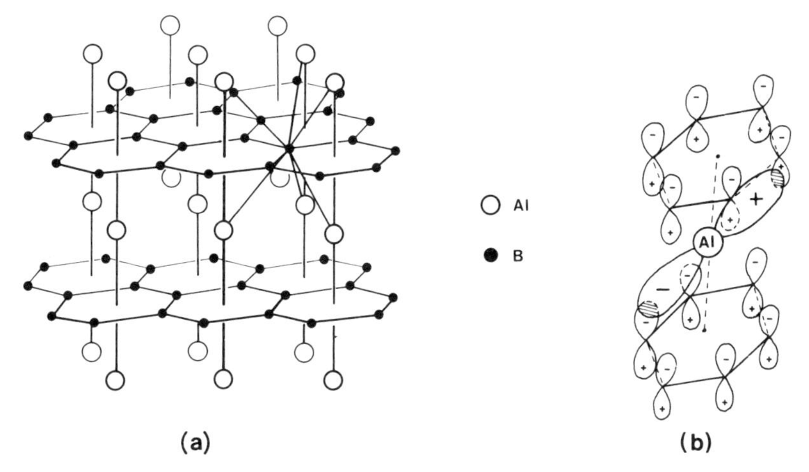

Figure 7-3 The AlB_2 structure (a), showing, (b), one overlap scheme (see text)

Figure 7-3. This structure is adopted by diborides of Mg, Al, Ti, Zr, Hf, V, Nb, Ta, Cr, Mo and U, as well as by some phases of USi_2, $PuSi_2$ and $ThSi_2$. The bonding is quite straightforward: the sheets themselves are constructed using sp^2 hybrids on boron (just as in graphite). Bonding to the metal atoms is by overlap of boron p_z-orbitals with *s*, *p* or *d* functions, as shown in Figure 7-3 for the case of *p*-orbitals.

The layer structure of Li_3N, more correctly formulated $Li[Li_2N]$, bears formal relation to the graphite type of net, although the bonding is quite different. In these hexagonal dark-red semi-transparent crystals each nitrogen has two co-linear nearest neighbours (1·94 Å) and six more at 2·11 Å completing a hexagonal bipyramid, Figure 7-4a. Nitrogen uses *p*-orbitals only. p_z

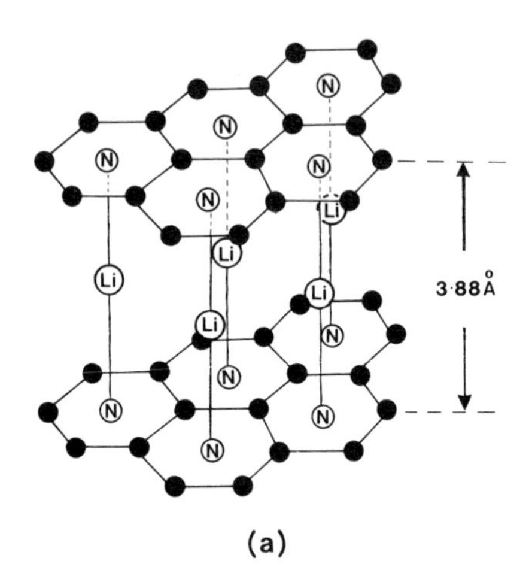

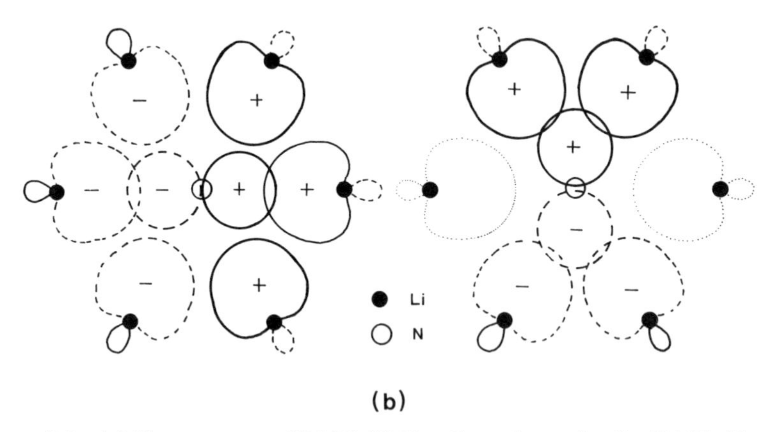

Figure 7-4 (a) The structure of Li_3N. (b) Bonding scheme for the $[Li_2N^-]$ layers, (From *Fundamentals of Inorganic Crystal Chemistry* by H. Krebs. © 1968 McGraw-Hill Book Company (UK) Ltd. Used with permission)

is used to overlap with *sp* hybrids on the Li atoms which lie *between* the $[Li_2N]$ layers. The degenerate p_x-, p_y-orbitals on N form two multicentre orbitals with sp^2 hybrids from the lithium atoms within the $[Li_2N]$ plane, Figure 7-4b.

7.1.2 Structures Based upon Buckled Three-connected Nets

Topologically there is no distinction between plane and buckled (or non-planar) versions of the same net, but the valence considerations requiring the non-planar version are quite different. In a formal sense (and in no other) **diamond** could be regarded as composed of buckled three-connected nets as shown in Figure 7-5a, the basic structural unit being a C_6 ring in the chair

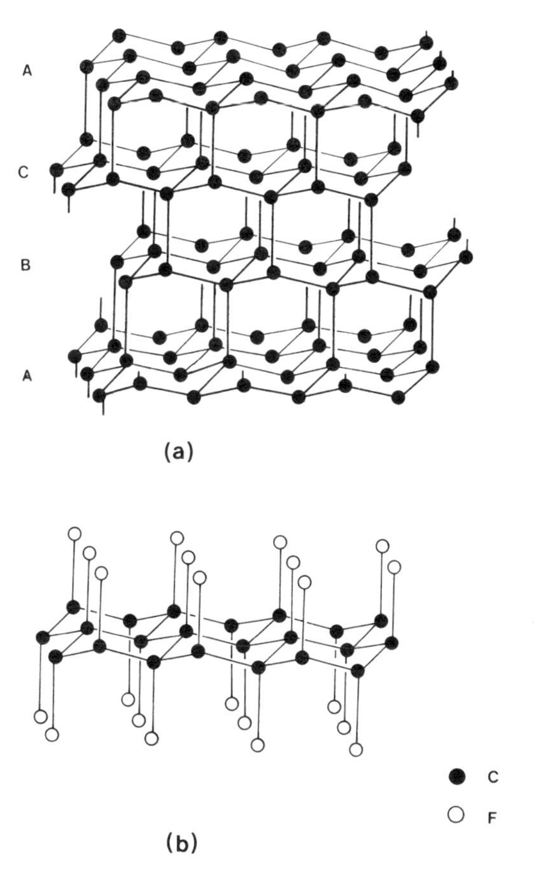

Figure 7-5 (a) The structure of diamond considered as constructed from buckled three-connected nets; cf. Figure 4-1a. (b) The layers found in CF

conformation. Single carbon (sp^3-hybridized) layers of this type are present in the grey CF, prepared by fluorination of graphite at 400–450°C, Figure 7-5b. Their stacking is irregular. More important is the rhombohedral allotrope of **arsenic** (the same structure is also adopted by antimony and bismuth) in which the same type of layer is formed. Each arsenic has three nearest neighbours trigonally disposed at distances $d_1 = 2{\cdot}51$ Å, but the layers are stacked in such a way that a distorted octahedron is completed by three next-nearest neighbours at 3·15 Å, shorter than the sum of van der Waals radii. This structure is usually related to the NaCl lattice, Figure 7-6, to emphasize the distorted octahedral coordination about each arsenic atom.

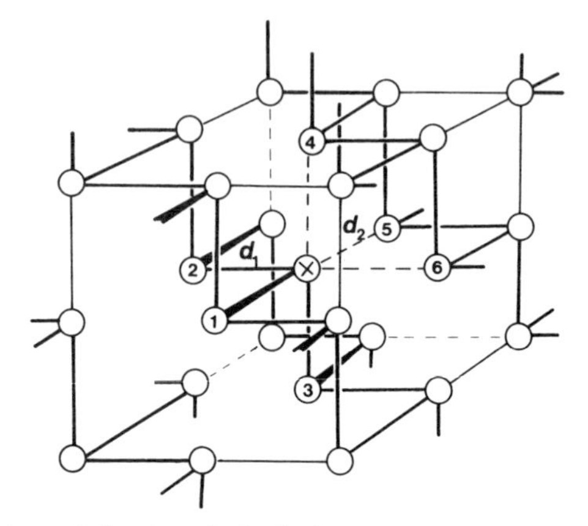

Figure 7-6 Relation of the rhombohedral arsenic layer structure to that of rock salt (schematic). Note the distorted octahedron formed about X by three intra-layer bonds (to atoms 1,2,3) and three interlayer bonds (to atoms 4, 5, and 6)

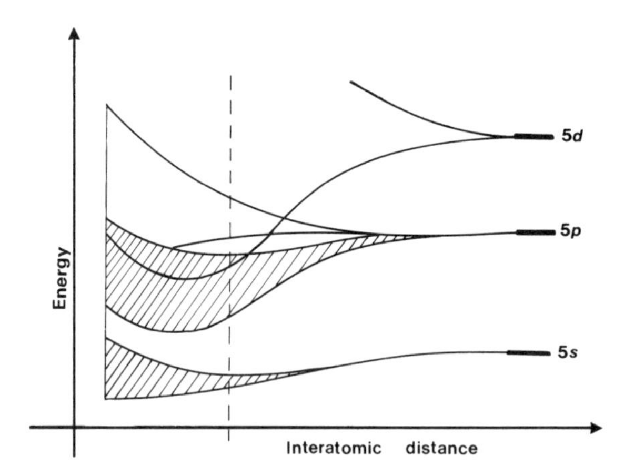

Figure 7-7 Band structure (schematic) of rhombohedral arsenic. (Mooser and Pearson in *Progress in Semiconductors*, **5**, Heywood and Company, 1960)

Bonding within the layers occurs via hydrid orbitals. If the interbond angles were 90°, bonding would be pure *p* type: arsenic is [core]s^2p^3. The angle is nearer 97° in arsenic (and close to 95° 30′ in antimony and bismuth) requiring admixture of some *s* character to give hybrids with mainly *p*

Table 1. Distances (Å) in the rhombohedral arsenic structure. (After Pearson (1972))

	Intralayer distance, d_1	Interlayer distance, d_2	van der Waals diameter	d_2/d_1
As	2·51	3·15	3·70[a]	1·25
Sb	2·87	3·37	4·00[a]	1·17
Bi	3·10	3·47	4·40	1·12
Simple cubic				1·00

[a] From Table 2, Chapter 2.

character, whilst accommodating the non-bonding pair in an orbital which is mainly *s*. Bonds between the layers are via sp^3d^2 hybrids which form a conduction band that is only slightly populated (at the expense of the intralayer bonding functions). Pearson (1972) has pointed out that this accounts for the known high electrical conductivity and metallic appearance of As, Sb and Bi. The situation is represented by the band scheme of Figure 7-7. As the Group is descended dehybridization increases, the conduction band is more highly populated, electronic conductivity rises and the distinction between nearest-next-nearest neighbour distances decreases, see Table 1. Pearson's proposal is

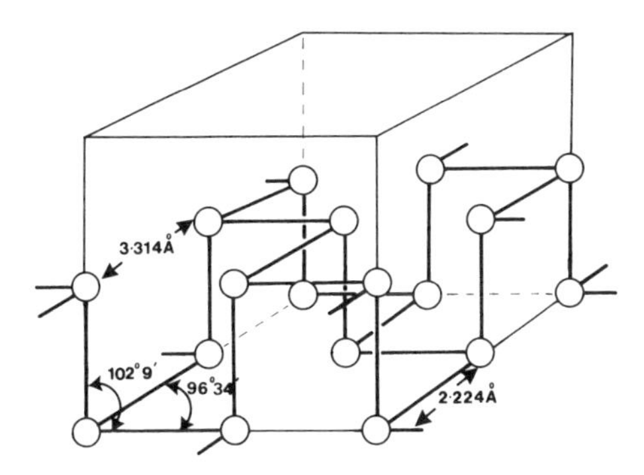

Figure 7-8 Formal relation of the layer structure of black phosphorus to that of rock salt. (The interbond angles are not shown correctly)

supported by the disappearance of electronic conductivity (semiconduction—by a different mechanism—remains) in the amorphous forms of As and Sb, which have small portions of the double-layer structure stacked in random fashion, with the next-nearest neighbour distance increased to 3·75 Å (effectively the van der Waals distance).

An alternative way of satisfying the same three-connectedness valence requirement is seen in the black phosphorus structure, which can also be, related formally to the NaCl lattice as shown in Figure 7-8. The distance between layers corresponds to the van der Waals diameter of phosphorus; this allotrope therefore does not show electronic conductivity.

The nets considered so far have had six-membered rings in the chair conformation. The boat conformation is seen in As_2O_3 (the mineral is known as claudetite) and a similar structure is adopted by As_2S_3 (orpiment). In the vapour phase, the so-called trioxides of phosphorus, arsenic and antimony, exist as tetramers P_4O_6, etc., in which phosphorus atoms at the vertices of a tetrahedron are bridged by oxygen atoms across each tetrahedral edge.

7.1.3 Structures Based upon Simple Square Nets

This class of structure is best exemplified by SnF_4 in which the metal is in octahedral coordination (Figure 7-9); PbF_4 and NbF_4 are isostructural with it. The layers are stacked in the only possible way, with the vertices of one layer fitting into the hollows of the next but, since the net is a very open one,

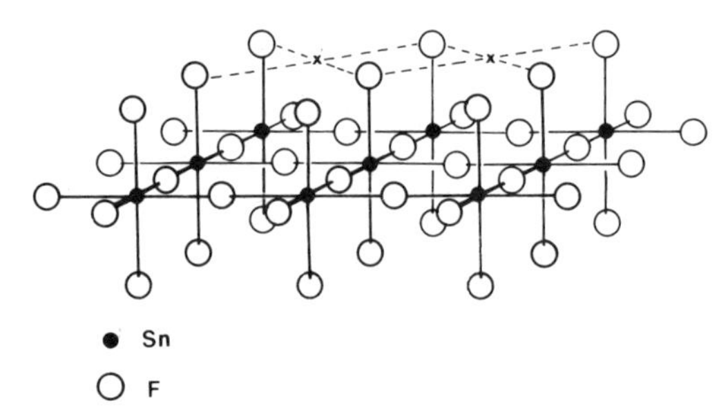

Figure 7-9 The layer structure of SnF_4. Fluorines from adjacent layers are located in positions x

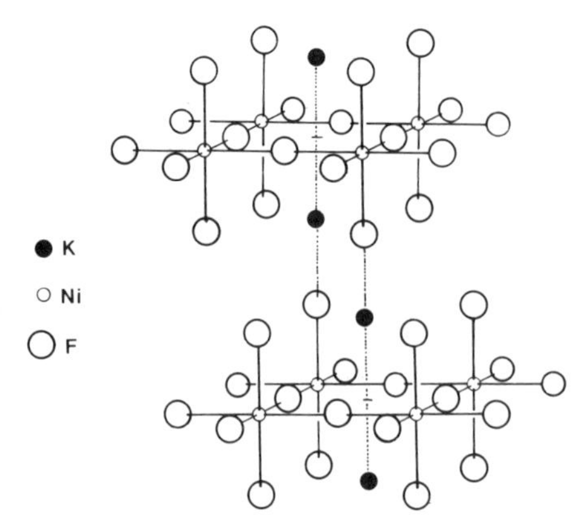

Figure 7-10 The K_2NiF_4 structure, showing the packing of layers of the SnF_4 type. (From *Fundamentals of Inorganic Crystal Chemistry* by H. Krebs. © 1968 McGraw-Hill Book Company (UK) Ltd. Used with permission)

this means that the two sets of fluorines are almost co-planar. A group of related compounds K_2MF_4, where M = Mn, Fe, Co, Ni, and oxides A_2BO_4 such as Sr_2TiO_4, Ca_2MnO_4 and Ba_2SnO_4, consists of similar MF_4 or MO_4 layers separated by cations which thereby achieve formation of a double layer of the rock salt type, Figure 7-10. In both these and SnF_4 there is doubtless considerable Madelung energy contributing to stabilization of the structure. Indeed, they are commonly related to perovskite (Wells, 1962; Krebs, 1968).

7.1.4 Sandwich-type Layer Structures

The $CdCl_2$ and CdI_2 structures were outlined on p. 57 where it was noted (Table 4, Chapter 3) that the hexagonal CdI_2 form is strongly favoured by the heavier anions. We now seek to explain why this should be.

The $CdCl_2$ and CdI_2 structures differ only in their layer-stacking sequence ($\overline{ABC}$ and $\overline{AB}$ respectively). The true situation is much more complex than has been indicated so far. CdI_2 is actually trimorphic. The known structures, all of which differ in the layer-stacking sequence, are summarized in Table 2.

Table 2. Types of layer stacking in dihalide sandwich structures. (After Wells (1962))

N	Stacking sequence	Example
2	AB (h.c.p.)	CdI_2 (C6)[a]
3	ABC (c.c.p.)	$CdCl_2$ (C19)
4	ABCB or ABAC	CdI_2 (C27)
6	ABCACB	CdI_2
9	ABCBCACAB	CdBrI
12	ABCBCABABCAC	CdBrI

N = No. of layers in repeat unit.
[a] All common structure types are assigned a *Strukturbericht* number for convenient reference.

The structure in which CdI_2 crystallizes depends upon conditions of preparation; a form is also known in which the stacking sequence is random. PbI_2 can be obtained in both C6 and C19 forms, and similar behaviour is almost certainly exhibited by many other dihalides of the sandwich-structure type.

Molybdenite, MoS_2, has a structure that is very closely related to those of the above dihalides. The individual sandwiches are constructed from *superimposed* close-packed sulphur layers with metal atoms in the interstitial sites. Due to the stacking sequence (AA) these holes are not octahedral but trigonal prismatic. Individual sandwiches are stacked in the usual offset fashion, see Figure 7-11. Synthetic MoS_2 has been prepared with the C19

($CdCl_2$) structure, whilst TaS_2 is known to adopt no less than four structures: C6, C19, C27 and the twelve-layer CdBrI structure (Wells, 1962).

It is evident that very subtle energetics control the adoption of all these layer structures, and it may turn out that the true experimental facts have not yet been established. Ideas on the forces responsible for a twelve-layer repeat (for example) are, at best, vague. It does seem clear that within any one sandwich XMX, the bonding can vary from only slightly to quite highly directional. Indeed, these two structures are essentially transitional in type between rock salt and CsCl on the one hand, and the more molecular structures which begin to appear among trihalides (e.g. Al_2Br_6). Compounds with the C6 and C19 structures fall in a middle domain on Mooser–Pearson $\bar{n}$ vs $\Delta\chi$ plots, see Figure 5-16. For transition metals d^2sp^3 hybrids are available for intralayer bonding although in PbI_2, with its inert pair of s-electrons, bonding is via

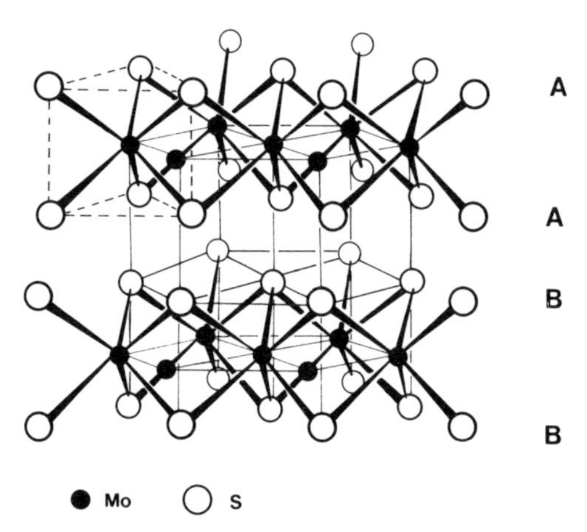

Figure 7-11 The layer structure of molybdenite, MoS_2, showing trigonal-prismatic coordination at molybdenum

two resonating p-electrons. The trigonal-prismatic coordination in MoS_2 is highly unfavoured electrostatically (Section 4.3.1) and clearly implies the presence of strongly directional bonding. This symmetry requires the use of s, p_z, d_{yz}, d_{zx} and either p_x, p_y or $d_{x^2-y^2}$, d_{xy} orbitals or some combination of them. Energetically, d^4sp hybrids seem the most likely; Krebs noted that these are directed towards the vertices of a trigonal prism. Similar coordination is known in some transition-metal complexes with thio ligands, e.g. $Mo\{S_2C_2(CN)_2\}_3$. Since MoS_2 can crystallize both in the molybdenite and $CdCl_2$(C19) structures, it is evident that the C19 structure is also compatible with a high degree of directed bonding. Transition from MoS_2 to $CdCl_2$(C19) structures with its accompanying change of metal six-coordination from trigonal-prismatic to octahedral is no more surprising than the occurrence of Mo in these geometries in coordination complexes: the energy difference between the two arrangements is evidently not large.

In **summary**, bonding within the X—M—X sandwiches of both $CdCl_2$ and CdI_2 structures varies from fairly non-directional (i.e. largely ionic) to quite highly covalent. $CdCl_2$ has a higher Madelung constant (4·489) than CdI_2 (4·383); the electrostatic part of the bonding is therefore best accommodated in its lattice, which is adopted by chlorides rather than bromides or iodides. The reason why dichlorides are found with the $CdCl_2$ structure is best understood by considering the main alternative AB_2 lattices, fluorite and rutile. The fluorite structure is only adopted by the larger cations and requires radius ratios larger than those associated with compounds having the $CdCl_2$ structure, whereas the rutile lattice only allows the anion to be planar three-coordinate, a situation implying use of sp^2 hybrids. We note that rutile and $CdCl_2$, CdI_2, phases are well separated on Mooser–Pearson diagrams (Figure 5-16).

The real difficulty with the $CdCl_2$, CdI_2, structures comes in understanding the interlayer X---X bonding. There is no doubt that this is largely van der Waals in type although it is also clear that additional interactions are of great importance in explaining (*a*) the conduction properties, and (*b*) the layer-stacking orders.

For the most part, compounds with the C6 and C19 structures are semiconductors but some show metallic conductivity, in particular, the dichalcogenides of Co, Rh, Ni, Pd and Pt; indeed, $PdTe_2$ is a superconductor. These same compunds are also distinguished by their X---X distances which are distinctly shorter than van der Waals radius sums and comparable with the interchain separations in Se and Te. Rather more surprisingly (because their X---X distances are normal), the sulphides, selenides, and tellurides of Ti and V Groups of the Periodic Table also show high electronic conduction. In the latter group of compounds *d*-orbitals are rather large so that there is the possibility of weak direct-bonding overlap between cation layers through the intervening X---X layers as in oxides MO (Section 5.5.4). However, another mechanism cannot be dismissed and may also be important in the Ni and Co groups of metallic conductors. Most, if not all, of these dichalcogenide phases are non-stoichiometric. If they are metal rich (i.e. MX_n with n slightly less than 2) the additional cations will enter the 'vacant' layers of octahedral sites; the same kind of distribution is also obtained by disordering the cations of both stoichiometric and cation-poor phases. The cation sites in an h.c.p. lattice are sufficiently close to allow metal–metal interaction (see discussion of NiAs and β-$ZrCl_3$) thereby permitting metallic conduction normal to the layers. The substantial X---X compression of the Co and Ni group chalcogenides is more likely to be due to a direct overlap band scheme. Krebs (1968) noted that in CdI_2 three approximately straight lines can be traced oblique to the layers, and that mesomeric bonding is therefore possible since *p*- (and, where appropriate, *d*-) orbitals are aligned with these directions. Thus:

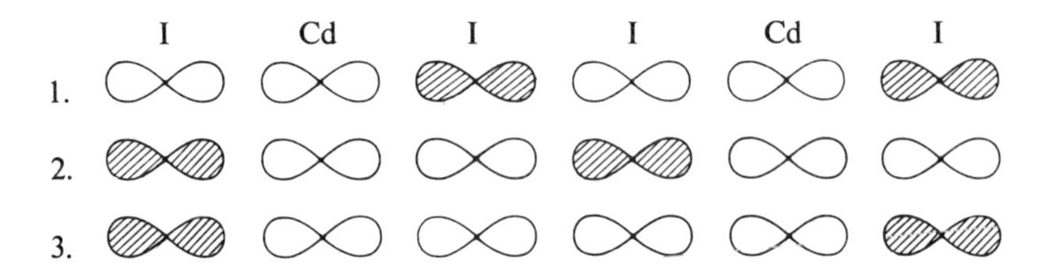

where the boundary surfaces represent filled (shaded) and half-filled *p*-orbitals. It is the resonance hybrid 3 which provides the bonding between the layers.

Similar mesomeric bonding is possible in the $CdCl_2$ structure but chlorine shows less tendency to enter into this type of bonding scheme (see discussion of halogen structures, Section 6.1.3); it seems that the higher Madelung constant therefore makes this structure marginally preferable.

Finally, we consider briefly the question of the complex stacking sequences associated with some of the related phases (see Table 2). Since it is improbable that any force will ensure a particular stacking sequence as long as twelve or even six layers high, we must consider explanations based upon shorter-range ordering. One means of achieving this would be to have a proportion of cations in the 'vacant' layers but *ordered* such that particular sequences were favoured. In particular, the partial sequence ABA would be favoured as this brings two sets of octahedral sites into fairly close proximity (as in NiAs). Considering, for example, the sequence of the nine-layer CdI_2 structure we see the regular occurrence of hexagonal repeats of this type (indicated by brackets).

7.2 STRUCTURES BASED UPON LINKED POLYHEDRA

The requirements of both the simple ionic theory (Section 4.3.1) and of directional-bonding schemes lead to formation of coordination polyhedra around atoms. Thus, in NaCl each ion is surrounded octahedrally by six of the other kind, whilst in ZnS tetrahedral coordination is found because of the need to accommodate highly directional bonding. Given that these nearest neighbours are disposed at the vertices of polyhedra in accord with principles which we understand, the next problem, the arrangement of more distant neighbours, is one of topology (see Section 4.2.2). From the geometrically possible arrangements of polyhedra, one or more will occur in nature, the selection being made on the basis of the most favourable energetics. For close-packed highly ionic materials it is not particularly helpful to emphasize the interconnections of the various coordination polyhedra, but with more

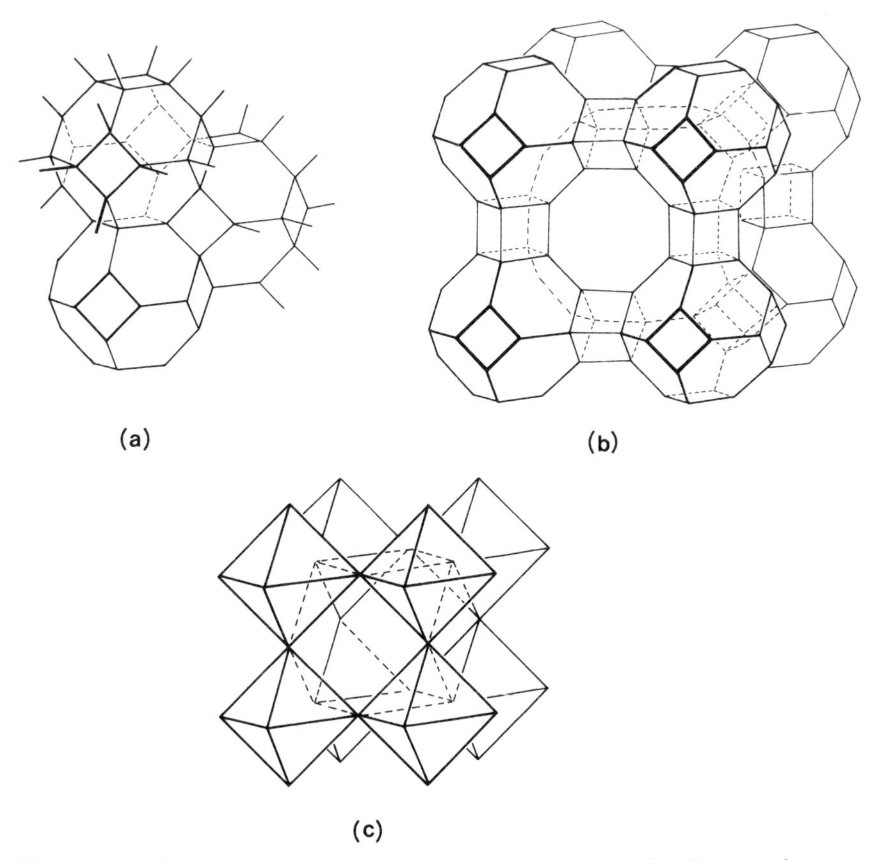

Figure 7-12 Some space-filling polyhedral arrangements. (a) Truncated octahedron. (b) Truncated octahedron, cube, and cuboctahedron. (c) Octahedron and cuboctahedron. (From A. F. Wells, *Structural Inorganic Chemistry*, The Clarendon Press, Oxford, 1962)

directed bonding it is sometimes fruitful to consider structures in this way, especially if the structures are complex. The advantages are twofold: (*a*) visualization of structures is much easier, (*b*) systems of classification develop naturally on this basis. All this may seem to be rather academic; in fact such considerations are of very real value in dealing with some problems in non-stoichiometry (Chapter 9) and in silicate chemistry. Part of the object of this

Table 3. Relative distances between two atoms centering polyhedra joined as shown

	Common vertex	Common edge	Common face
Octahedra	1·00	0·71	0·58
Tetrahedra	1·00	0·58	0·33

section is to lay a basis for that work. Of course, polyhedra can be described equally well by nets: both descriptions of a given structure are equally valid but it is often easier to visualize a structure in terms of a recognizable polyhedron.

In the representation of crystal structures by polyhedra, two general cases have been considered:

(*a*) the polyhedra fill all space,
(*b*) more open networks are formed.

(*a*) Fedorov (1904) showed that if space is to be filled by identical convex polyhedra, all having the same orientation, then only five shapes are admissible: cube, hexagonal prism, truncated octahedron, rhombic dodecahedron, elongated dodecahedron. If the orientation restriction is lifted, other polyhedra (such as the trigonal prism) are allowed. If more than one kind and orientation is permitted, a very large number of possibilities arise. Of all the above ways of filling space, only a few are of any real concern in crystal chemistry: Figure 7-12 illustrates some of them.

Table 4 Compositional formulae resulting from linking octahedra or tetrahedra

	Elements shared per polyhedron		
No. of units joined	2 Vertices	2 Edges[a]	2 Faces
Octahedra			
1	MX_6	MX_6	MX_6
2	M_2X_{11}	MX_5	M_2X_9
3	M_3X_{16}	M_3X_{14}	MX_4
∞ chain or a ring of any size	MX_5	MX_4	MX_3
Tetrahedra			
1	MX_4	MX_4	
2	M_2X_7	MX_3	
3	M_3X_{10}	M_3X_8	
∞ chain or a ring of any size	MX_3	MX_2	

[a] Not adjacent.

(*b*) An infinite number of possibilities exist for the open packing of polyhedra. Our discussion is restricted to consideration of two of the most important, the octahedron and the tetrahedron.

In principle, polyhedra can be linked by sharing vertices, edges or faces or some combination thereof. Sharing faces is generally unfavourable since the atoms at the polyhedron centres are brought close to each other, increasing electrostatic repulsion; this situation can sometimes be stabilized by formation of a covalent bond as in β—$ZrCl_3$. Sharing of edges is more common than sharing faces, but both tend to reflect the electrostatic repulsion between central atoms by a slight distortion which allows them to move a little further apart. Thus, a shared edge is often found to be shorter than unshared edges. Relative central-atom distances are shown in Table 3. Table 4 shows the formulae which result from various schemes of polyhedron linkage.

7.2.1 Open Packing of Octahedra

Sharing a common vertex between two octahedra MX_6 yields M_2X_{11}. Continuing this process gives M_nX_{5n+1} which, in the limit of a long chain, becomes MX_5 since this is the repeat unit of the structure, see Figure 6-25*a*. Given three or more octahedra per ring, MX_5 can be formed. Several examples of these possibilities have been discussed in Section 6.1.4, viz. the chains of α-UF_5, Na_2AlF_5; the same chain type is also found in $SrPbF_6$, more correctly written as $Sr(PbF_5)F$. Examples of cyclic polyhedral structures are the tetrameric fluorides $(MF_5)_4$, p. 186.

Sharing two non-adjacent edges yields MX_4 chains such as those found in NbI_4 and $TcCl_4$ (Section 6.1.4); dimers M_2X_{10} are found among the chlorides and bromides of Nb, Mo and Re. Face-sharing occurs in $Fe_2(CO)_9$, $M_2Cl_9^{3-}$ and in chains such as β-$ZrCl_3$ (p. 56).

If more than two vertices or edges are shared, sheet structures are possible. For example, if each octahedron shares all four vertices in a square-plane the composition MX_4 results. This structure is exemplified by SnF_4, K_2NiF_4 (p. 228) and M^IAlF_4 (M^I = NH_4, K, Rb, Tl). All six vertices shared gives the ReO_3 structure, Figure 7-12c. In this figure, the octahedral vertices represent oxygen atoms whilst Re is at the centre of each one. This should be compared with the description (p. 66) based upon close packing: the figure clearly displays the incomplete nature of the oxygen lattice and makes the linear twofold coordination of oxygen clear.

If each octahedron shares three edges, layer structures such as $CrCl_3$ and BiI_3 result (compare with Figure 3-18a), Figure 7-13, while a particular form of sharing four edges is present in chain structures such as that of NH_4CdCl_3, Figure 7-14.

Sharing some vertices and some edges in the same structure leads to many possible arrangements. Two of especial relevance are rutile and MoO_3.

Rutile (TiO_2) can be described as chains of edge-sharing octahedra of composition (TiO_4) condensed by sharing the remaining vertices as illustrated in Figure 3-32. In Figure 3-31 is shown a projection of the rutile structure which is readily related to that shown in Figure 3-32. Using this nomenclature the essential features of many other structures are readily displayed, Figure 7-15 a to c. MoO_3 has a layer structure in which some oxygen atoms are three-coordinate whilst others are two-coordinate or are terminal.

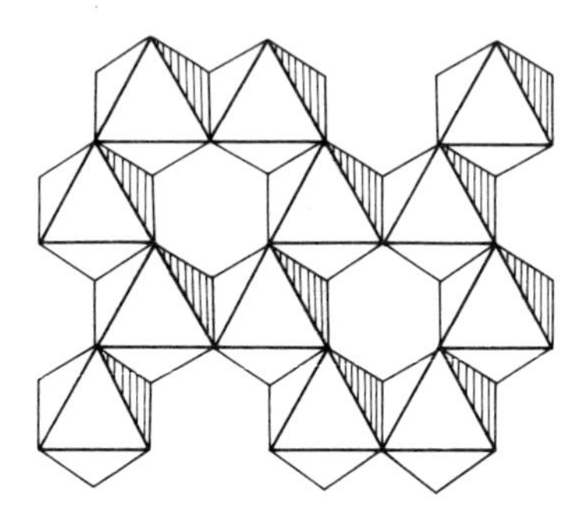

Figure 7-13 Layers of the type found in $CrCl_3$ and BiI_3 represented as MX_6 octahedra sharing three edges

None of the above structures are so complex that their description by normal unit-cell drawings presents any difficulties although, in some cases, description in term of octahedra is perhaps more readily grasped. But this approach really helps in dealing with groups of related structures of an involved nature such as the iso- and hetero-polyacids of molybdenum and tungsten.

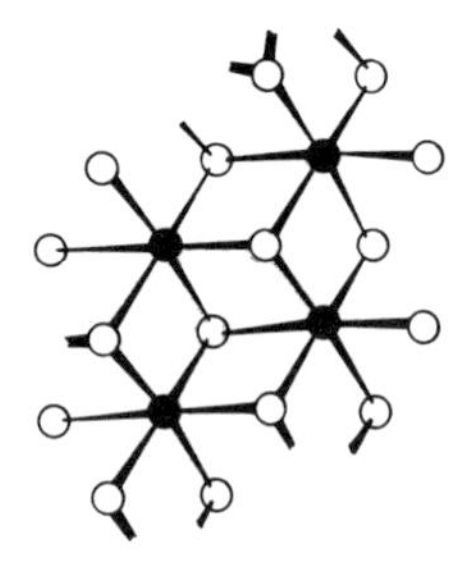

Figure 7-14 The structure of the $\{CdCl_3^-\}_n$ chains, in NH_4CdCl_3

Iso- and hetero-polyacids of V, Nb, Ta, Mo and W

A small group of elements in the Periodic Table are notable for the extent to which their oxyanions may be polymerized, both in solution and as solids. They are: borates (based upon three-coordinate boron), phosphates,

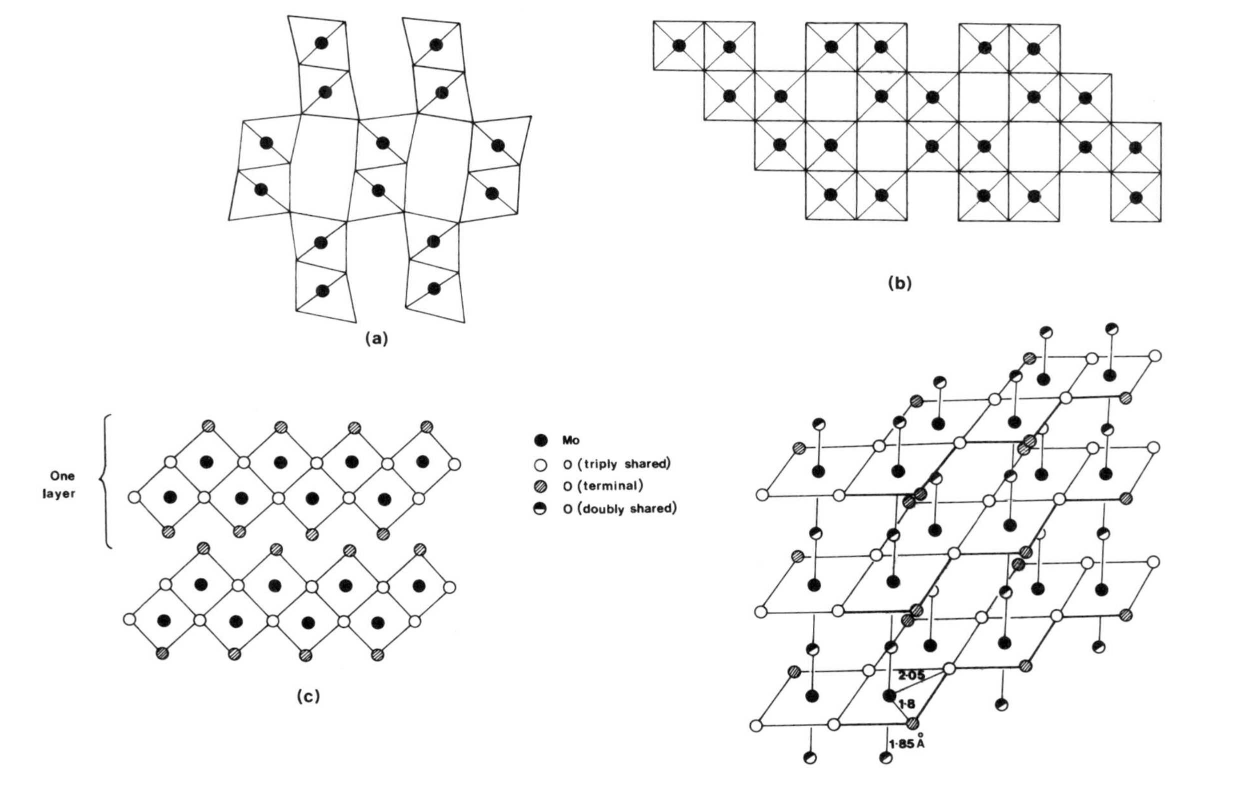

Figure 7-15 The structures of various oxides in which the metal has octahedral or distorted octahedral coordination, represented schematically in terms of their coordination polyhedra. (a) One form of MnO_2, ramsdellite. (b) V_2O_5 (idealized; see also Figure 4-3). (From A. F. Wells, *Structural Inorganic Chemistry*, The Clarendon Press, Oxford, 1962). (c) MoO_3. (d) Perspective view of one layer of the MoO_3 structure

aluminates and silicates (based upon tetrahedrally coordinate atoms), and the oxyanions of vanadium (V), niobium (V), tantalum (V), molybdenum (VI) and tungsten (VI), in which the metal is invariably octahedrally coordinated. We now outline the main structural features of the oxyanion polymers formed by the five transition metals. These polymers are collectively known as iso-

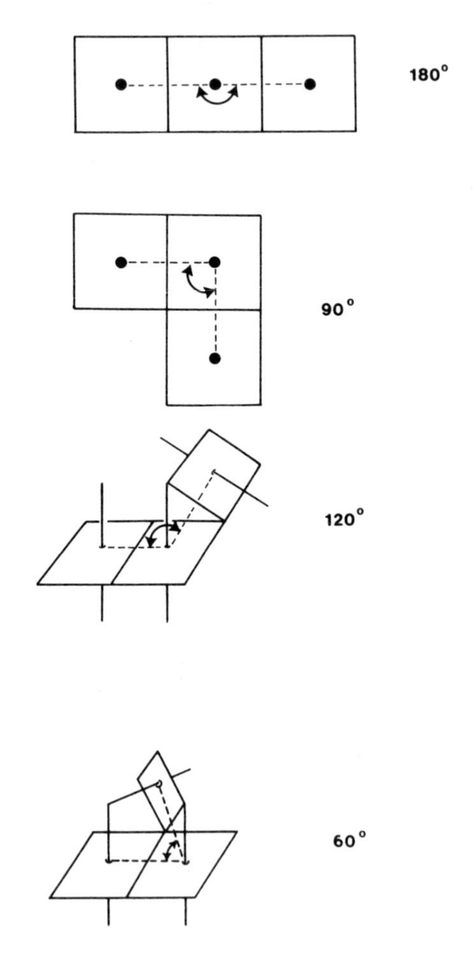

Figure 7-16 Four ways of joining three octahedra by edge sharing

polyacids. A further series, the heteropolyacids, can be made incorporating other elements which also form oxyanions.

Polymerization of one of these metal oxyanions proceeds stepwise but, under given conditions, always results in formation of a polymeric species of definite molecular weight rather than the complex mixture that would normally be expected. Thus, in the most basic solutions of vanadium (V), tetrahedral vanadate, VO_4^{3-}, is present. Lowering the pH causes dimerization, then trimerization, resulting finally in formation of $V_{10}O_{28}^{6-}$. The most stable polymers for other metals are $M_6O_{19}^{8-}$ (M = Nb, Ta), $Mo_7O_{24}^{6-}$, $Mo_8O_{26}^{4-}$ and $HW_6O_{21}^{5-}$.

Copolymerization with other oxyanions such as phosphate, arsenate, silicate, and many metallic species, yields heteropolyacids, also of definite composition; for example, the phosphotungstate $PW_{12}O_{40}^{3-}$. Others are

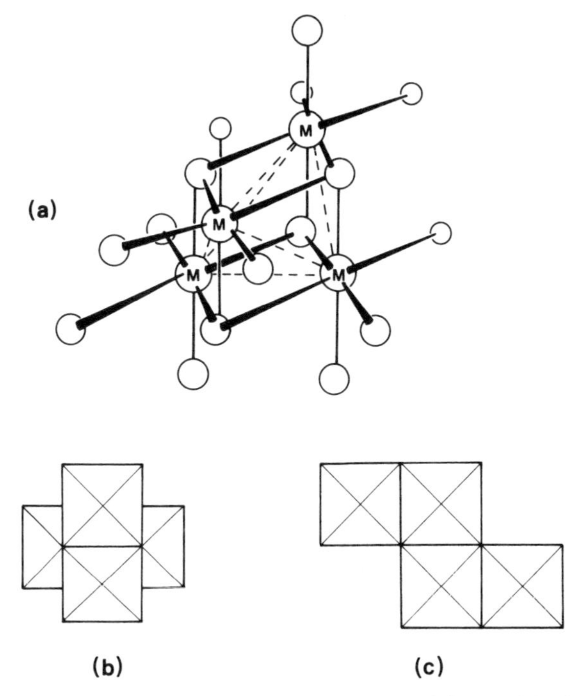

Figure 7-17 (a) and (b) show the structure of the M_4O_{16} polyhedral cluster. (c) Represents a further way of joining four edge-sharing octahedra

$TeMo_6O_{24}^{6-}$, $M^{III}Mo_6O_{24}^{9-}$ (M^{III} = Al, Cr, Fe, Co, Rh, Ga), and $M^{IV}Mo_9O_{23}^{6-}$ (M^{IV} = Mn, Ni). Many of the reactions in which these complex structures are formed are of analytical value, but of greater importance is their activity as catalysts for certain reactions of hydrocarbons.

The main features of polyacid chemistry have been recognized for a long time, but not understood. Why are polymers of definite molecular weight formed? Why does the extent of polymerization differ amongst the metals? Why are these specific structures formed? Answers to these questions have been offered recently by Kepert (1969), based upon an ionic model: we follow his arguments. As noted above (p. 235) sharing of octahedral edges introduces unfavourable electrostatic repulsion between the central metal ions, but this can be relieved to some extent by an off-centre displacement which will be easier the smaller the cation. As the size of the polyanion increases, Coulom-

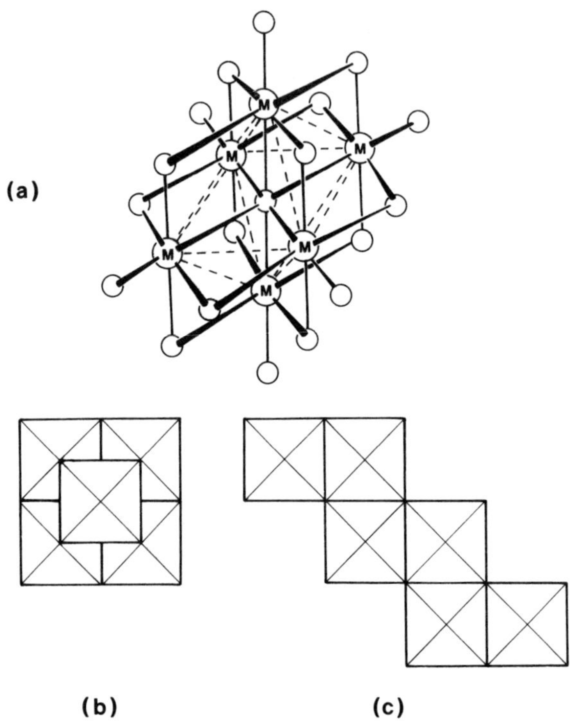

Figure 7-18 (a) Structure of $M_6O_{19}^{8-}$, (M = Nb, Ta). (b) and (c) show two favourable ways of linking six edge-sharing octahedra, (b) being equivalent to (a). (Kepert (1969))

bic repulsion rises as off-centre displacements become restricted by the arrangement of neighbouring octahedra and will eventually form a barrier to further polymerization; this situation will be reached earlier for the larger ions. The facts broadly support this contention.

Crystal radius of cation, Å (from Table 4, Chapter 2)	Complex ion
0·68	$V_{10}O_{28}^{6-}$
0·74	$Mo_8O_{26}^{4-}$, $Mo_7O_{24}^{6-}$
0·72	$HW_6O_{21}^{5-}$
0·78	$M_6O_{19}^{8-}$ (M = Nb, Ta)

The shapes of the polyanions are also suggested on the same basis. Addition of a third octahedron to a pair of edge-shared octahedra can be done in four ways giving respectively angles of 60°, 90°, 120° or 180°, depending upon the edges shared by the central octahedron (Figure 7-16). The 120° and 180° configurations are least favourable since the metal in the central polyhedron is subjected to opposed or nearly opposed forces. In fact, 60° edge sharing is very common in heteropoly-molybdates and -tungstates. For four octahedra joined by common edges, the structures of Figure 7-17 are most favourable.

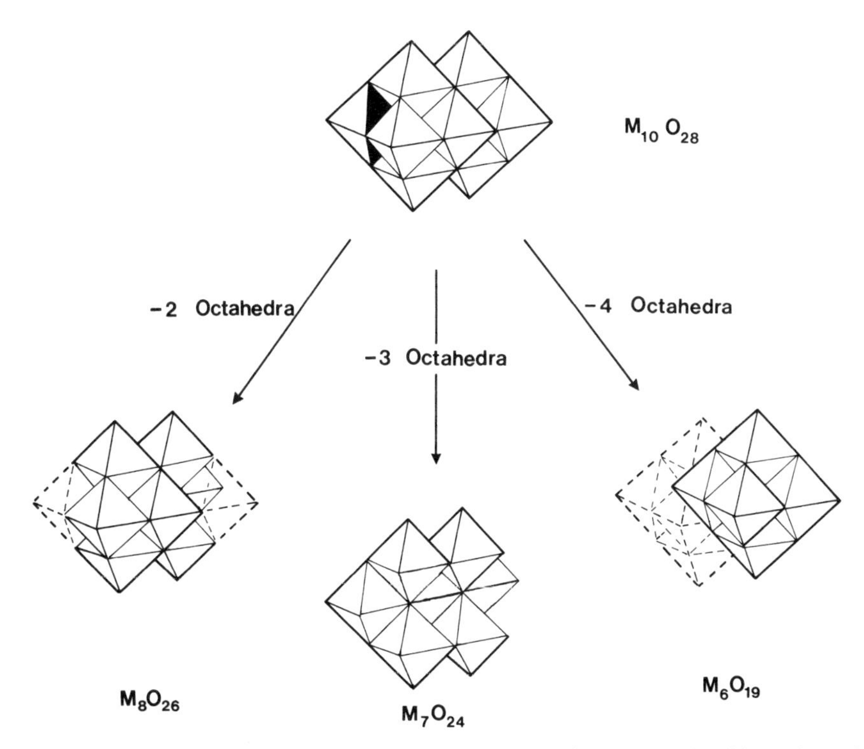

Figure 7-19 The relationships of M_8O_{26}, M_6O_{19}, and M_7O_{24} polyacids to that of $M_{10}O_{28}$. (Kepert (1969))

M_4O_{16} consists of a tetrahedron of octahedra with 60° M—M—M angles: this is the structure present in $Li_{14}(WO_4)_3(W_4O_{16})\cdot 4H_2O$. Only two combinations of six edge-shared octahedra avoid 120° or 180° interactions; these are shown in Figure 7-18 in which (a) represents the structures of the Nb and Ta polyanions. (c) In Figures 7-17 and 7-18 is a structure type which would seem to allow indefinite polymerization since the Coulombic barrier is independent of chain length. Since it is not observed it is presumed that it is unstable with respect to M_4O_{16} or M_6O_{19} in which further elimination of water is accompanied by a favourable entropy term.

When more than six octahedra share edges, some of the unfavourable 120° or 180° interactions are inevitable. Two 180° interactions occur in $V_{10}O_{28}^{6-}$, the largest known isopolyanion; its structure is considerably distorted by mutual repulsion of metal atoms resulting in a range of V—O distances from 1·59 to 2·22 Å and reduction of the central V—V—V angles from 180° to 175°. The structures of the Mo_8 and Mo_7 species can be regarded as derived from $M_{10}O_{28}$ by successive removal of fragments to relieve electrostatic strain (see Figure 7-19).

Heteropolyanions nearly always have the heteroatom in tetrahedral or octahedral coordination. The structures of a range of heteropolyacids with octahedrally coordinate heteroatoms have been rationalized along lines similar to those above; the interested reader is referred to Kepert (1969) for further details.

7.2.2 Open Packing of Tetrahedra

A system of structures based upon tetrahedra can be developed analogous to that for octahedra. Sharing of faces never occurs (cf. filling all tetrahedral sites in an h.c.p. array). Edge sharing brings the central atoms very close together, an electrostatically unfavourable situation relieved to some extent,

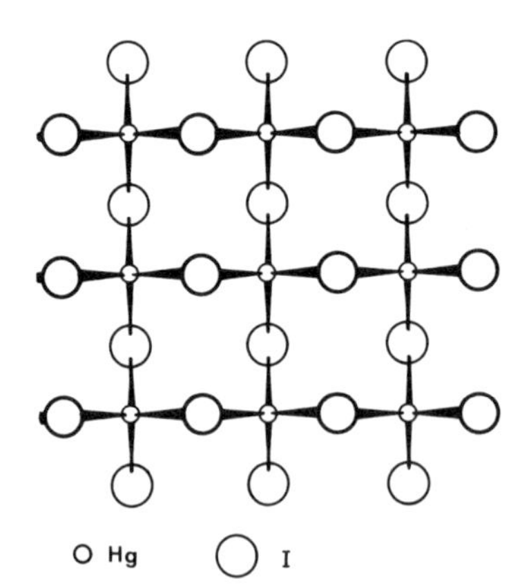

Figure 7-20 The layer structure of red HgI_2. (From A. F. Wells, *Structural Inorganic Chemistry*, The Clarendon Press, Oxford, 1962)

as above, by distortion shortening the shared edge. Examples are: (*a*) the dimeric halides M_2X_6 (Fe_2Cl_6, Al_2Br_6 and similar halides of gallium and indium); (*b*) chains such as $BeMe_2$, SiS_2 described in Chapter 6.

By far the commonest way of linking tetrahedra is via common vertices.

Sharing *one* vertex yields finite pyroanions such as $Cr_2O_7^{2-}$, $S_2O_7^{2-}$, $P_2O_7^{4-}$ but sharing *two* vertices per tetrahedron allows both rings (e.g. γ-SO_3, $P_4O_{12}^{4-}$) and chains (e.g. α-SO_3, $(SiO_3^{2-})_n$, etc.) to be formed, see Figure 6-22. Sharing of *three* or *four* vertices allows many types of structure including layers and three-dimensional frameworks. If all four vertices are shared the

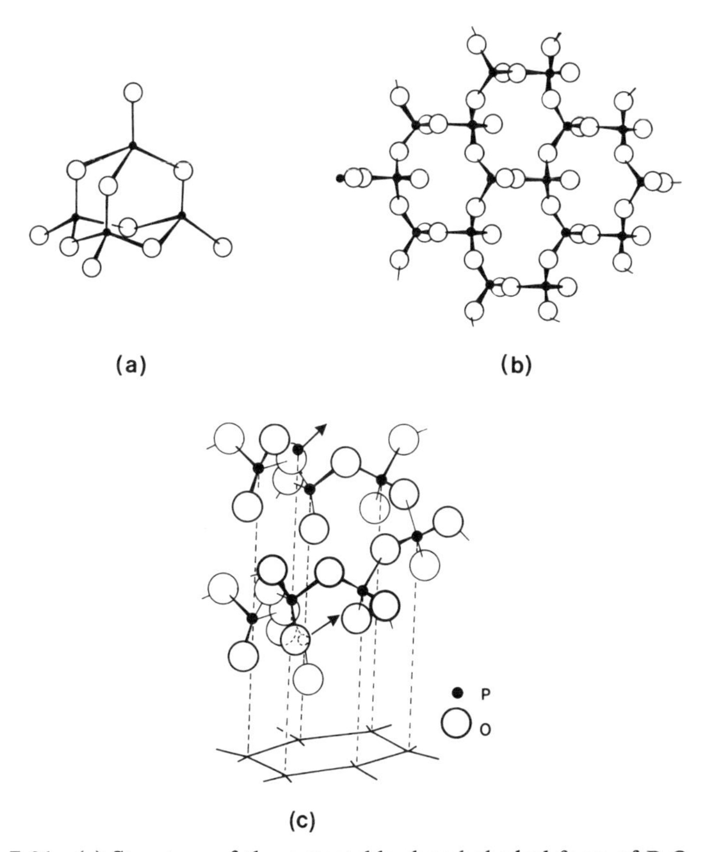

Figure 7-21 (a) Structure of the metastable rhombohedral form of P_2O_5. These molecules P_4O_{10} are also present in the vapour. (b) Structure of the orthorhombic layer form of P_2O_5; it is based upon a hexagonal net. (c) Structure of the stable orthorhombic form of P_2O_5. (*Structure Reports*, **8**, 144 (1940–41))

formula becomes AB_2; two examples are silica, SiO_2, and red mercuric iodide, HgI_2. The latter has the layer structure of Figure 7-20; silica has several polymorphs but two of the commonest (cristobalite and tridymite) have the tetrahedral framework structures already discussed on p. 79.

Phosphorus pentoxide provides an instructive case: it has three polymorphs with the structures shown in Figure 7-21. Wells (1968) has discussed these as an example of different topological ways of accommodating the same tetrahedral unit (PO_4), the structure of which is determined by valence theory. (a) is the simplest polyhedral configuration; (b) has the form of the simplest three-connected plane net (see Figure 4-2a) and is thus a layer structure. The stable orthorhombic form has a three-dimensional structure in which three vertices of each (PO_4) are shared; the P atoms are at the vertices of one of the two simplest three-connected three-dimensional nets, having rings of ten (PO_4) units with the oxygen atoms approximately close packed. These structures represent three totally different ways (of approximately equal energy) of packing tetrahedra whilst allowing each to share three vertices. What is of especial interest is not so much that this happens, but that the three structures are among the simplest *possible* configurations associated with three-connectedness.

7.2.3 The Mineral Silicates

The most extensive series of linked polyhedra occurs in silicon chemistry. The earth's crust is composed very largely of silica and silicates (rocks, sand, clay, soil) as are many important structural materials (bricks, cement, ceramics) and the majority of semiprecious gemstones. Natural silicates were formed from a complex molten magma; it is therefore not surprising that they are of variable composition. It is preferable to refer to them as minerals rather than compounds. To each mineral is ascribed an 'ideal' composition—the composition that it would have if it were homogeneous. Isomorphous replacement is common, cations being replaced by others of similar *size*, although often not of the same charge. Thus, Na^+, Mg^{2+}, Ca^{2+}, Fe^{2+}, Mn^{2+}, Fe^{3+}, are readily interchangeable. Anions also suffer isomorphous replacement; O^{2-} can be replaced by F^- or OH^-.

Aluminium occupies a special role in silicate chemistry; the unravelling of its behaviour was one of the earlier successes of X-ray crystallography. Due to its size, Al^{3+} can replace Si^{4+} in silicates and does so in a random manner and to an indefinite extent. In order to retain electrical neutrality, every time Al^{3+} replaces Si^{4+} there must be a compensating replacement such as Fe^{3+} for Fe^{2+}, Ca^{2+} for Na^+, or Al^{3+} for Mg^{2+}, in which aluminium can take part because its size makes it suitable for six-coordination as well. In other words, aluminium is found both in tetrahedral coordination as part of the basic silicate framework and in octahedral coordination as a compensating cation.

These points are well illustrated by the analysis figures of a sample of the mineral hornblende, ideal composition $Ca_2Mg_2(Si_4O_{11})_2(OH)_2$. Approximately 25% of the silicon was replaced by aluminium; most of the Mg^{2+} was replaced by Fe^{2+}, with smaller amounts of Fe^{3+}, Mn^{2+} and Ti^{4+}; about 30% of the Ca^{2+} was replaced by Na^+ and K^+.

The structural unit in silicate chemistry is the SiO_4^{4-} tetrahedron, as was recognized and used by W. L. Bragg in his pioneering X-ray work on silicates. We can therefore classify silicates in terms of the ways in which tetrahedra link together.

Oxygen requires a total of two electrostatic bonds to it; see p. 85 for definition of the electrostatic bond. This may be satisfied by: 2 Si^{4+}; Si^{4+}, 2 Al^{3+} (in octahedral holes); Si^{4+}, 3 Mg^{2+} (in octahedral holes); etc. Many such combinations can be deduced from Table 5. When Al^{3+} replaces Si^{4+}

Table 5 Electrostatic bond strengths for some ions commonly found in silicates

Ion	C.n.	Electrostatic bond strength
Li^+	4	$\frac{1}{4}$
Na^+	6, 8	$\frac{1}{6}$, $\frac{1}{8}$
K^+	6–12	$\frac{1}{6}$–$\frac{1}{12}$
Cs^+	12	$\frac{1}{12}$
Be^{2+}	4	$\frac{1}{2}$
Mg^{2+}	6, 8	$\frac{1}{3}$, $\frac{1}{4}$
Ca^{2+}	8	$\frac{1}{4}$
Al^{3+}	4, 6	$\frac{3}{4}$, $\frac{1}{2}$
Si^{4+}	4	1
Cr^{3+}	6	$\frac{1}{2}$
Mn^{2+}	6, 8	$\frac{1}{3}$, $\frac{1}{4}$
Fe^{2+}	6, 8	$\frac{1}{3}$, $\frac{1}{4}$
Fe^{3+}	6	$\frac{1}{2}$
Zn^{2+}	4	$\frac{1}{2}$
Ti^{4+}	6	$\frac{2}{3}$

in a silicate, charge compensation occurs as described above. In aluminosilicates, which are based upon networks of (SiO_4) and (AlO_4) tetrahedra, the additional electrostatic bond to oxygen ($Al \overset{\frac{3}{4}}{—} O \overset{1}{—} Si$) is satisfied most commonly by Ca^{2+} or Mg^{2+} in eight-coordination.

The Structures of some Mineral Silicates

1. The simplest type of silicate (in structural terms) contains discrete $[SiO_4]^{4-}$ tetrahedra. These are typified by *olivine* and related minerals of ideal formula $M_2^{II}SiO_4$ ($M^{II} = Fe^{2+}, Mn^{2+}, Mg^{2+}$). The structure of olivine was determined by W. L. Bragg. It may also be regarded as an h.c.p. array of oxygen with Si^{4+} in one-eighth of the tetrahedral sites, Mg^{2+} (partly substituted by Fe^{2+}) in half of the octahedral sites. Due to its c.p. structure it is dense and difficult to fracture. It is widely used as a constituent of refractory bricks and cements. The gemstone *zircon*, $ZrSiO_4$, also contains discrete tetrahedra.

2. Joining two tetrahedra yields $[Si_2O_7]^{6-}$; minerals containing this pyro anion are rare.

3. If each (SiO_4) tetrahedron shares two vertices, rings and chains $(SiO_3^{2-})_n$ can be formed:

Rings	$n = 3$	e.g. benitoite, $BaTi(Si_3O_9)$
	$n = 6$	e.g. beryl, $Be_3Al_2(Si_6O_{18})$
Chains (pyroxenes)		e.g. diopside, $CaMg(Si_2O_6)$

There are shown in Figures 7-22 to 7-24. Note in *benitoite* the way in which coordination of oxygen (Ti^{4+}(6), Si^{4+}(4), Ba^{2+}(6)) satisfies the electrostatic bond strength rule: $(\frac{2}{3} + 1 + \frac{1}{3})$.

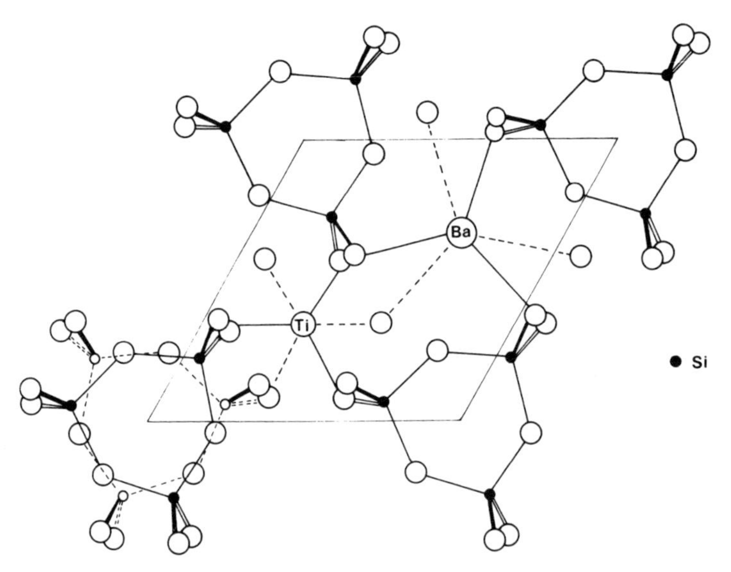

Figure 7-22 The structure of benitoite, $BaTi(Si_3O_9)$, in projection

In *beryl* the $[Si_6O_{18}]^{12-}$ rings are stacked one above the other and bound together by the cations in appropriate coordination arrangements as required by the radius ratios (see Chapter 3). Again, the electroneutrality rule is satisfied. There are small channels running through the structure which can accommodate small atoms (but not ions as the electrostatic requirements of the oxygen atoms are already satisfied). Helium is often found occluded in beryl and can be removed, without damaging the crystal, by gentle heating.

The *pyroxene* group of silicates has long chains of (SiO_4) tetrahedra, which are bound together by the cations, as shown in Figure 7-24c. Cleavage of the crystals parallel to the chain axes takes place at an angle characteristic of the mineral class. Thus, for all pyroxenes the angle is 89°; for all amphiboles (see below) the value is 57°.

Many other ways of forming chains of the same composition are illustrated in Figure 7-25.

4. Double-chain silicates (the *amphiboles*) are formed when alternate (SiO_4) tetrahedra, all with the unique vertex pointing in the same direction, share two or three vertices (Figure 7-26). The structure is equally well described as a condensation of two chains of the diopside variety. A typical member of this class is *tremolite*, $Ca_2Mg_5(OH)_2(Si_8O_{22})$; an essential difference from the pyroxene class is the presence of OH^-, which cannot be coordinated to silicon as this would completely saturate the electrostatic

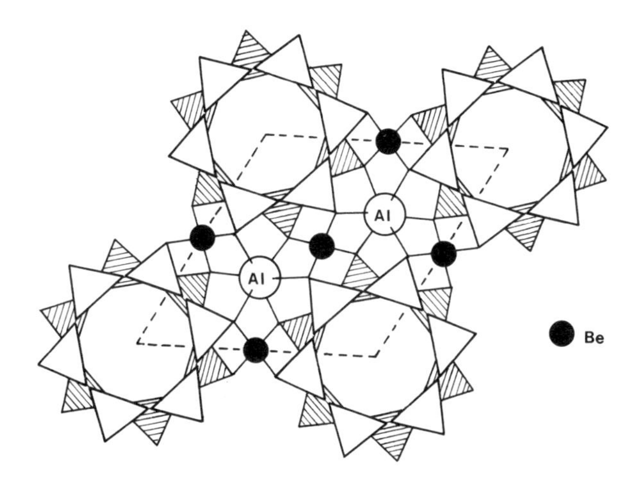

Figure 7-23 The structure of beryl. (SiO_4) groups are represented as triangles, the shaded ones being at a lower level. Al and Be lie in a plane halfway between the two sets of rings. (From, R. C. Evans, *Crystal Chemistry*, 2nd Ed., Cambridge University Press, 1964)

bond requirement of Si^{4+}, leaving it unattached to the rest of the framework.

As in all silicates, there is isomorphous substitution in these chain minerals: F^- can replace OH^-. Cation replacement has already been discussed for *hornblende* which falls in this class.

Amphiboles such as tremolite are part of the group of fibrous silicates sold under the designation 'asbestos'. Commercial asbestos however is largely *chrysotile*, a sheet silicate.

5. When each (SiO_4) tetrahedron shares all four vertices either (*a*) layer structures, or (*b*) three-dimensional frameworks are formed. We consider two important examples, talc and muscovite.

Talc, $Mg_3(OH)_2(Si_4O_{10})$ is built up from a 'sandwich' of two sheets of

(a)

(b)

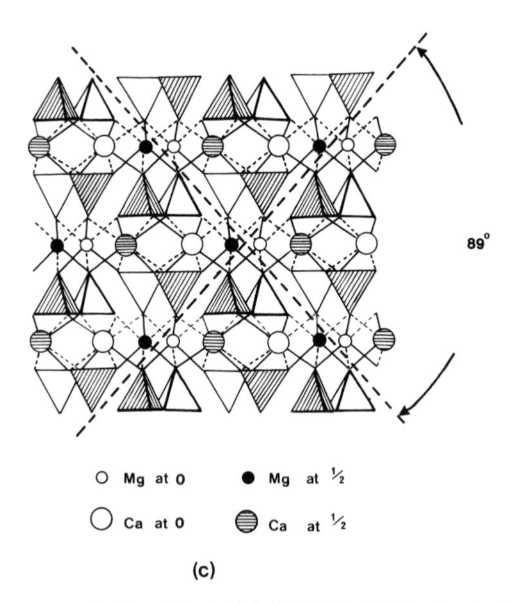

Figure 7-24 The structure of diopside, $CaMg(Si_2O_6)$. (a) The basic chains, showing equivalent ways of representing the structure: in both, the tetrahedron vertices are above the plane of their bases. (b) Various equivalent ways of representing the chains as seen in cross-section (i.e. the view in the direction indicated by the arrow in (a). (c) Projection of the idealized diopside structure, showing planes of cleavage. (Figure 7-24c from R. C. Evans, *Crystal Chemistry*, 2nd Ed., Cambridge University Press, 1964)

tetrahedra with the bases forming the outside of the sandwich; the two layers of sheets are strongly bound together by Mg^{2+}, in octahedral coordination with two oxygen atoms from each sheet (and two OH^- groups), Figure 7-27. The 'sandwich' layers are weakly bound by van der Waals forces only, which therefore allows cleavage between the planes. Talc is extremely crumbly, is the softest mineral known, and is therefore used as a lubricant, in medical and cosmetic preparations, as a filler in paper, textiles, etc.

In *muscovite*, $Al_2K(OH)_2(Si_3AlO_{10})$, (AlO_4) tetrahedra are introduced into the network, the charge deficit being compensated by K^+ or Na^+. This is one of the micas and although it still has the basic sheet 'sandwich', these are held together more tightly than in talc by K^+ ions (Figure 7-27). Although still easily cleaved, the micas are nevertheless harder than the talcs. There are several related micas in which different cations are used to compensate for the charge deficit. Thus, in *paragonite* sodium is used, in *margarite*, calcium. Subtle variations in hardness and ease of cleavage result from these changes.

Framework structures are formed when (SiO_4) tetrahedra are linked by all four vertices forming infinite three-dimensional arrays. The polymorphs of silica—tridymite, β-cristobalite, and quartz—are the simplest examples of this type, and represent three of the various ways in which tetrahedra may be disposed with respect to one another when four-connected. Clearly, in these polymorphs the electrostatic bond requirements of both atoms are satisfied without the need to introduce compensating cations. It is, of course, only a formalism to talk of an 'electrostatic' bond strength in such compounds, i.e. a suitable means of book-keeping. Structure is determined by a combination of valence and topological requirements, as related in Chapter 4.

When Al^{3+} replaces some of the silicon, the aluminosilicate minerals result; viz. feldspars, zeolites, and ultramarines. The framework is, of course, negatively charged and requires other cations for electrical neutrality. The majority of aluminosilicates are consequently less dense than the silica polymorphs and form a variety of rigid open frameworks of considerable flexibility of composition. They are of great industrial importance.

Feldspars are the major constituent of igneous rocks. Granite, for example, consists of quartz, feldspars and micas. Feldspars are basically constructed of rings of four tetrahedra with alternate pairs of vertices pointing in opposite directions. The rings are then joined in layers via the apices giving structures such as that of Figure 7-28. Typical members of this group are:

Orthoclase, $K(AlSi_3)O_8$, in which one in four silicon atoms has been replaced by aluminium, together with a compensating K^+ ion.

Celsian, $Ba(Al_2Si_2)O_8$, and *anorthite*, $Ca(Al_2Si_2)O_8$, in both of which the same sort of charge compensation occurs. Note that in feldspars the small first-row transition-metal ions do not occur (cf. chain and layer silicates), probably due to the difficulty of closing around them. The cations K^+, Ca^{2+}, Ba^{2+}, are in ten-coordination with oxygen (MO_{10});

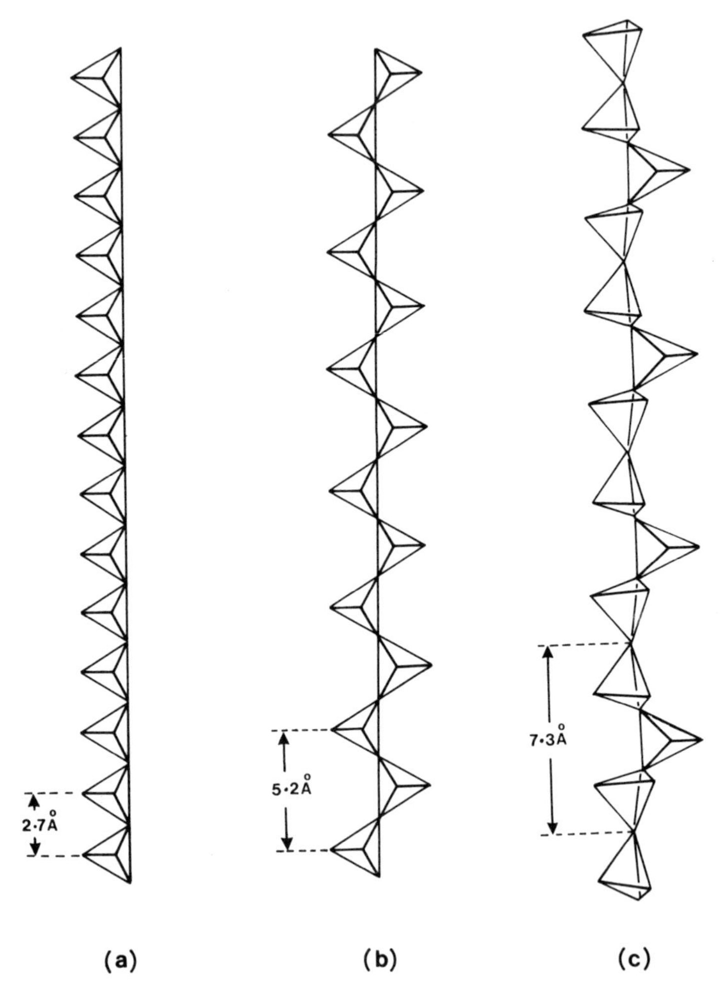

Figure 7-25 Various ways of coupling chains of AB_4 tetrahedra (Libau, (1959)). Examples of compounds adopting these structures are as follows. (a) $CuGeO_3$. (b) Diopside. (c) $CaSiO_3$; $NaBeF_3$. (d) $AgPO_3$. (e) Rhodonite, $MnCa(SiO_3)$. (f) Pyroxymangite, $(Mn, Fe, Ca, Mg)SiO_3$

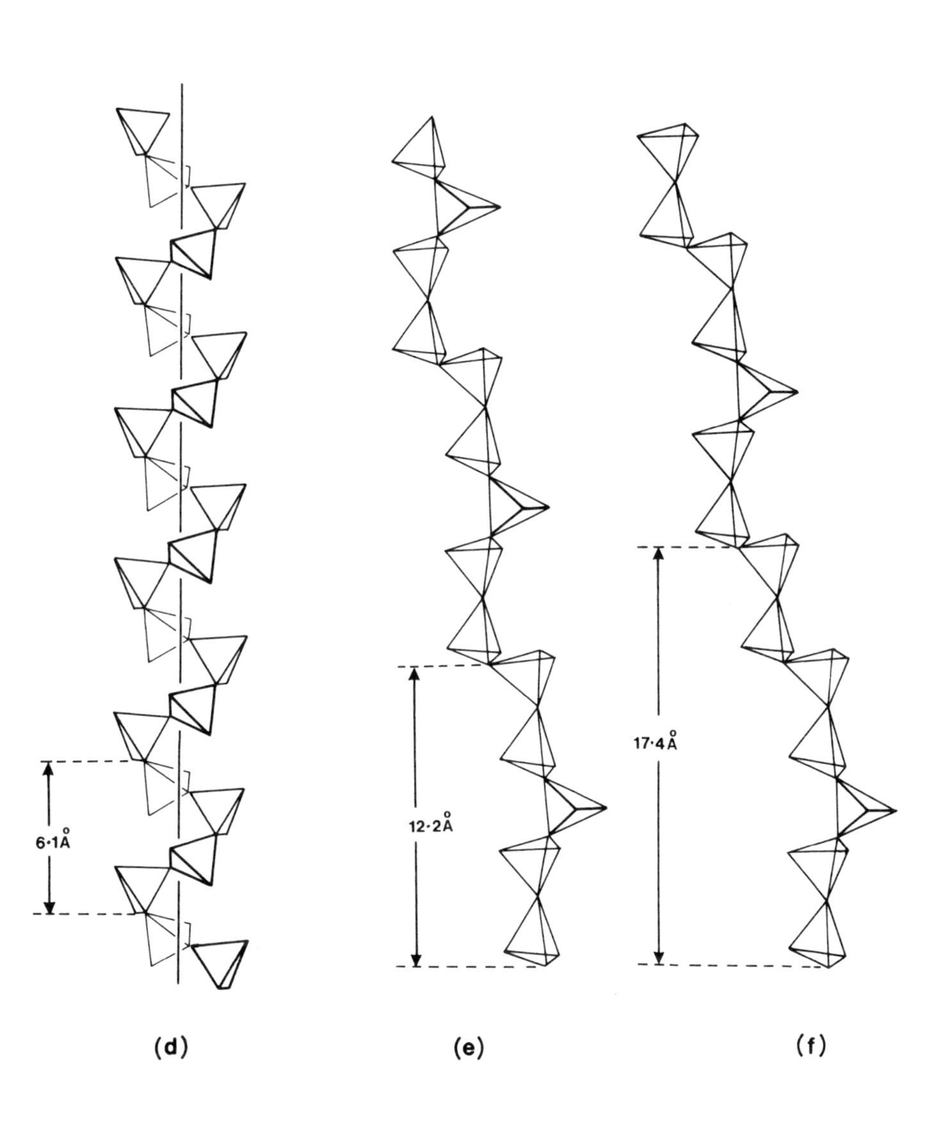
6·1Å
12·2Å
17·4Å
(d)
(e)
(f)

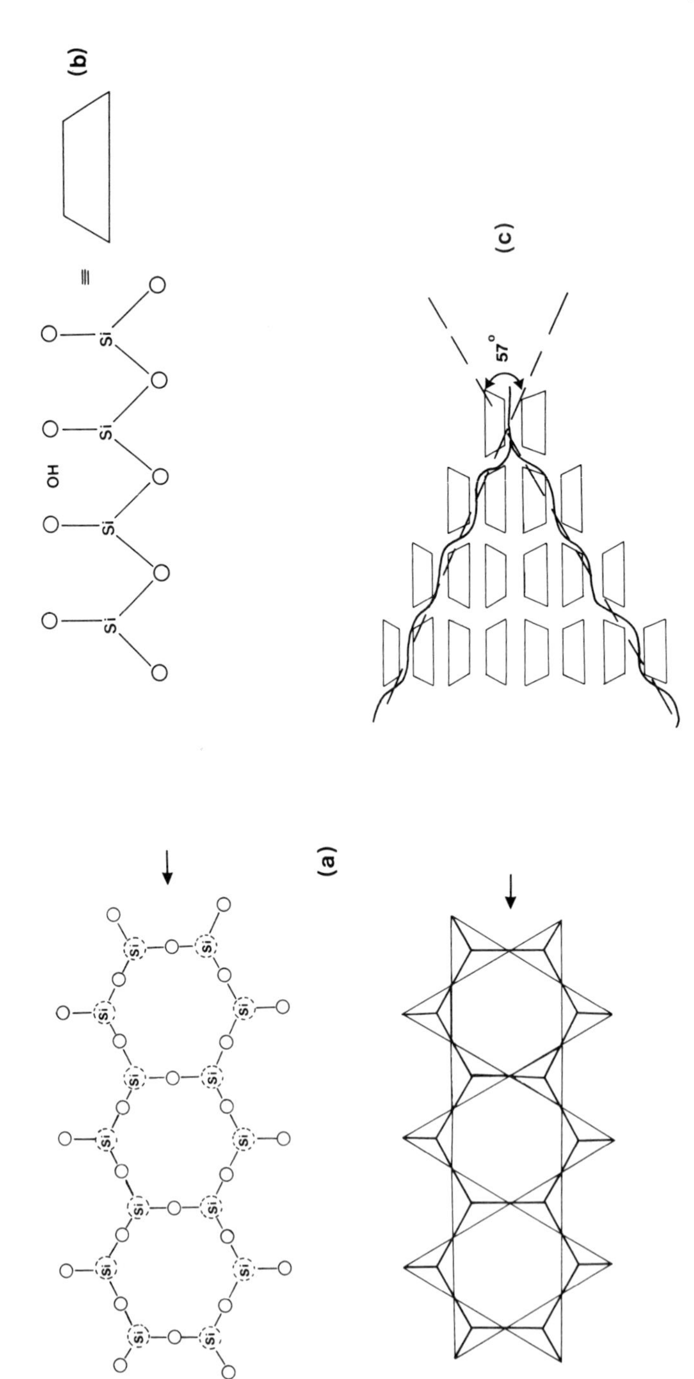

Figure 7-26 (a) The double chain structure of the amphiboles, shown in two equivalent ways. Cross-sections (along the direction of the arrows) are shown in (b) in two equivalent forms; (c) illustrates the mode of chain packing (anions are omitted) and the characteristic angle of cleavage

feldspars cleave parallel to layers of these cations, which are also quite readily leached out.

The *ultramarines* form another class of aluminosilicate mineral differing from those above in having anions such as Cl^-, SO_4^{2-}, CO_3^{2-} or S^{2-}, which are readily exchanged for other anions. These minerals find use as pigments; lapis lazuli is of this type. The origin of the intense colour (which can be varied from red to blue) is uncertain, although S^{2-} seems to be implicated. Sodium can also be exchanged (e.g. for Li^+, Ag^+, Tl^+, Ca^{2+}) giving a series of artificial ultramarines of colour varying from yellow to red to blue. Their structures are all based upon a framework which is most readily described as a truncated octahedron with silicon atoms at all of the vertices, Figure 7-29. This corresponds to one of the completely space-filling arrangements discovered by Fedorov. Typical minerals of the ultramarine class are of ideal formula

$$Na_8(Al_6Si_6O_{24}).X_2$$

where $X = \frac{1}{2}S^{2-}$, ultramarine
$X = Cl^-$, sodalite (this is colourless)
$X = \frac{1}{2}SO_4^{2-}$, noselite.

The *zeolites* are without doubt the most important and interesting of the framework silicates. They have much more open frameworks than the other silicates discussed above and these are responsible for their technologically valuable properties.

Zeolites are of three structural types:

(*a*) a class typified by *analcite*, $Na(AlSi_2O_6)H_2O$; *chabazite*, $CaNa_2(Al_2Si_4O_{12})6H_2O$; and *faujasite*, $NaCa_{1/2}(Al_2Si_5O_{14})10H_2O$; in which there is strong bonding in three dimensions;

(*b*) layer zeolites, of which little is known, and in which the strongest bonding is within the layers;

(*c*) fibrous zeolites.

Zeolites differ from nearly all other silicates in being hydrated; this water is not essential to the structure, and can be removed reversibly; they are therefore of use as desiccants.

The range of natural silicates has been considerably extended by the preparation of synthetic zeolites, which often have different types of structure from the natural ones; for example, the so-called Zeolite A, $Na_{12}(Al_{12}Si_{12}O_{48})27H_2O$, which has the framework shown in Figure 7-13b, in which silicon atoms are at all of the polyhedral vertices. Note the channels which run through the structure. Similar channels are found in all of the type (*a*) zeolites, although the actual arrangement of (SiO_4) and (AlO_4) tetrahedra is much more complex in analcite and chabazite. The arrangement in faujacite is shown in Figure 7-30.

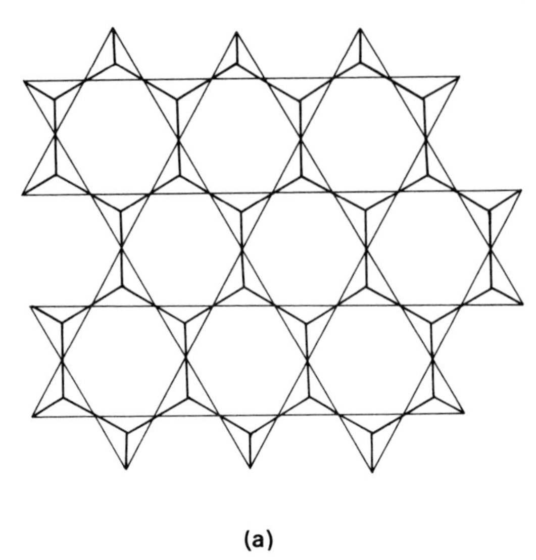

(a)

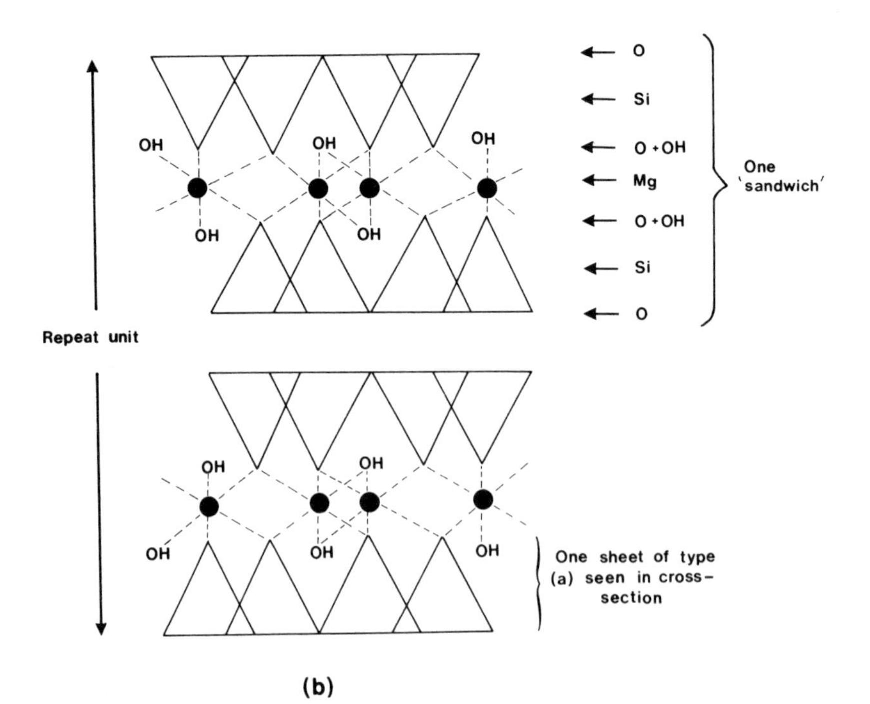
O
Si
O + OH
Mg
O + OH
Si
O
One 'sandwich'
OH
OH
OH
OH
OH
OH
Repeat unit
OH
OH
OH
OH
OH
OH
One sheet of type (a) seen in cross-section

(b)

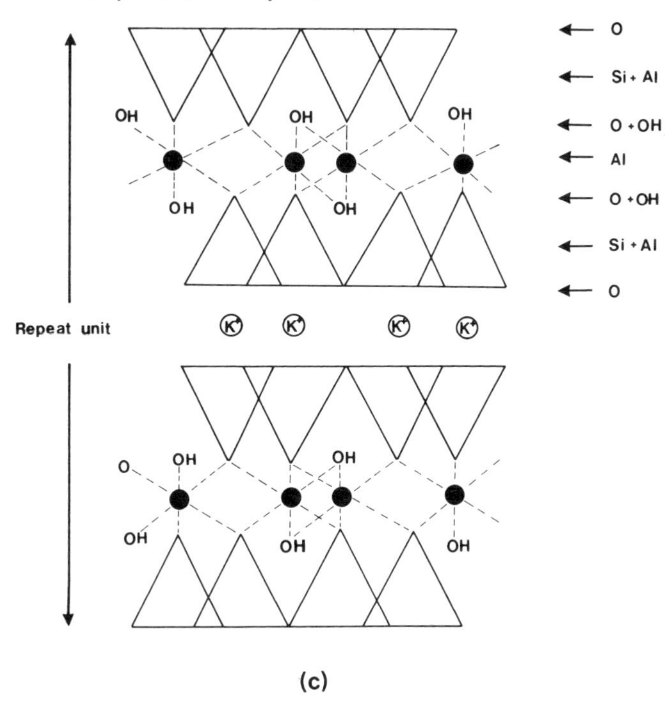

Figure 7-27 (a) The basic silicate sheet structure of the type found in talc, muscovite, and many related minerals. (b) and (c) show schematic representations of talc, $Mg_3(OH)_2(Si_4O_{10})$, and the mica muscovite, $Al_2K(OH)_2$ (Si_3AlO_{10}). Each has a two-layer repeat consisting of four sheets of type (a), but in the micas cations bind the 'sandwiches' together; in muscovite, K^+ is twelve-coordinated to basal oxygens of the sheets. Note the octahedral coordination (O + OH) of Mg in talc; the central Al atoms are similarly coordinated in muscovite

The channels in zeolites are of sizes similar to the diameters of many small molecules. Dehydrated zeolites will usually absorb small neutral molecules such as H_2, CO, NH_3, CH_4, etc. They can be used to separate chemically similar molecules differing only in size; one zeolite will readily pass ethane but not propane; another will pass small straight-chain hydrocarbons but not branched-chain ones. Due to the variability of the basic structural framework, it has proved possible to tailor-make zeolites for specific functions. As 'molecular sieves' they find many applications in industry, especially in the petrochemicals field.

A further important property of zeolites is the ease with which cation exchange takes place in aqueous solution. Due to the size of the channels, equilibrium between the incoming and outgoing ions is rapidly achieved. This process forms the basis of the *Permutit* water-softening process in which

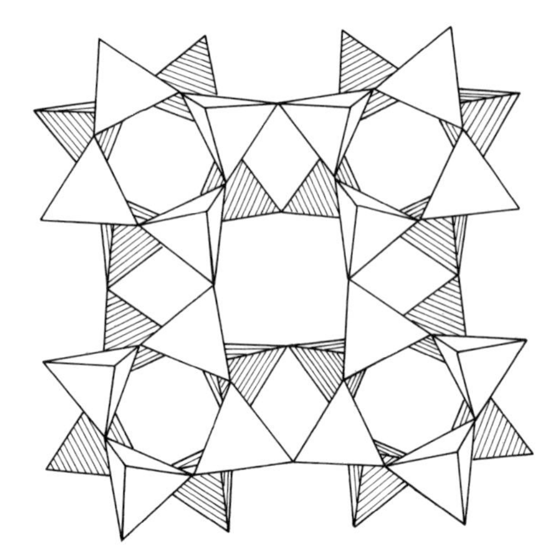

Figure 7-28 The structure of the feldspars. Each (SiO_4) tetrahedron shares all four vertices with neighbours thereby forming an extremely rigid lattice

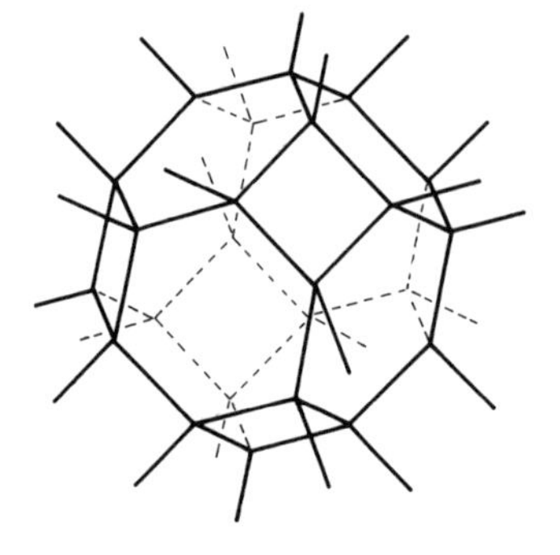

Figure 7-29 The framework of linked (SiO_4) tetrahedra which forms the basis of the ultramarines, and of the zeolite chabazite. The geometry is shown more clearly by using the silicon atoms only to depict the vertices of a truncated octahedron. (From A. F. Wells *Structural Inorganic Chemistry*, The Clarendon Press, Oxford, 1962)

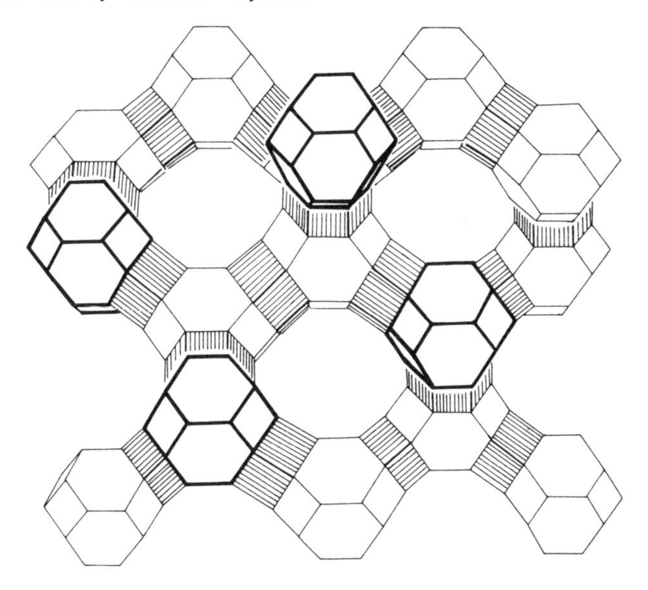

Figure 7-30 The structure of faujasite, showing the aluminosilicate framework: silicon or aluminium atoms are at the apices of the truncated octahedra. (Wells (1962))

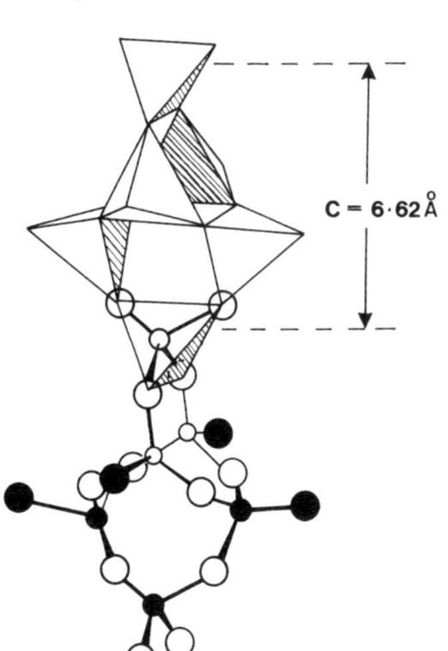

Figure 7-31 The basic chain structure of the fibrous zeolites. (Modified from A. F. Wells *Structural Inorganic Chemistry*, The Clarendon Press, Oxford, 1962)

Ca^{2+} is removed from 'hard' water by exchange for sodium. The spent silicate is simply regenerated by treatment with strong brine.†

Finally we mention the fibrous zeolites. They consist of chains with a characteristic repeat unit of five tetrahedra, Figure 7-31, crosslinked by the oxygen atoms indicated. Although these crosslinks are stronger than the interchain bonds in (say) diopside, they are relatively few in number and form the planes along which the needle-like crystals cleave, giving them their fibrous nature.

In our survey of silicates we have looked at only one or two examples from each class. There are many hundreds of others all differing in detail. Some are representative of other ways of linking or orienting tetrahedra, or other polyhedra. We have, however, illustrated the more important structural principles operative in this enormous class of minerals.

BIBLIOGRAPHY

Eitel, W. *Physical Chemistry of the Silicates*, University of Chicago Press, 1954.

Evans, R. C., *An Introduction to Crystal Chemistry*, Cambridge, 1964.

Kepert, D. L., *Inorganic Chemistry*, **8**, 1556 (1969).

Krebs, H., *Fundamentals of Inorganic Crystal Chemistry*, McGraw Hill, London, 1968.

Libau, F., *Acta Cryst.*, **12**, 180 (1959).

Mooser, E. and W. B. Pearson, *Progr. Semiconductors*, **5**, 105 (1960).

Pearson, W. B., *The Crystal Chemistry and Physics of Metals and Alloys*, Wiley, New York, 1972.

Wells, A. F., *Structural Inorganic Chemistry*, 3rd edn., The Clarendon Press, Oxford, 1962.

Wells, A. F., *Acta Cryst. B*, **24**, 50 (1968).

† The many industrial uses of zeolites have developed from 'pure' research into their physical chemistry. This is an excellent illustration of the truth that good technology derives from fundamental research.

CHAPTER 8

The Metallic Elements

All but twenty six of the chemical elements are metals. Although some of them have unique and idiosyncratic structures, the overwhelming majority have one of three simple forms: body-centred cubic, hexagonal close packed, cubic close packed. The energy differences between these structures are usually small; consequently, quite modest changes of temperature or pressure (or both) can induce transformation to one of the other structures (see Section 5.3.4). The particular structure adopted at n.t.p. depends primarily upon the electronic configuration of the element. Figure 8-1 shows the distribution of structure types across the Periodic Table. An important qualification must be attached to the description of a metal as 'h.c.p.'. For true h.c.p., the axial ratio $c/a = \sqrt{\frac{8}{3}} = 1{\cdot}633$. Examination of the ratios given in Figure 8-1 shows that only Mg and Co come close to this value; all other metals have $c/a < 1{\cdot}633$, with the notable exception of Zn and Cd for which $c/a > 1{\cdot}633$. We must now consider why this pattern of structures occurs.

8.1 STRUCTURES OF THE METALLIC ELEMENTS

The non-metallic elements, when binding either to themselves or to other atoms, make maximum use of their valence orbitals. Typically an octet is completed through formation of more or less localized electron-pair bonds. In contrast, metals have too few valence electrons for completion of an octet by electron pairing in any reasonable structure. Electrons are therefore shared among many bonds of partial order to neighbours. The situation can be described in more or less equivalent ways, using the 'resonating-bond' concept of valence-bond theory, or by band theory. In the so-called 'simple' metals (*s* and *p* block) the bonding electrons form a nearly free electron (n.f.e.) gas: the picture is then one of a periodic array of positively charged metal atoms embedded in a sea of itinerant electrons (which can be discussed in terms of plane waves). In contrast, the majority of valence electrons in transition metals are much more tightly bound and cannot then be described as 'nearly free'.

Considered from the viewpoint of volume occupation alone, the existence of metals with c.c.p. and h.c.p. structures requires no explanation; these lattices represent the two simplest and most economical ways of packing

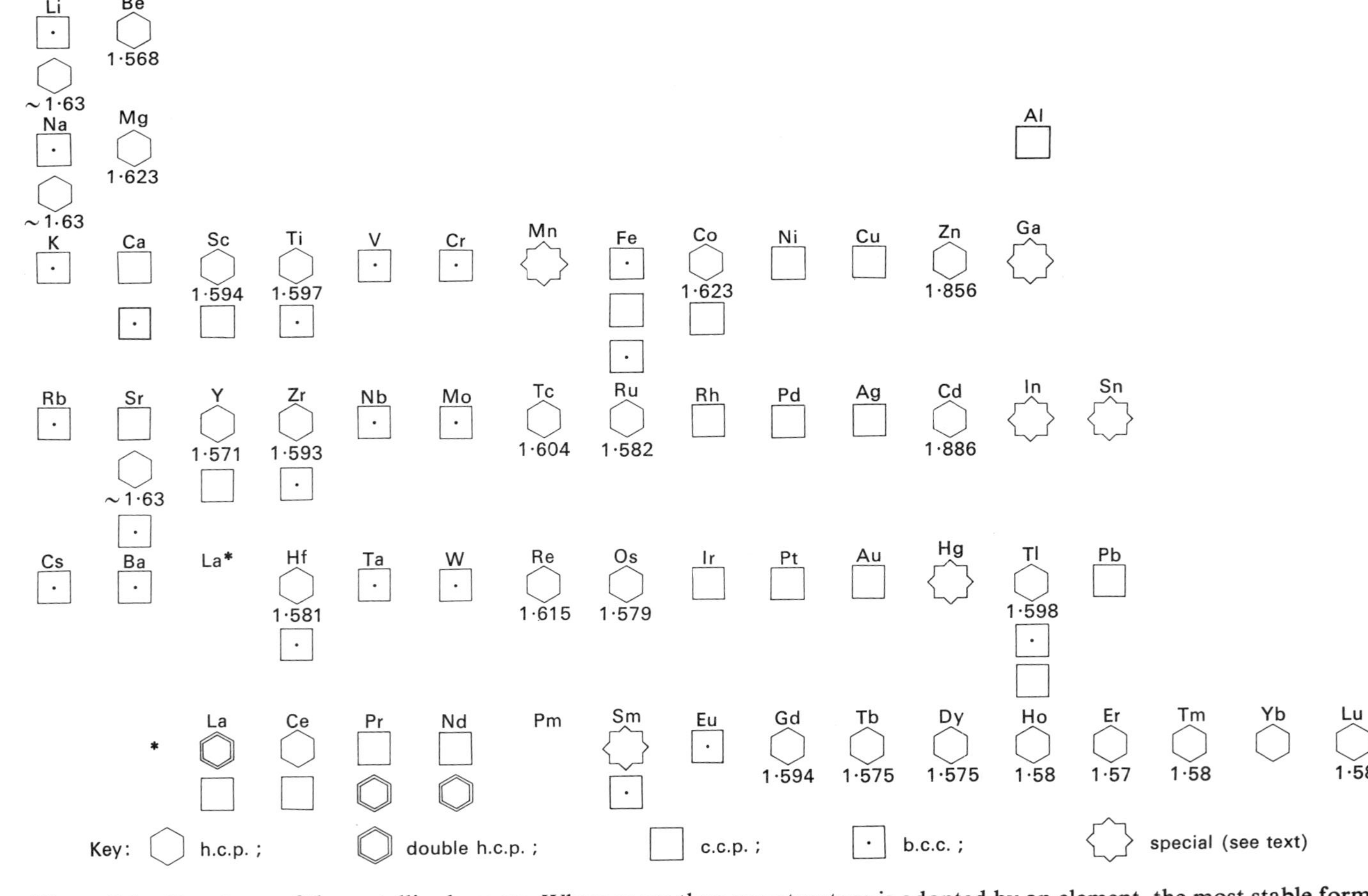

Figure 8-1 Structures of the metallic elements. Where more than one structure is adopted by an element, the most stable form is given first and less stable ones underneath. For hexagonal structures the *c/a* ratio is also given

identical spheres to give structures of high symmetry in which a large percentage of volume is occupied (74%), see Section 3.1. The b.c.c. structure is only marginally less efficient in the use of volume (68%) and confers the advantage of bringing eight next-nearest neighbours very close to a given atom, allowing the coordination number to approach fourteen (Section 3.4).

We should be cautious, however, about the meaning of facile statements such as those above. One of the distinctive features of metal and alloy structures is that the atoms appear to be more or less compressible; i.e. distances to neighbours within the first coordination shell may vary. For this reason it is preferable to think more in terms of their atomic volumes than of their atomic radii. It follows then, that in terms of the *actual* occupation of volume, the differences between b.c.c. and c.c.p. or h.c.p. structures of a given metal may be of little significance. *What we do need to explain* is why some elements show a preference for one, rather than another, of these three structures.

8.1.1 Bonding in Metals

In talking about bonding in metals, it is sometimes convenient to use the language of valence-bond theory, and at others to use band theory (i.e. MO theory as applied to periodic solids). The different strengths and emphases of these approaches are too well known to be rehearsed here, but we note that if both 'are treated as approximations to a complete treatment, their conceptual framework may be regarded as complementary.' We shall use either, as appropriate. It is often stated that the metallic bond is non-directional in character, implying that the cohesive forces in metals are altogether different in type (and more mysterious) from those in normal valence compounds. This is only a partial truth. Electron energies in metals are quantized just as in any other solid. One-electron orbitals are available to electrons in metal atoms, as in any other kind of atom, and are freely used when discussing bonding between metal and non-metal atoms both in lattices and in coordination complexes. In essence, bonding in metals is to be understood in terms of the same orbitals, and their hybridization and overlap schemes, as are used in other areas of application of valence theory, although for the n.f.e. 'simple metals' this approach is less meaningful than for transition metals. In metals, as opposed to normal valence compounds, differences arise because of the paucity of valence electrons. It follows that if bonding in metals is described, at least in part, in terms of overlap of partially populated appropriate hybrid orbitals, a measure of directionality is associated with such bonding. This is generally not so great as in (say) grey tin (diamond structure), but it is true that bonding, at least in transition metals, is probably more directional than has commonly been admitted. In support of this view, Altmann, Coulson and Hume-Rothery (1957) point to the comparative rarity of stacking faults in metals crystallizing with h.c.p. or c.c.p. structures; these two structures differ principally in their stacking sequences (AB or ABC respectively). The fact that these stacking sequences are maintained implies that the forces linking a given layer to those both above and below it are at least partially directional. This statement in no way conflicts with adoption by metals of c.c.p. and h.c.p. structures. These are the simplest, highest-symmetry ways of packing identical atoms which bond via non-directional forces (e.g. argon). But it is illogical to state that because a metal adopts these structures, the bonding must be non-directional: it *may* indeed be non-directional, but that point cannot be deduced from the observation that the metal crystallizes with such a structure. As we shall see below, what happens is that these high-symmetry structures are *also* compatible with hybridization schemes which allow a measure of directionality to be added to the bonding. In a formal sense, an analogy may be drawn with the rock salt structure; it is a very simple, high-symmetry, lattice which is compatible with ionic, metallic and covalent, bonding requirements.

The accurate description of the electronic structure of metals, and its correlation with structure type, is a matter of very great difficulty. Energy differences between h.c.p., c.c.p. and b.c.c., structures for any one metal are often small; prediction of which structure is the most stable for a given electronic configuration is therefore extremely difficult as the various energy terms involved in the summation are large compared with the physical differences which it is desired to calculate. There is no generally agreed theory of the structure of metals; each of the several approaches has its strengths and weaknesses, its successes and its failures. In part, the differences are semantic in that what the chemist says, talking in valence-bond language, may not really be so very different from what the physicist says using band theory or pseudopotentials, although at first sight problems of communication may obscure the common ground. Nevertheless, there are very basic differences of view which currently divide metallurgists. The situation was aptly described in 1967 by Lomer at an interdisciplinary conference devoted to this subject. The major problem to be solved is that of calculating the total energy of any solid system. 'The formal path is difficult, and the chemical/empirical thinkers try to improve things by injecting "commonsense" variables like size, electronegativity, and electronic configuration into the discussion, and most physicists keep trying out little shortcuts. All parties are setting out from a low-lying plain to scale a high mountain; the formal path lies up a tricky shoulder ridge, and is hard going from the start; the chemist's valley goes far into the mountains, giving many perspectives of the peak but never providing a really good climbing base; the physicists have not so much a valley as a set of little terraces where they play with one factor or another in isolation, and they greatly enjoy the game and get to know the boundaries of each terrace rather well. The formal path may still turn out to be the only one which runs from bottom to top: it is not often easy to join it partway up.'†

Despite this assessment of the chemical/empirical approach, it is the one which we now follow, partly because the language in which it is couched is familiar to chemists, but also because to go far into other theories requires introduction of more theoretical groundwork than is appropriate in a treatment at the level of this book. What we can do is to note the kind of correlation which has been observed between metallic structure and electron configuration, and discuss it in simple terms.

8.1.2 Electronic Configuration and Crystal Structure: The Simple Metals

Consider the first three elements of the third row: sodium, magnesium and aluminium. Sodium, with a single $3s$ valence electron, can only form one pair bond but this is spread out over the six nearest, the eight next-nearest, and still more distant, neighbours of the b.c.c. structure. In its electronic ground configuration, $3s^2$, magnesium cannot form any bonds and would therefore be a gas, or perhaps form a monatomic 'molecular' crystal as do the inert gases. The heat of sublimation of metallic magnesium is a couple of orders of magnitude above those of the inert gases, leaving no doubt that the bonding is altogether different. Magnesium can use the $3s3p$ configuration and hence make an average of two electron pair bonds spread over the twelve nearest (and more remote) neighbours of its h.c.p. structure. The promotion energy necessary to reach the $3s3p$ configuration is more than offset by the binding energy. A similar situation is found for aluminium; promotion from the $3s^2 3p$ ground state to $3s3p^2$ allows use of all three valence electrons in bonding; the c.c.p. structure is adopted. The number of bonding electrons per metal atom is clearly reflected in the heats of atomization of these elements:

Na	108 kJ/g atom
Mg	149 kJ/g atom
Al	326 kJ/g atom

The important point to note is: that *each valence electron configuration appears to be associated with a different and characteristic solid-state structure.* This was first pointed out by Engels and has since been much elaborated by Brewer (1968).

It is evident from Figure 8-1 that the simple story outlined above requires elaboration. All the alkali metals adopt the same (b.c.c.) structure, but in Group IIa there is a switch from h.c.p. at Mg to more complex behaviour for succeeding elements of the Group which is not completely understood. The transition metals exhibit an interesting progression of structures, whilst the

† From *Phase Stability in Metals and Alloys*, edited by Rudman, Stringer and Jaffee.

B-metals (i.e. the members of the B sub-Groups of the Periodic Table) are chiefly notable for their high proportion of idiosyncratic forms. We consider these groups of elements in turn.

8.2 STRUCTURES OF THE *d*-BLOCK TRANSITION METALS

8.2.1 Cohesive Energies of the Transition Metals

Use of *d*-orbitals profoundly increases the strength of bonding in metals. This is true both for the crystalline solids and for liquid metals: crystalline transition metals have binding energies in the range 300 to 800 kJ mol^{-1} but their heats of fusion are only c.a. 20 to 42 kJ mol^{-1} implying that the major cohesive forces are present also in the liquid state and do not depend for their explanation upon periodic properties of the lattice. Figure 8-2 shows how the binding energies vary for each *d* series. These rise to a pronounced maximum around the middle of each series and thereafter decrease; in the first row there

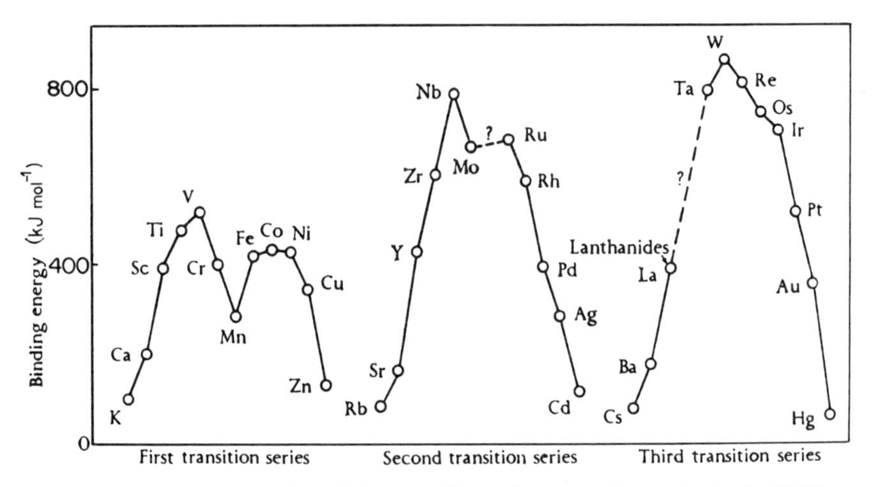

Figure 8-2 Binding energies of the transition elements. (From C. S. G. Phillips and R. J. P. Williams, *Inorganic Chemistry*, The Clarendon Press, Oxford 1965)

are complicating features associated with the spin configurations of Cr, Mn and Fe. A number of physical properties correlate loosely with the order of binding energies; the variation with atomic number of atomic diameter, compressibility and coefficient of thermal expansion are illustrated in Figure 8-3. It is clear, therefore, that in transition metals valence electrons cannot be described as n.f.e.; the appropriate model is one of 'tight binding'. Alternatively expressed, in transition metals Fermi surfaces are markedly different from the nearly spherical shapes shown by simple metals.

A simple outline explanation of these features follows from band theory.

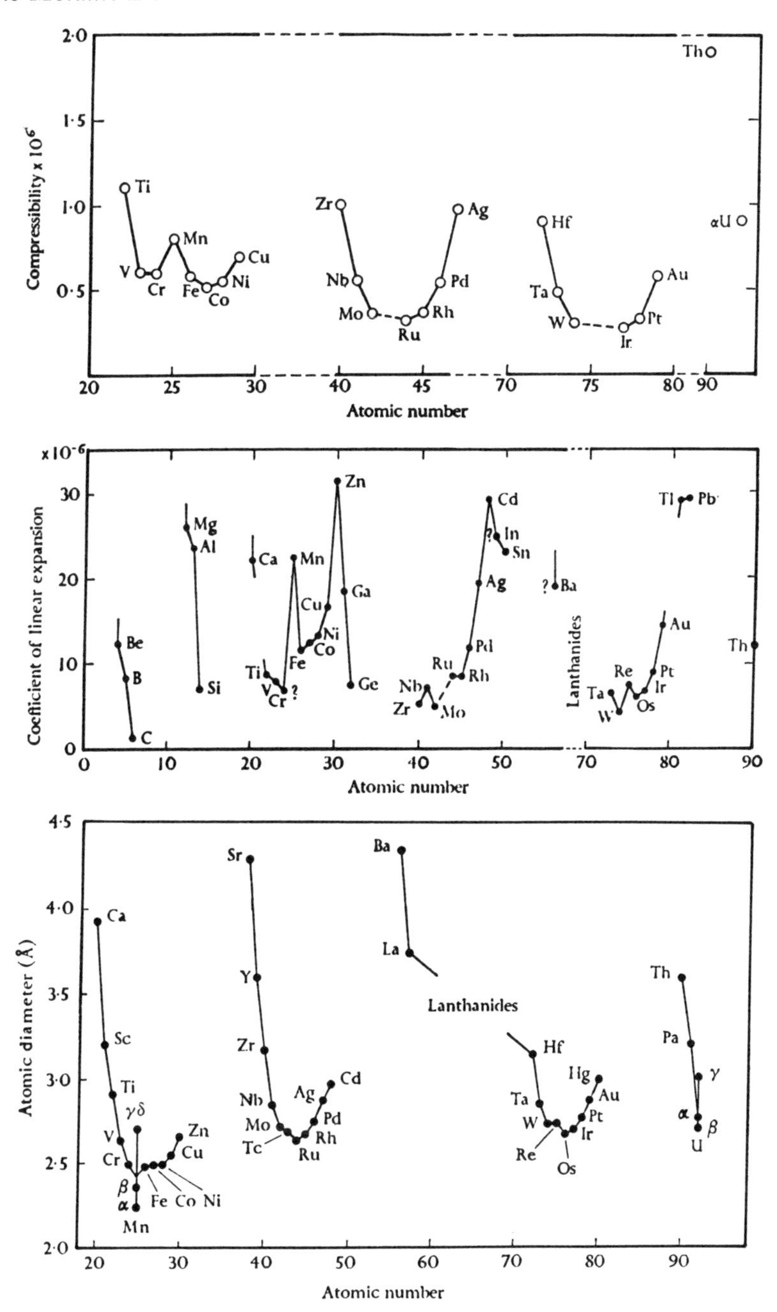

Figure 8-3 Variation with atomic number of atomic diameter, compressibility, and coefficient of thermal expansion for the transition metals. (From C. S. G. Phillips and R. J. P. Williams, *Inorganic Chemistry*, The Clarendon Press, Oxford, 1965)

In a metal crystal the *s*, *p* and *d* atomic orbitals are broadened into bands of finite width. A distinction is usually drawn between bands of predominantly *d* character and those of *s* or *s*/*p* type, although there is mixing between them which contributes further to stability of the crystal. From left to right across each *d* series there is a progressive contraction of the *d*-orbitals, due to increasing nuclear charge and the poor shielding effect of other *d*-electrons, which results in narrowing of the *d* bands. In contrast the *s* band is much broader and there is considerable overlap between neighbouring *s* functions due to their greater extension. However, the density of states associated with the *s* band is low because it can only accommodate two electrons per atom, whereas the *d* bands can accommodate a much greater density of states. A representation of this situation is given in Figure 8-4: at Cu the *d* band is

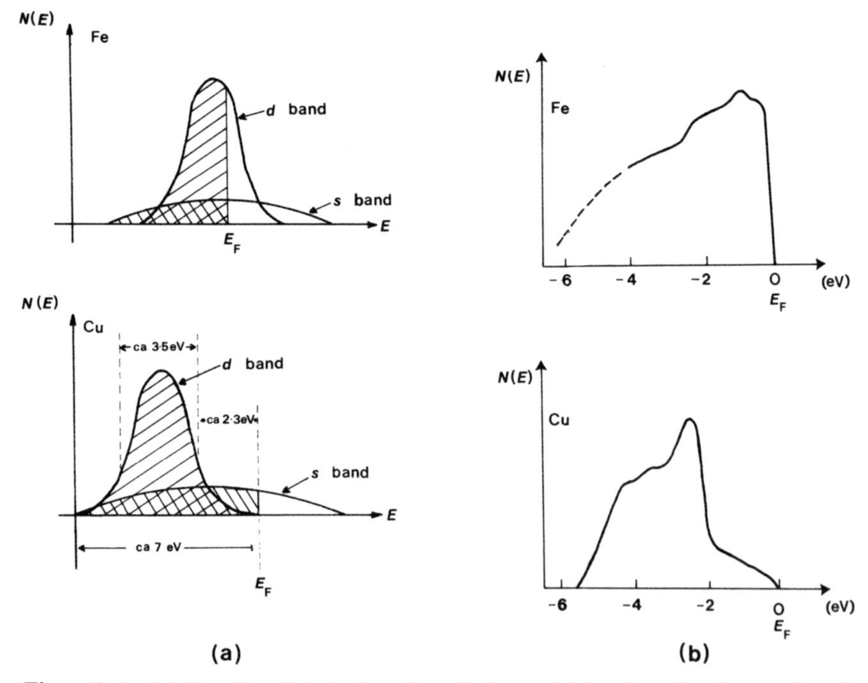

Figure 8-4 (a) Density of states functions (schematic) for *d*- and *s*-electron bands of iron and copper. E_F is the Fermi level. (b) Experimental density of states functions from photoemission studies. (Eastman (1972))

full and so tightly bound as almost to form part of the atomic core. Towards the middle of each series the *d* and *s* bands are of comparable energy.

The stabilizing effect of a half-filled or less than half-filled *d* band is understood in terms of the types of function of which the band is composed. Just as combination of two atomic orbitals (with respect to phase) yields a bonding and an antibonding level with the bonding level lower in energy than the original combining levels, so in the formation of a band the upper levels have antibond character and, for half-filled or less than half-filled situations, the overall energy of the system is lower than that of the original atomic orbitals. Thus, as a *d* band is progressively filled and electrons enter the lowest, most strongly bonding part of it, the cohesive energy increases (nearly parabolically) with atomic number. The maximum comes near the half-filled band position (Cr, Mo, W). The actual value will depend also upon the exact form of the density of states function, $N(E)$ vs E, which will be modified by s/p band mixing and by spin-correlation effects. Beyond the middle of each series the *d* band continues to contract in width as the *d*-electrons become more tightly bound to their nuclei; occupation of higher, antibonding, levels of the band (and, in some cases, of non-bonding levels) leads to decrease in cohesive energy.

8.2.2 Structures of the Transition Metals

Having noted the broad lines along which the trends in cohesive energy are explained by band theory, we now consider in more detail the structures concerned. There is a clear progression of structure types across each *d* series. The triads Sc–La and Ti–Hf are h.c.p., V–Ta and Cr–W all adopt the b.c.c. structure, and so do Ti–Hf at elevated temperatures. The slightly more open b.c.c. structure allows lattice vibrations of greater amplitude and energy, giving a higher entropy term; thus, when the c.p. and b.c.c. structures of an element have similar energies, the one with the greater entropy will be the more stable at high temperatures. There is one important exception to this general principle: in iron there is a magnetic contribution to the entropy at lower temperatures which helps to stabilize it in the b.c.c. form. On raising the temperature this is succeeded by a non-magnetic c.c.p. structure but above ca. 1400°C the b.c.c. again becomes the more stable due to vibrational entropy.

There now follow four elements (Tc, Re, Ru, Os) with h.c.p. lattices, after which the remaining members up to and including Cu, Ag and Au, all crystallize with the c.c.p. lattice. This is the *same* progression as was found for Na (b.c.c.), Mg (h.c.p.) and Al (c.c.p.): Brewer (1968) considers that this is significant and suggests that the crystal structure adopted by each transition metal is determined primarily by the number of *s*- and *p*-valence electrons. The *d*-electrons are considered to play an indirect role. Consider the example of tungsten. It has the ground-state configuration $5d^4 6s^2$ but promotion to $d^5 s^1$ makes six electrons available for bonding. The major contribution to the cohesive energy comes from the *d*-electrons which are in the inner shell and form more or less localized bonds to nearest neighbours, but the more extensive s-electrons, which are responsible for most of the electronic conductivity, are considered to determine the structure: the presence of only one s-electron per atom (as in Na) correlates with adoption of the b.c.c. lattice. In the case of metals with two or three s/p-electrons effectively, the h.c.p. or c.c.p. structures respectively are found. As in tungsten, the s- and p-electrons are highly delocalized (effectively becoming an electron gas); their distribution within each Brillouin zone is given experimentally by determinations of the Fermi surface. However, although Brewer's theory has much in its favour, it also has many opponents. We give below a theory due to Altmann, Coulson and Hume-Rothery (1957); although Brewer has claimed that this supports his own ideas, there *are* differences of emphasis.

8.2.3 Relation Between Bond Hybrids and Metal Structures

A certain amount of conceptual support, and more precise insight into bonding, comes from consideration of the symmetry-allowed hybrid orbitals which may be used in the three common metal structures. The symmetry classification of atomic orbitals is common ground to all quantitative theories of metals and therefore forms a good bridge between band and valence-bond theories. A model involving hybrid orbitals is most likely to succeed with metals having many bonding electrons, because this situation most closely resembles bonding in normal covalent compounds.

In both the b.c.c. and c.c.p. structures there is a centrosymmetric arrangement of nearest neighbours around any given atom, see Figures 3-13 and 3-3a. In the h.c.p. lattice the six neighbours in the c.p. plane are also arranged centrosymmetrically about a given atom, but the two sets of three atoms which are respectively above and below that plane are at the corners of a trigonal prism (Figure 3-3b) and do not have inversion symmetry. Similarly, the s- and d-atomic wave functions both have inversion symmetry and are termed *gerade* (g), even, whereas p-orbitals are *ungerade* (u), uneven. In relation to bonding in metals there are three important things to note about hybrids of g type.

(a) Formation of hybrids from mixtures of orbitals of both g and u symmetry results in an antisymmetrical distribution of charge with strongly directional character, well suited to formation of strong localized bonds. In contrast, combinations of g-type orbitals always retain g character with an even charge distribution on either side of the atom, Figure 8-5; bonding via such orbitals will be weaker than via the u type, and corresponds to a greater degree of delocalization. Owing to this delocalization property, g hybrids are especially suited to bonding in electron-deficient systems, such as metals, forming directional bonds of partial order.

(b) Hybrids of g and u symmetry have different energies. For the case of a linear chain, calculation shows that their energies vary with the lattice $\boldsymbol{k}$

vector ($=2\pi/\lambda$) as in Figure 8-6. The point to note is that at the Brillouin-zone centre ($\boldsymbol{k} \approx 0$) *g* hybrids have their minimum energy and *u* hybrids their maximum. Now, to a first approximation, the energy of a crystal can be written as the sum $E_A + E_0 + E_F$, where E_A = atomic energy of the free atoms, E_0 is the potential energy, and E_F the mean Fermi energy. E_F will be

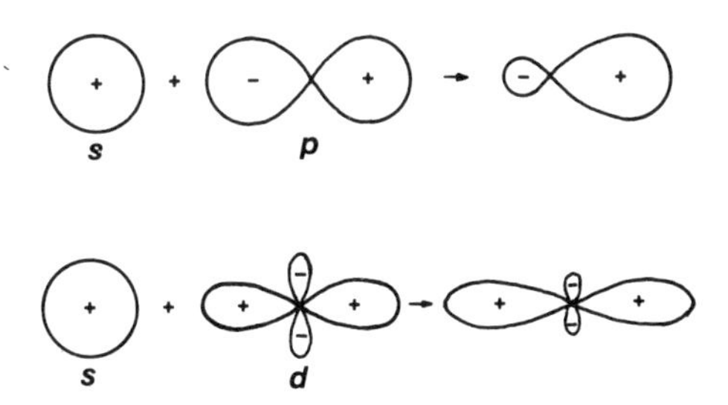

Figure 8-5 Hybrids formed from *s*-, *p*- and *d*-atomic orbitals

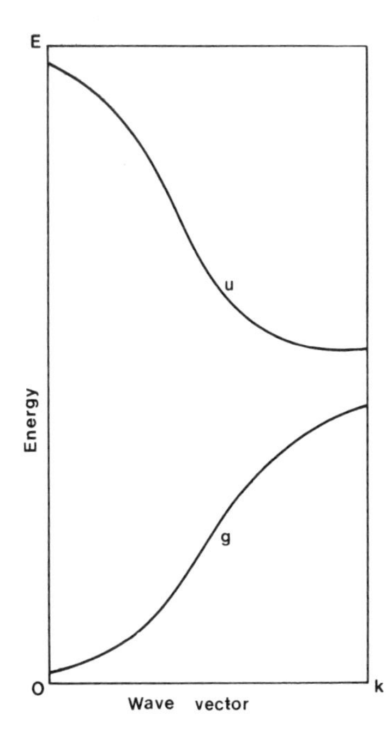

Figure 8-6 Variation of the relative energies of *g* and *u* hybrids with wave vector. (From S. L. Altman, C. A. Coulson, and W. Hume-Rothery, *Proc. Roy. Soc. A* **240**, 145 (1957))

low if the density of states is *high* for low values of $\boldsymbol{k}$. In other words, energetic arguments strongly favour the use of *gerade* hybrids in a crystal. This is a conclusion of considerable importance. *Any* mixing in of *u* functions destroys the *g* nature of a function.

It is of particular interest to note that this argument makes one of the rare quantitative points in support of Laves's postulated (Section 1.2) 'symmetry principle'.

(*c*) In a lattice of N atoms held by normal covalent bonds there must be $2N$ binding electrons and these would fill the first Brillouin zone, giving an insulator. But if *gerade* hybrids only are used, corresponding to half bonds, only N binding electrons are necessary and the material will be a metallic conductor.

Hybrids compatible with the three main structures of metals are as shown in Table 1.

Table 1. Hybrid orbitals compatible with metal crystal structures (Altmann, Coulson, and Hume-Rothery, 1957)

Crystal structure	Hybrids	
Body-centred cubic	sd^3, d^4	Nearest neighbours (pure g)
	d^3	Next nearest neighbours (pure g)
Cubic close packed	p^3d^3	(mixed g and u)
	sd^5	(pure g)[a]
Hexagonal close packed	sd^2	Neighbours in c.p. plane (pure g)
	pd^5, spd^4	Neighbours in planes above and below (mixed g and u)

[a] Altmann, private communication, 1973.

Body-centred Cubic Structure

As discussed earlier (p. 101), eight hybrids of cubic symmetry cannot be formed from *s*-, *p*- and *d*-orbitals alone. However, the four hybrids of tetrahedral symmetry, sd^3 (i.e. d_{xy}, d_{yz}, d_{zx}), are *gerade* and hence have eight equivalent directions for binding. A similar situation obtains for the set of d^4 hybrids of tetragonal symmetry. The situation is shown diagramatically in Figure 8-7.

In the b.c.c. lattice the six next-nearest neighbours are only 15% more distant than the nearest neighbours and bonding to them must therefore be considered. The only *gerade* hybrids possible are d^3, which form a trigonal pyramid of perpendicular bonds, but which can provide octahedral bonding if account is taken of their *gerade* nature. Note that insistence upon use of g hybrids rules out the usual d^2sp^3 type for bonding to next-nearest neighbours. If these were used, it would be logical to assume that sp^3 hybrids should be involved along with those of sd^3 type, thereby destroying the *gerade* nature of the latter. Thus, the bonding in the b.c.c. lattice is due to resonance between sd^3, d^4 and d^3 hybrids, and may be represented as $(sd^3)^a(d^4)^b(d^3)^c$. The relative weight of, say, the sd^3 hybrid is $a/(a + b + c)$, but we cannot write $(a + b + c) = 1$ since the total hybrid is also in resonance with other structures such as those of ionic nature.

Cubic Close-packed Structure

The hybrids p^3d^3 have lobes which point to the corners of a trigonal antiprism (Figure 8-8). Resonance of four such hybrids about the threefold axes of the crystal covers all of the twelve nearest neighbours; sd^5 hybrids are also suitable.

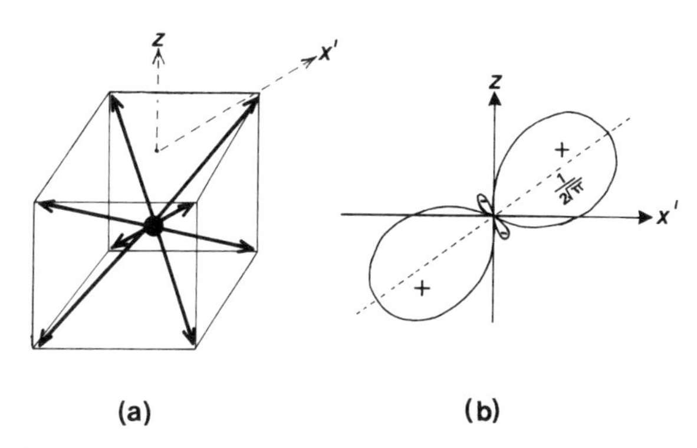

Figure 8-7 The lobes of sd^3 and of d^4 hybrids point towards the vertices of a cube and are represented in (a) by arrows. (Altmann, Coulson, and Hume-Rothery (1957)). An sd^3 hybrid is shown to scale in (b), the *d*-orbitals being from the t_{2g} set

Hexagonal Close-packed Structure

Bonding to the six in-plane neighbours is via sd^2 hybrids: these are *gerade*, and there are three of them at angles of 120° (just like sp^2 hybrids), see Figure 8-8. The arrangement of the two sets of three atoms above and below the central c.p. plane is not centrosymmetric as the atoms are at the vertices of a trigonal prism. The hybrids spd^4 and pd^5 have the appropriate symmetry. The bonding is therefore described as $(spd^4)^a(pd^5)^b(sd^2)^c$.

Correlation of Electronic Configuration with Crystal Structure

The hybridization schemes listed above and in Table 1 are those which symmetry theory shows to be admissible. But it cannot tell us anything about the *extent* to which each possible hybrid is populated because that depends upon the relative energies of the *s*-, *p*- and *d*-electron bands. The closer they are in energy, the more readily they will hybridize. In particular, the 'weight' of *d*-orbitals in the hybrids might reasonably be expected to vary with the energy of the *d* band from metal to metal. In each *d* series there is a gradual transition of *d*-orbital levels from high (i.e. outer orbitals) at the left-hand side, through bonding to core at the Group II end, where they are now too deep to be used for bonding. We may reasonably conclude that the contribution of *d*-orbitals to the hybrid schemes of Table 1 will increase up to the middle of each series (i.e. to Cr, Mo, W) but will fall off on either side. In short, that it mirrors the order of cohesive energies (Figure 8-2).

Since the hybrids are only partly occupied, what matters is the proportion of *d*-orbital contribution to each. This is simply defined by:

$$\text{c.c.p.} \quad (p^3d^3)^a(sd^5)^b \qquad \therefore\ P_c = \frac{3a + 5b}{6a + 6b}$$

$$\text{h.c.p.} \quad (spd^4)^a(pd^5)^b(sd^2)^c \qquad \therefore\ P_h = \frac{4a + 5b + 2c}{6a + 6b + 3c}$$

$$\text{b.c.c.} \quad (sd^3)^a(d^4)^b(d^3)^c \qquad \therefore\ P_b = \frac{3a + 4b + 3c}{4a + 4b + 3c}$$

Of course, values of a, b and c, are unknown and not easily estimated. Taking $a = b = c$, $P_c = 0{\cdot}66$ for c.c.p., $P_h = 0{\cdot}73$ for h.c.p. and $P_b = 0{\cdot}91$ for the b.c.c. structure. Although the values of P_c, P_h and P_b are dependent upon the indices a, b and c, their formulae show, without ambiguity, that the *order* of increasing *d*-electron participation is b.c.c. > h.c.p. > f.c.c. The correlation with structure may therefore be summarized:

Weight of *d*-electrons per orbital	Crystal structure
0.6	c.c.p.
0.7	h.c.p.
0.9	b.c.c.

Even taking values of the indices varying by factors as large as five only results in weights $\pm$ 0·1 from the above figures.

Consider now the sequence of structures shown by the second and third transition series, and the elements preceding them. The most stable structure for Ca and Sr is c.c.p., which has the lowest *d* weight of the three. This is reasonable as the *d*-orbitals are relatively high in energy for these elements. However, their presence is enough to cause a switch from the h.c.p. structure of Be and Mg, which must be determined by the *s*- and *p*-electrons alone. Proceeding across the series, as the weight of *d*-electrons increases, the sequence h.c.p. and then b.c.c. is observed, followed by reappearance of h.c.p. structures (Tc, Re, Ru, Os) and then, finally, c.c.p. exactly in accord with the theory outlined above. The reappearance of h.c.p. and c.c.p. structures along with the progressive stabilization of the *d* band is especially pleasing. This story is mirrored by the behaviour of the first transition series, although Mn,

Fe and Co, require comment. It is probable that these metals exhibit high-spin behaviour in contrast to the second and third rows for which low-spin behaviour is always found. Thus, iron is ferromagnetic (lowest temperature b.c.c. phase only) with a saturation moment corresponding to two unpaired electrons.

Comparison of the Altmann–Coulson–Hume-Rothery theory with that of Brewer shows a measure of compatibility. The order of increasing *s*-, *p*-electron contribution is the same in both, viz. c.c.p. > h.c.p. > b.c.c. But Brewer insists that it is the *s*- and *p*-electrons alone which determine the

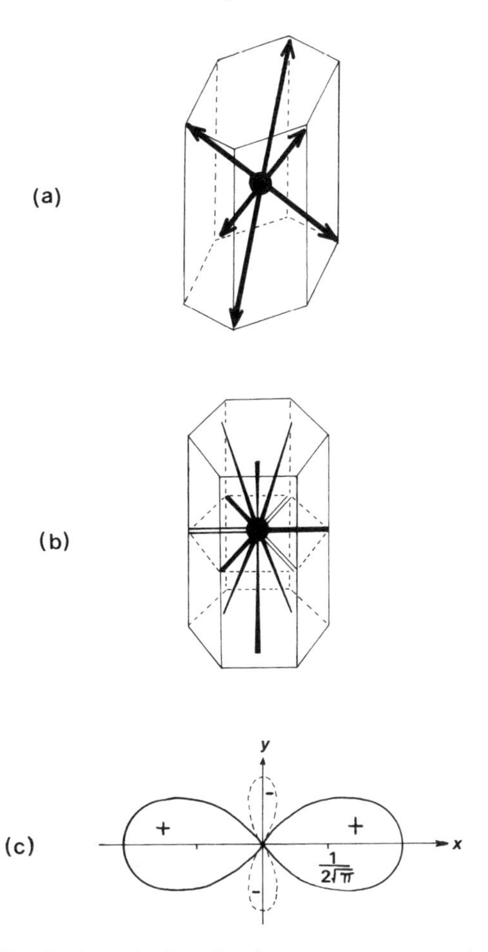

Figure 8-8 (a) The basic unit for the f.c.c. structure, a trigonal prism with p^3d^3 or sd^5 hybrids. (b) The h.c.p. structure showing the disposition of mixed pd^5, spd^4 hybrids (above and below central plane), and sd^2 hybrids (from $s, d_{x^2-y^2}$ and d_{xy}) in the plane. (Altmann, Coulson, and Hume-Rothery (1957)). (c) Detail of an sd^2 hybrid, drawn to same scale as Figure 8-7b

structure adopted, whilst recognizing that much of the cohesive energy originates in *d*-orbital overlap. It must be admitted that it seems a little unrealistic to allow the *d*-orbitals to provide most of the cohesive energy without reference to the particular orbitals used by various members of each transition series. The Altmann–Coulson–Hume-Rothery theory also accounts rather well for the reappearance of h.c.p. and c.c.p. structures late in each series, and for the c/a ratios in h.c.p. metals. Although this is deeply satisfying, we must conclude this paragraph by recalling that the stability of a structure is related to its free energy, $G = H - TS$, and the entropy term has not been explicitly included.

c/a Ratios in Hexagonal Close-packed Transition Metals

A particular strength of the Altmann–Coulson–Hume-Rothery theory is that it provides an explanation for the puzzling changes in c/a ratio exhibited by metals. The bonding in the h.c.p. structure is described in terms of sd^2 hybrids within each c.p. layer, but the interlayer bonds are of spd^4 or pd^5 type. Since the interlayer bonds are localized, and those within each layer are more delocalized and hence weaker and longer, it follows that $c/a < 1{\cdot}633$ if the occupation of bonds within the c.p. plane is the same as those between planes. Although such occupation numbers are not known, we *can* say that the c/a ratio is determined by the relative populations of these two groups of hybrid orbitals. Two factors affect this competition. (*a*) The total binding energy will tend to favour occupation of interlayer hybrids as these form the stronger bonds. (*b*) The Fermi energy favours the weaker intralayer bonds (i.e. the purely *g* hybrids, since these are of lower energy).

8.3 STRUCTURES OF THE B-METALS

Zinc and cadmium both crystallize in hexagonal lattices which are distinguished by having $c/a > 1{\cdot}633$, nearer to 1·9 in fact. Distances to nearest neighbours (within the c.p. layers) and next-nearest neighbours (in adjacent layers) are:

Zn	2·659	2·906 Å
Cd	2·973	3·286 Å

No satisfactory account of the bonding is available. Mercury, which of course is liquid at room temperature, has a unique rhombohedral structure, Figure 8-9, with six nearest neighbours at 3·00 Å and six others at 3·47 Å. It may be regarded as formed by a trigonal distortion of the f.c.c. lattice. Below 79 K it transforms to β-Hg, which has a tetragonal structure.

Gallium also has a curious structure, chiefly notable for the occurrence of one very short nearest neighbour distance, 2·44 Å, and an exceptionally low melting point, 30°C. Other neighbours are at 2·71(2), 2·74(2) and 2·80(2) Å. The tendency to have one nearest neighbour is retained in the liquid state, and in a variety of compounds of the element such as γ-Ga_2S_3 and $Ga_2Cl_6^{2-}$. Gallium is isostructural with iodine; both structures have the same space group and unit cell occupancy, Figure 6-10. It may, not unreasonably, be described as a molecular crystal of metallic character. This fact appears not to have been remarked upon before and could have some importance theoretically in helping to form a bridge between the descriptions advanced by theoretical physicists in terms of pseudopotentials and band theory on the one hand, and the valence-bond language more familiar to chemists on the other. Particular interest also attaches to this problem because the basic structure of iodine is well described on the basis of Kitaigorodskii's packing theory, once it is agreed that the crystal is composed of molecules, not atoms. In contrast, if it were described by pseudopotential theory, it would be necessary to begin with the assumption that it is atomic and then show that it could not be.

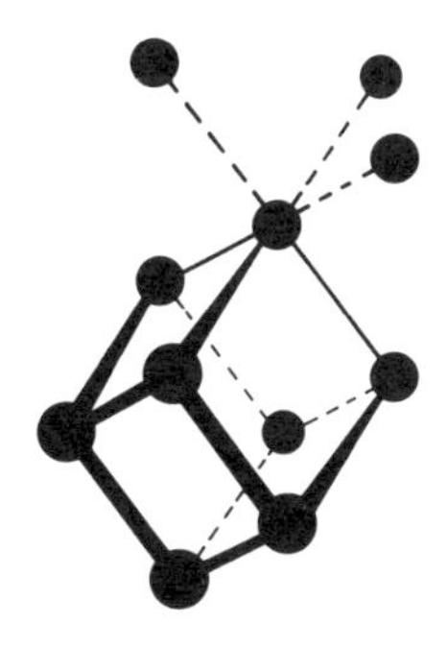

Figure 8-9 The rhombohedral structure of α-mercury. (From W. Hume-Rothery, *Atomic Theory for Students of Metallurgy*, 3rd edn., Institute of Metals, London, 1960)

Indium, and the isostructural γ-phase of manganese, have a tetragonal structure which is a very slightly distorted form of the f.c.c. lattice. The axial ratio is only 1·08. Tin is diamorphic. Grey tin crystallizes in the diamond lattice; white tin has the form shown in Figure 5-1, and is related to the diamond lattice by compressing the latter along one axis.

8.4 PSEUDOPOTENTIAL THEORY AND THE STRUCTURES OF THE SIMPLE METALS

8.4.1 Pseudopotential Theory

The mid-1960s witnessed the appearance of a new theoretical approach to the study of metals. It has found a measure of success in dealing with the so-called simple metals (i.e. non *d* or *f* block), and with some of their alloys in which the valence electrons are effectively itinerant so that their conduction bands can be described as containing 'nearly free electrons' (n.f.e.); i.e. for metals having an approximately spherical Fermi surface. The essential difference from n.f.e. theory, which treats electrons in terms of plane waves is, that by a simple ruse, the effect of atomic cores is included.

The central idea is that the true wave function of a valence electron in an atom, Ψ, which has a variety of radial nodes, can be replaced by a pseudowave-function, Φ, which is identical with Ψ outside the atomic core but inside the core is a smooth function with all the radial nodes removed; Figure 8-10. The two are related by an exact transformation of the Schrödinger equation from

$$[-(\hbar^2/2m)\nabla^2 + V(\boldsymbol{r})]\Psi = E\Psi$$

to

$$[-(\hbar^2/2m)\nabla^2 + V_{ps}]\Phi = E\Phi$$

where V_{ps} is a pseudopotential. The eigenenergy, E, remains the same, and both equations are written in one-electron form. The pseudopotential of each atom is thus a screened potential with long-range variations only. It is mainly

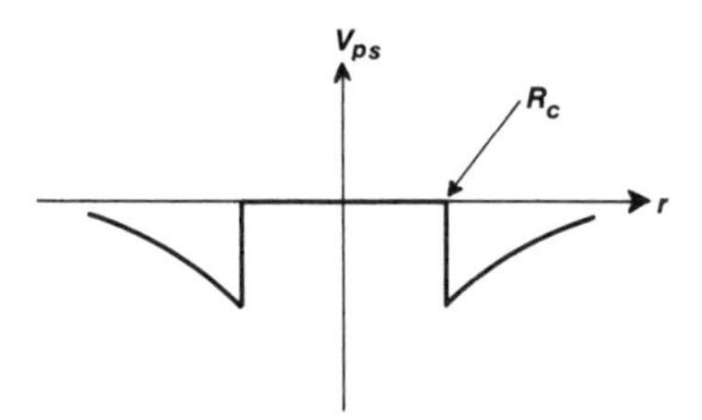

Figure 8-10 Schematic representation of a pseudopotential, cancelled off to approximately zero inside the ion-core radius R_c, and equal to $-z/r$ outside. (From B. J. Austin and V. Heine, *J. Chem. Phys.*, **45**, 928 (1966))

determined by (*a*) the ratio of atomic radius to ion-core radius, and (*b*) the energy difference between the lowest *s*- and *p*-valence states of the atom. There exists, therefore, a conceptual link between discussions of structure in terms of pseudopotentials, and those involving dehybridization. Because the radial oscillations of Ψ have been cancelled against the strong attractive potential $V(\boldsymbol{r})$ inside the core, V_{ps} is a relatively weak potential function. Physically, this statement corresponds to recognition that the scattering power of atoms for electrons is weak (as revealed, for example, by the low electrical resistivities of metals and their thermal dependence). The cancellation within the core, represented diagrammatically in Figure 8-10, is perhaps visualized more readily in terms of the relation (for $l = 0$ states),

$$V_{ps} \approx V - \sum_c \left(\int \Psi_c^* V \,\mathrm{d}v \right) \Psi_c$$

where Ψ_c are core orbitals.

The key assumption is then made that V_{ps} is sufficiently weak that it may be treated in terms of second-order perturbation theory: it is just this assumption that is at the root of the most serious conceptual objections to pseudo-potential theory. It has been claimed that the two basic assumptions of the theory are incompatible, that the conditions for applicability of second-order perturbation theory are grossly violated, and that any agreement with experiment is fortuitous. Even the proponents of pseudopotential theory agree that it still has major deficiencies, one of the most serious being adequate definition of V_{ps} at larger distances, but it has a large and enthusiastic following and it does give another way of discussing the relation between band structure and lattice stability. We therefore report, in outline, some of the conclusions reached as a result of pseudopotential calculations, leaning especially on the work of Heine and Weaire.

A major difficulty facing anyone attempting to write descriptively about pseudopotential theory is that it is cast, for the most part, in terms of reciprocal space. Whilst this presents no problems for crystallographers, other

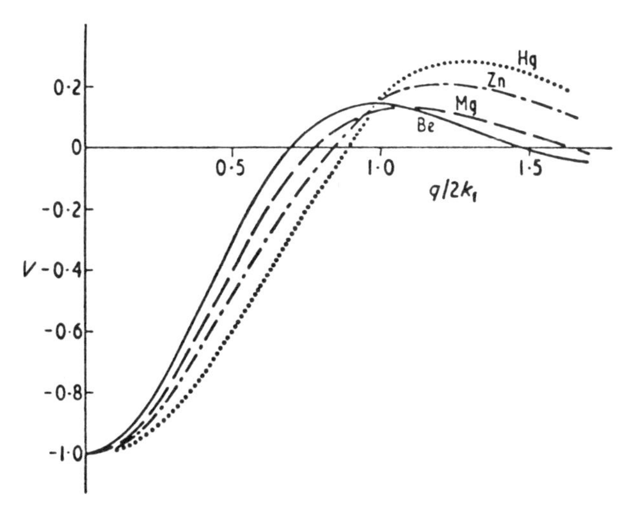

Figure 8-11 The form of the model potential, *V(q)*, for some divalent simple metals. The curve for Cd is almost identical with that for Zn. (From D. Weaire, *J. Phys. C*, Series 2, **1**, 210 (1968) © Insititute of Physics)

readers may wish to refresh their understanding of this simple concept by reference to a standard text. The use of reciprocal space necessitates expression of V_{ps} in terms of its Fourier transform, $V(q)$, where q is a reciprocal lattice vector.

There are two major approaches to determining the form of $V(q)$. (*a*) It may be calculated *a priori*, using procedures which need not concern us here. (*b*) Since a variety of different properties of metals are related to the pseudo-potential, its form may be deduced by fitting a set of experimental data, and then used to account for other properties. For example, the $V(q)$ values for both antimony and indium were obtained by varying the theoretical band structure of InSb until it fitted the observed optical spectrum. The $V(q)$ so obtained for indium was found to provide a basis for understanding the curious structure adopted by that metal. The pseudopotentials of most metals are surprisingly similar, taking the form shown in Figure 8-11. The essential difference is that the point at which $V(q)$ passes through zero, q_0, varies from metal to metal. It is plotted in terms of the ratio $q/2k_F$, where k_F is the radius of the Fermi sphere (see p. 261), because this keeps the reciprocal lattice vectors fixed.

8.4.2 Crystal Energy in Terms of the Pseudopotential

The total internal energy, U, of a metal may be expressed as the sum

$$U = U_0 + U_E + U_{BS}$$

where U_0 is the structure-independent energy. The other terms, which are our concern, are structure dependent. U_E is the Ewald energy, the energy of an assembly of point metal ions, at the lattice positions, in a uniform electron gas. It corresponds to the Madelung energy of an ionic crystal (Section 5.1.1). U_{BS} is the band-structure energy of the electron gas as it reacts to the crystal potential, and only occurs separately from U_E when expressed in terms of reciprocal space. Harrison has shown that when transformed to real space

$$U_E + U_{BS} = \frac{1}{2} \sum_{n,m} \Phi(|\boldsymbol{R}_n - \boldsymbol{R}_m|)$$

where $\boldsymbol{R}_n$ and $\boldsymbol{R}_m$ denote atomic positions. (Strictly speaking, Φ is a pair potential only if terms higher than second order in U are neglected.)

The Ewald term, U_E, always favours high-symmetry structures; its variation with departure from some symmetrical structure has been demonstrated in a number of cases by explicit calculation. The real advances of late have come in understanding the behaviour of U_{BS} and its relation to structure. We must now examine its form, which may seem complex at first sight, but is really quite straightforward.

$$U_{BS} = W(g)E(g) = W(g)[V(g)]^2\varepsilon(g)f(g)$$

$W(g)$ is the structural weight, defined by

$$W(g) = n(g)\Big|\sum_n \exp(ig\boldsymbol{R}_n)\Big|^2 = n(g)|S(g)|^2$$

where $S(\mathbf{g})$ is the structure factor of the Brillouin zone plane $\mathbf{g}$, and $n(g)$ the number of equivalent planes. $\varepsilon(g)$ is a screening factor which does not vary greatly from unity and may be ignored in so far as structural determination is concerned. $f(g)$ is a logarithmic function arising from use of perturbation theory; it varies with g as shown in Figure 8-12a. $E(g)$, the 'characteristic energy' of a structure, being the product of $\varepsilon(g)$, $f(g)$, and the square of the pseudopotential, varies in a complex manner; Figure 8-12b. The essential features to note are the existence of a zero value, q_0, and its position in relation to $2k_F$. $W(g)$ is an invariant for any given lattice. It is shown for the three simple metal lattices in Figure 8-13.

The relation between structure stability and pseudopotential should now become clear. Since $E(g)$ has zero or very low values in the region of q_0, U_{BS}

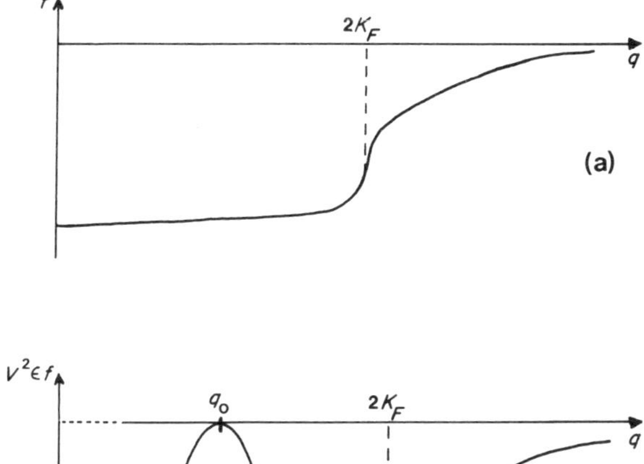

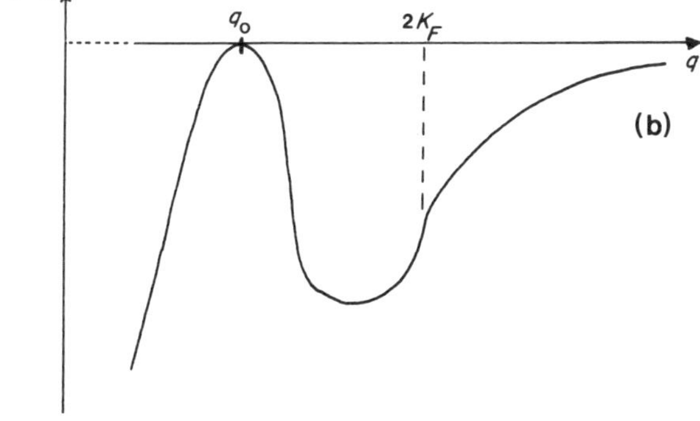

Figure 8-12 Schematic representation of: (a) the function f, (b) the characteristic energy $E(g) = [V(g)]^2\varepsilon(g)f(g)$, which occur in pseudopotential theory. The scale in (b) between q_o and $2k_F$ has been greatly expanded. (Heine and Weaire (1966))

will be small; hence, any structure with weights $W(g)$ near q_0 will be unstable. This is a fundamental point, so let us put it in other ways as well.

If a particular metal has a structure of relatively low symmetry (e.g. Ga) and not one of the high-symmetry ones (b.c.c., h.c.p. or c.c.p.), this is to be understood in terms of a low value of U_{BS} overcoming the Ewald energy, U_E (which always favours the high-symmetry case). However, U_{BS} can assume a more favourable value, and hence stabilize a structure, if the weights $W(g)$

can be redistributed, i.e. if the (hypothetical) high-symmetry form of the metal can be transformed by shifting the atoms to new positions thereby defining a new lattice with $W(g)$ values away from the region of q_0. Examples are discussed below. By similar arguments, a choice can, in principle, be made between simple structures by noting which has the distribution of $W(g)$ which best avoids q_0. Finally, we note that the shape of $f(g)$, Figure 8-12a, implies that a lattice plane has a greater stabilizing influence if g is just *less* than $2k_F$ than if it is greater. There is, of course, *a major qualification* to this argument. Structure stability depends upon minimizing the *free energy*; in pseudopotential calculations, such as those now described, the entropy term is neglected.

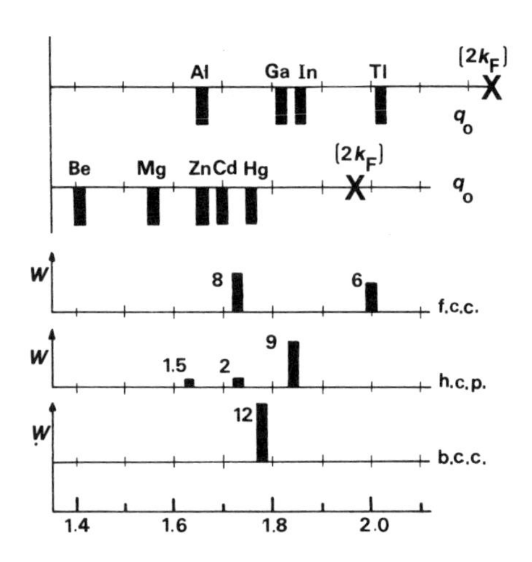

Figure 8-13 Values of q_o for several metals, and of the reciprocal lattice vector weights $W(g)$ for f.c.c., h.c.p. and b.c.c. structures. (Heine and Weaire (1966))

The above correlation (due to Heine and Weaire, 1966) between the value of q_0 and structure stability, clearly requires that we focus attention upon the relation of q_0 to fundamental atomic properties if we are to lay claim to any real understanding of simple-metal structures. The value of q_0 depends upon the valence of the metal and the atomic number, see Figure 8-13. Outside the core, R_c, the pseudopotential $V_{ps} = -z/r$ is a function of r/R_c only, if we take V_{ps}/z. Hence $q_0 \propto R_a/R_c$, where R_a = atomic radius. R_a/R_c, and hence q_0, increases with z, but within a group of constant z, R_a/R_c decreases slightly with increasing atomic number. This is, however, offset by a greater change down each group in Δ_{sp}, the energy difference between s and p states in an atom. Δ_{sp} increases, relative to the energy levels themselves, down a group. Thus, for example, the s-electrons of Hg, Tl and Pb, are rather tightly held (the 'inert-pair' effect). It is worth noting, in passing, that q_0 of the mean pseudopotential in alloys varies with the electron/atom ratio, thereby giving a fundamental reason for Hume-Rothery's classic discovery of the correlation between alloy structure type and electron/atom ratio.

8.4.3 Crystal Structure and Pseudopotential

As noted previously, the energy differences between the simple metal structures are small, making it extremely difficult to predict which will be adopted by a given metal. There is, in any case, a certain degree of arbitrariness in definition of the pseudopotential (what is R_c, for example?), and there is as yet no satisfactory way of defining it at large distances. Further, the entropy term is neglected. Any agreement between the structure adopted by a metal and that predicted by this theory is most probably fortuitous. What *does* appear, is that the q_0 values are at least seen to be compatible with the weight distribution, $W(g)$, for the high-symmetry structures of the alkali metals, and for those of Be, Mg and Al. A particular success of pseudopotential theory is a rationalization of the distorted structures exhibited by the B-metals (see Section 8.3.).

Comparing the structure weight distributions of Figure 8-13 with the q_0 values (Figure 8-11) of Zn and Cd, we see that the c.c.p. lattice is unfavourable since one peak comes near q_0. H.c.p. is clearly the most stable of the three lattices since its main $W(g)$ peak is well removed from q_0. For Be and Mg, which have smaller values of q_0, the balance between the possible structures is more even and no clear prediction emerges.

The distinctive feature of Zn and Cd, their high c/a ratios, was explained on p. 274 in terms of competition between two hybridization schemes. In pseudopotential language, these facts are explained as follows. Although the Ewald energy, U_E, is a minimum for the h.c.p. lattice at the ideal c/a value (1·633), there is no requirement that U_{BS} be minimized at that point. Consequently, the total energy, U, of an h.c.p. metal may be expanded in terms of deviation from the ideal value of the axial ratio:

$$U = A + B\lambda + C\lambda^2 + \cdots$$

where $\lambda = (c/a)-(c/a)_{ideal}$. The coefficient B arises almost entirely from the band structure term. Explicit calculation of U_E and U_{BS} as a function of c/a showed that U minimized at different c/a values for each metal, Be, Mg, Zn and Cd, and that these minima are near the observed values.

Whilst the two available explanations of these c/a ratios are apparently well founded, the relation between them is not as clear as is desired.

The structures of aluminium (f.c.c.), gallium and indium, have all been accounted for on the theory outlined above, viz. that pseudopotential q_0 values must not fall close to major lattice-weight terms. For aluminium, q_0 is close in value to those of zinc and cadmium but, nevertheless, has the f.c.c.

structure, apparently because the weights for that latice now bear a different relation to $2k_F$. The indium structure is best viewed as a slightly tetragonally distorted f.c.c. lattice with $c/a = 1{\cdot}08$. In principle, the f.c.c. structure may be distorted either rhombohedrally (along a cube diagonal) or tetragonally (along one axis). It turns out that for indium it is the value of the Ewald term that swings the balance in favour of the observed tetragonal distortion.

Mercury occurs in two forms, both being viewed as distortions of f.c.c. α-Hg corresponds to the rhombohedral distortion and β-Hg to the tetragonal one. The Ewald term opposes both distortions, but not to the same extent, although the U_{BS} term is favoured by the rhombohedral modification since it splits the lattice weights better with respect to q_0.

8.5 CONCLUSIONS

It is evident that our understanding of the factors which determine the structures of the metallic elements is far from satisfactory. In this field the problems of computing lattice energy are probably more severe than for any other type of crystalline solid. Nevertheless, within their respective limitations, the extant theoretical approaches can all lay claim to some successes. Perhaps the most important task now is to show their equivalence more clearly.

BIBLIOGRAPHY

Altmann, S. L., C. A. Coulson and W. Hume-Rothery, *Proc. Roy. Soc.*, A **240**, 145 (1957).

Austin, B. J. and V. Heine, *J. Chem. Phys.*, **45**, 928 (1966).

Brewer, L. *Science*, **161**, 115 (1968).

Eastman, D. E., in *Experimental Techniques in Solid Metals*, Pergamon, Oxford, 1972.

Harrison, W. A., *Pseudopotentials in the Theory of Metals*, Benjamin, New York, 1966.

Heine, V. and D. Weaire, *Phys. Rev.*, **152**, 603 (1966).

Heine, V., *J. Phys. C.*, Series 2, **1**, 222 (1968).

Rudman, P. S., J. Stringer and R. I. Jaffee (Ed.), *Phase Stability in Metals and Alloys*, McGraw Hill, New York, 1967.

Weaire, D., *J. Phys. C.*, Series 2, **1**, 210 (1968).

CHAPTER 9

Degrees of Order

As we have progressed in our study of solids through the preceding chapters, the data and our understanding of them have gradually come into focus. We now add a further stage of refinement by considering a variety of imperfections and defects, from some form of which no crystal is exempt. Many of the topics we now approach are of enormous complexity and currently absorb the efforts of many large research teams. The aim, here, is to present in summary some of the main structural facts as they appear today, together with current theories which seek to account for them.

A brief introduction to this area of solid-state chemistry was given in Section 1.4. In addition to the faults and defects enumerated there, further types are recognized. (*a*) Electrons and positive holes; phonons; excitons. Phonons are the quasiparticles of energy associated with excitation of one of the modes of vibration of the perfect crystal, whilst excitons are semi-localized electrons produced by electron excitation in insulators and semiconductors. (*b*) Three types of *transient* imperfections are also classified: electromagnetic quanta (of any frequency), charged and uncharged radiations such as α- and β-particles and neutrons. The groups (*a*) and (*b*) will not be considered further as they fall within the classical content of solid-state physics and are clearly described in a wide range of texts.

The remaining classes of defect may be broadly classified as inherent and structure specific. *Inherent* (Schottky and Frenkel) defects are present in all crystals by virtue of the energy fluctuations which occur above absolute zero. They are considered as *point defects*; they occur randomly and in very low concentration and may therefore be handled theoretically by statistical mechanical procedures analogous to those developed for describing dilute solutions. The Schottky and Frenkel defect concentration, even just below the melting point of a solid, is several orders of magnitude below the concentration of defects present in non-stoichiometric materials with a chemically significant range of composition (0·1 to 20%). One of the most important recent advances in solid-state chemistry is the recognition that non-stoichiometric compounds cannot be understood in terms of high concentrations of randomized point defects, as was held for a long time. We return to this theme in Section 9.2.

Schottky and Frenkel defects are basic to transport processes in solids; thus, explanation of ionic conductivity and of diffusion properties requires the existence of mobile defects. These aspects are fully discussed in many physics and chemistry textbooks and will not be considered further since we are principally concerned here with the description of compounds in which defect concentrations are at least 0·1 atom per cent and require additional concepts for their understanding.

This chapter therefore deals with imperfections and defects which are *structure specific*. We consider them in two main groups: 9.1 Stoichiometric Compounds; 9.2 Non-stoichiometric Compounds.

9.1 DISORDER IN STOICHIOMETRIC COMPOUNDS

In stoichiometric compounds, the constituent atoms are present in rational proportions understood in terms of the oxidation states available to them: i.e. the composition is constant. In addition to such imperfections as dislocations, stacking faults, and inherent defects, several types of structure-specific defects are known. They can be regarded as due to partial (occasionally complete) loss of order.

9.1.1 Orientational Disorder in Molecular and Ionic Crystals

Formation of crystals of carbon monoxide can take place with a more or less random orientation of the molecules with respect to each other since there is little difference between the two ends.

Because the X-ray scattering power of an atom is proportional to its atomic number, it is extremely difficult to distinguish atoms of similar atomic number by this technique. In particular, oxygen and carbon cannot be distinguished from each other. However, an excellent test of orientational order is comparison of the experimentally determined residual entropy with a calculated value. For complete randomness the entropy of the CO crystal would be $R \ln 2 = 5{\cdot}78$ J mol^{-1} K^{-1}; the experimental value is 4·61 J mol^{-1} K^{-1} indicating retention of some, but not much, order. Other examples are N_2O ($S_{exp} = 4{\cdot}86$ J mol^{-1} K^{-1}) and perchloryl fluoride, $FClO_3$; for the latter there are now *four* possible orientations of the Cl—F bond so that for complete randomness the entropy would be $R \ln 4 = 11{\cdot}52$ J mol^{-1} K^{-1}. The experimental value is 10·13 J mol^{-1} K^{-1}.

Another form of disorder is found in molecular and complex ionic crystals in which one component can undergo *rotation between equivalent positions*. The onset of rotation is revealed by a maximum in the heat capacity (C_p) versus temperature curve (i.e. a λ point) indicating the need to supply energy to support the new motion. Above such a transition point, the rotating entity appears more symmetrical than below it with the consequence that the higher temperature phase usually has higher symmetry. Thus, CsCN has the CsCl structure above 200 K, but below that adopts a deformed version of that lattice. Methane is known to behave as pseudospherical above ca 20 K, and many ions such as NH_4^+, CO_3^{2-} and NO_3^-, also show orientational disorder attended by various crystallographic and thermodynamic symptoms. For details of the ammonium halides see p. 209.

Chain crystals can also show transitions which are associated with specific types of motion. Wells quotes the example of the hydrocarbon $C_{29}H_{60}$, Figure 9-1. Any chain molecule composed of atoms that are not co-linear is allowed one degree of rotation freedom about the chain axis. In addition each chain undergoes translational vibrations in two mutually perpendicular directions in a plane normal to the chain axis. Combination of these two types of motion gives the zigzag hydrocarbon chains a pseudocylindrical

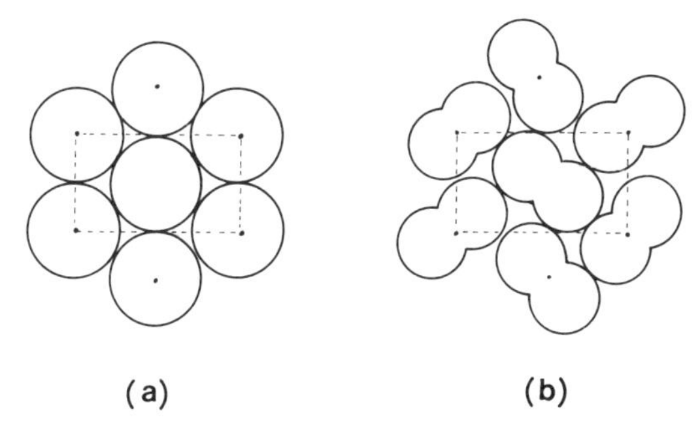

Figure 9-1 Modes of chain packing in $C_{29}H_{60}$. (a) High-temperature form. (b) Low-temperature form. The chains are seen in cross-section

appearance with the result that a more symmetrical crystal structure is adopted at higher temperature. In accord with close-packing principles each chain has six nearest neighbours. Wells states that in the low-temperature form the m.c.n. = 4 only, but this is almost certainly wrong; from the diagram there are clearly *six* nearest neighbours and correct drawing of van der Waals radii would indicate contact with all of them.

A number of *inorganic crystals* are known in which orientational disorder results from failure to distinguish between oxygen and fluorine. The earliest such example, studied by Pauling in 1924, is $(NH_4)_3MoO_3F_3$. If no distinction is made between oxygen and fluorine it has the same face-centred cubic lattice as the series of hexafluorometallates $M^I_3MF_6$, where M = Al, V, Cr or Fe, and $M^I = NH_4^+$ (and also Li^+–Cs^+ for the ferrates). The symmetry of an octahedral complex ion $[MoO_3F_3]^{3-}$, whichever isomer is involved, is incompatible with the cubic lattice in which it occurs. The structural data can only be interpreted in terms of randomly oriented octahedra, random in the sense that either oxygen or fluorine may be found along the octahedral axes. An equivalent statement, crystallographically speaking, is that oxygen and fluorine are disordered upon the set of lattice sites occupied by fluorine alone in the $(NH_4)_3AlF_6$ structure.

A similar situation occurs in $FeF_2.4H_2O$ and the analogous zinc salt. These compounds have the metal in octahedral coordination with two fluorines and four water molecules; due to the similar scattering powers of oxygen and fluorine, it is difficult to distinguish them but the fluorines are believed to be *trans* to each other. Since these octahedra can be aligned in three ways the crystals should have a residual entropy at 0 K of $R \ln 3$; the experimental value is in close agreement. It should be stated that this is not the only interpretation of the data, but it seems the most likely.

Choice of site occupation and valency disordering in spinels also come logically in this section. Their essential features were described in Sections 3.5.4 and 5.5.1.

9.1.2 AgI and 'Superionics'

The polymorphism of silver iodide is not fully understood in all respects, but in outline the situation is as follows.

	γ-AgI	β-AgI	α-AgI
Preparation	Precipitation	Crystallization from the melt	From β-AgI
Temperature range	Up to 136°C	136–146°	146–555° (m.p.)
Structure type	Zinc blende	Wurtzite	B.c.c.
Conductivity (ohm^{-1} cm^2)		$3{\cdot}4 \times 10^{-4}$	1·31

In fact, precipitation generally yields mixtures of γ- and β-AgI and the transition at 136° has been questioned. We are concerned here with the $\beta \rightarrow \alpha$ transition. In it the iodine lattice changes from h.c.p. to b.c.c. Concurrently, the electrical conductivity changes from a low value to one in the metallic range; it has been shown by transport experiments that all the current is carried by silver ions (i.e. the conduction is not electronic as in metals). Self-diffusion work with silver isotopes confirms that these ions move freely between the available sites in the iodine lattice. In the b.c.c. lattice there are the following interstitial sites (see Figure 3-13).

(*a*) Three highly distorted octahedral sites per packing atom (I in this case). They are so distorted that they are sometimes referred to as positions of linear two-coordination, since the distances from the centre of such a site to neighbouring atoms are $0{\cdot}154R$ and $0{\cdot}633R$ for the case in which the packing atoms first touch (R = radius of the packing atoms).

(*b*) Six tetrahedral sites per packing atom; these are larger than sites (*a*) and have a lower-limiting radius ratio $r/R = 0{\cdot}291$.

(*c*) Twelve sites per packing atom in which the centre is equidistant from three iodine atoms. All these sites are indicated in Figure 9-2.

Clearly, complete occupation of any one of these three sets of sites does not yield a compound of formula AB and, since there is a total of twenty one interstitial sites per iodine, there is a wide choice of positions open to each silver ion. The structure of α-AgI has therefore been described as a b.c.c. lattice of I^- with a 'molten' sublattice of Ag^+.

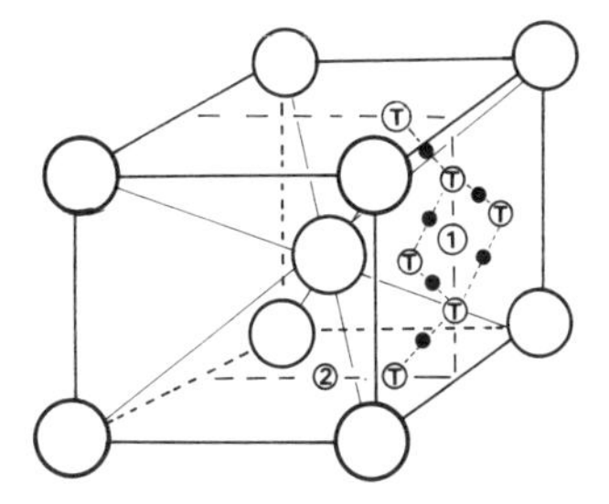

Figure 9-2 The body-centred cubic structure showing some of the distorted octahedral sites (1 and 2); tetrahedral sites, Ⓣ; and trigonal sites (●)

Why should the $\beta \rightarrow \alpha$ transition occur, and why should the product be a b.c.c. lattice? There is considerable disorder in the β form, although this is not nearly so great as in α-AgI. The silver ions may be in any one of five positions: the normal tetrahedral one and four equivalent sites at the centres of I_4 tetrahedron faces. With rising temperature and increasing entropy, the β structure decreases in stability until it becomes necessary to reconstruct the material in a new form compatible with the high entropy of the 'molten' silver sublattice. In other words, the directed bonding in γ- and β-AgI is not sufficiently pronounced in character to stabilize the structure via the enthalpy term.

If AgI is described in terms of band theory, it is interesting to note that we should expect a b.c.c. structure, on the basis of the Engels–Brewer correlation (Section 8.1.2), if we agree that only one valence electron per atom is involved; but this is probably misleading. The relative sizes of Ag^+ and I^- require the b.c.c. structure if the bonding is regarded as *less* directional than that which requires retention ofthe wurtzite structure at all temperatures, but directional enough to force AgI into the wurtzite form when the entropy term is not dominant.

The special electrical properties of α-AgI have led to the search for other materials exhibiting extremely high ionic conductivity. A small group of such materials is now known, commonly referred to as 'superionics', and a number of commercial applications have been developed (D. Greene, *New Scientist*, 11th May 1972). The basic approach is to seek compounds in which a fairly rigid anion sub-lattice exists through which smaller cations can move readily. The majority of superionic conductors contain both silver and iodine, but other stable frameworks have been constructed in which I^- is combined with sulphide, phosphate, or certain organic ions. Ag_3SI is based upon the CsCl structure (the SI sublattice) with Ag^+ distributed among interstitial positions as in α-AgI. An isostructural series $M^IAg_4I_5$ exists with $M^I = K^+$, Rb^+ or NH_4^+: Na^+ is too small to stabilize this framework and Cs^+ is too large. Superionic behaviour is also exhibited by an entirely different type of material, $NaAl_{11}O_{17}$, named β-alumina for historical reasons. This is most readily understood as slabs of spinel structure with Na^+ or K^+ ions interposed; it is, of course, the alkali-metal ions that are responsible for the ionic conductivity.

$RbAg_4I_5$ has been employed as the electrolyte in solid-state batteries in which energy is derived from the reaction Ag (at one electrode) and RbI_3 (at the other electrode) to yield more $RbAg_4I_5$. Such cells can operate over a wide temperature range ($-55°$ to $+200°C$), have a long shelf life and can withstand mechanical shock. Aerospace and other applications are envisaged.

β-Alumina is the electrolyte in a secondary (rechargeable) cell which has an unusually good energy-to-weight ratio. Its power derives from a reaction which produces sodium sulphide.

9.1.3 'Incomplete' Lattices

Many compounds may be considered, in a formal sense only, as having incomplete lattices. A classic example is ReO_3 which is commonly related to the perovskite structure simply as an aid to visualization. ReO_3 can be described as having an incomplete c.c.p. lattice of oxygen with Re in octahedral sites (see Section 3.5.4), although it is more realistically considered as the simplest topological solution to packing ReO_6 octahedra (required by valence considerations) so as to give stoichiometry AB_3. As a result, the structure has a certain openness. Incidentally, ReO_3 has an electronic conductivity comparable with that of copper; the reader is invited to account for this in terms of band theory.

Tetrahedral structures such as zinc blende are rather open for the same basic reason; the fact that only half of the tetrahedral sites in the c.c.p. array of sulphur are filled with zinc in no way justifies us in describing zinc blende as a structure with an 'incomplete' lattice. The zinc blende structure represents one of two topologically acceptable ways of accommodating atoms which form four tetrahedrally disposed covalent bonds to nearest neighbours. An entirely different set of reasons accounts for the fluorite structure in which *all* the tetrahedral sites in a c.c.p. array are filled.

A very large number of compounds are structurally related to zinc blende and have been the subject of much research by virtue of their semiconduction properties. General formulae for all possible binary and ternary compounds have been given by Goryunova (1965). We consider a very few illustrative cases. A similar but much more restricted range is based upon the wurtzite structure.

Either the metal or non-metal positions in the zinc blende lattice may be occupied by more than one type of element, and both variations can occur at the same time. A further feature of note (which justifies inclusion of these compounds in this chapter) is that available positions may be only partially occupied depending upon the valence of the substituting atoms; this occupation can be either ordered or random. Consider some examples.

(*a*) Replacement of zinc by equal proportions of copper and iron disposed regularly yields the chalcopyrite structure, $CuFeS_2$. Because the atoms at the corners of the original zinc blende unit cell are now of different kinds, the repeat unit is doubled in one direction, Figure 9-3. In the mineral stannite, Cu_2FeSnS_4, each sulphur is tetrahedrally coordinated to two copper atoms and one each of iron and tin.

(*b*) Variants of the chalcopyrite formula exist in which two metal atoms are distributed among the metal sites at *random*, e.g. $CdZnSe_2$ and $GaInSb_2$.

(*c*) Many compounds A_2B_3, where A is a B-metal or a rare earth, and B = S, Se or Te, adopt the basic zinc blende structure but, due to their stoichiometry, one third of the metal sites are left vacant. Such compounds are commonly written as $A_2\square B_3$, where $\square$ represents a vacant site, and are referred to as having 'defect' or 'incomplete' lattices. Typical examples are the γ-phases of Ga_2S_3, Ga_2Se_3, Ga_2Te_3 and In_2Te_3. In all these compounds the metal atoms are randomly distributed over the metal sites. Ga_2S_3 has a β-phase entirely analogous to the γ-phase but based upon wurtzite; the α-phase, also based upon wurtzite, has an *ordered* arrangement of metal atoms.

The use of the term 'defect' in this context is unfortunate and not recommended; we reserve it for use in describing Schottky, Frenkel and other departures from perfection. Description of a material such as Ga_2S_3 as having an 'incomplete' zinc blende lattice is valid but writing its formula as $Ga_2\square S_3$ is of doubtful value and no more meaningful than pointing out the formal relation of zinc blende to fluorite. The structure of Ga_2S_3 is different from that of ZnS for good reasons. The bonding in Ga_2S_3 is highly directional (Ga is a B-metal, and the Phillips electronegativity difference $\chi(S) - \chi(Ga)$ is only 0·74) and requires tetrahedral coordination. In this it differs from Ga_2O_3 ($\Delta\chi_{(Phillips)} = 1{\cdot}87$) which has two polymorphs (described below) both having the metal in some form of six-coordination compatible with the rather higher ionic contribution to the bonding. Ga_2S_3 therefore adopts the ZnS-based structures because none of the A_2B_3 structures is compatible with its bonding requirements; Ga achieves tetrahedral coordination to sulphur whilst each sulphur atom is surrounded tetrahedrally by four electron pairs which are either used in bonding to Ga neighbours or are lone pairs.

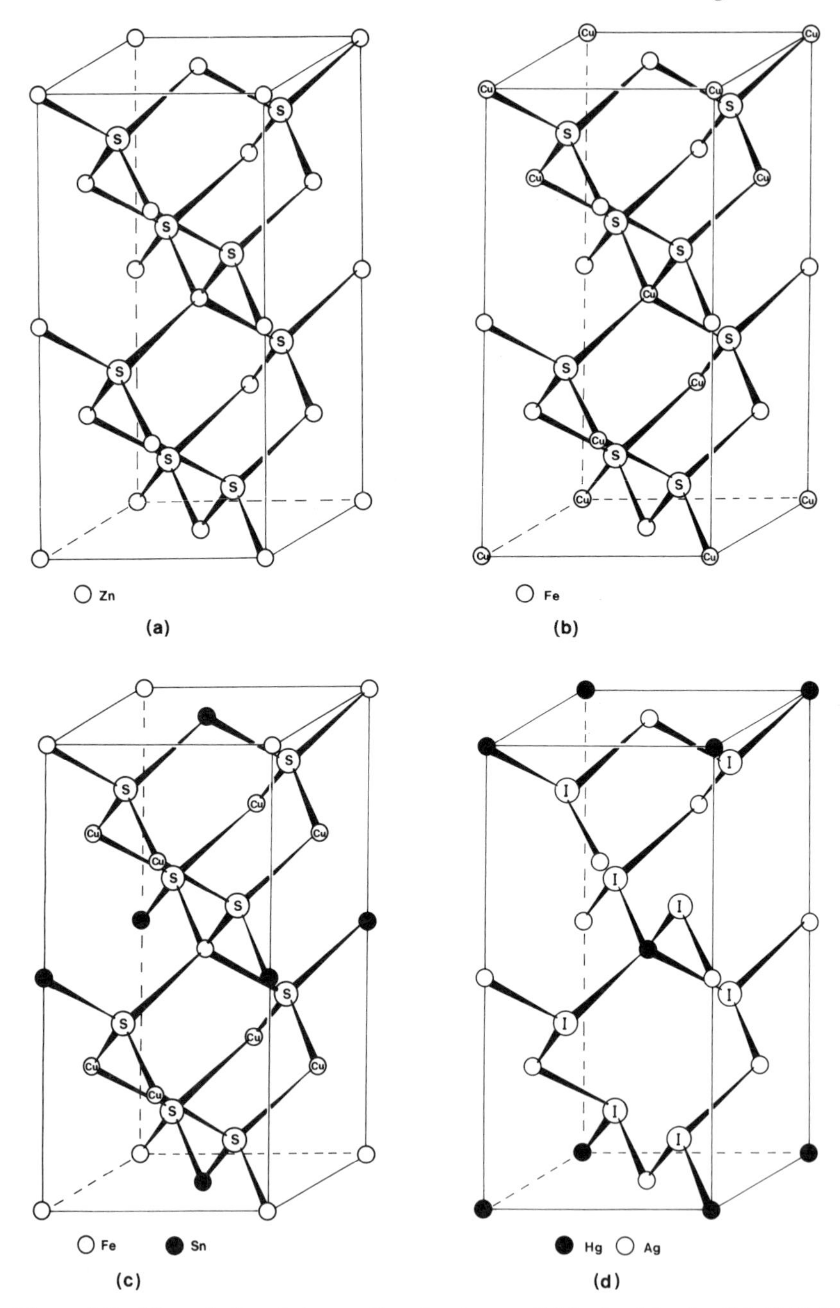

Figure 9-3 The relation between the structures of (a) zinc blende, (b) $CuFeS_2$, (c) Cu_2FeSnS_4 and (d) β-Ag_2HgI_4. (From A. F. Wells, *Structural Inorganic Chemistry*, The Clarendon Press, Oxford, 1962)

The less strongly directional bonding in Ga_2O_3 allows adoption of structures in which Ga achieves higher coordination. Thus, one polymorph has the corundum structure (p. 74) whilst the other crystallizes with the remarkable C–M_2O_3 structure which is most readily understood in terms of its relation to fluorite. In fluorite each M atom has eight nearest neighbours at the corners of a cube. In C–M_2O_3 each M is six-coordinate but this has been achieved in two different ways such that half of the metal atoms have a different environment from the others: the two arrangements can be considered as obtained by removal of two cubic neighbours as shown in Figure 9-4. The oxygen atoms are four-coordinated. This structure is also adopted by α-Mn_2O_3, In_2O_3, Tl_2O_3 and some M_2O_3 oxides of the lanthanides and actinides. The adoption of the C–M_2O_3 structure provides a further example of the trend to high coordination numbers for the more ionic compounds of

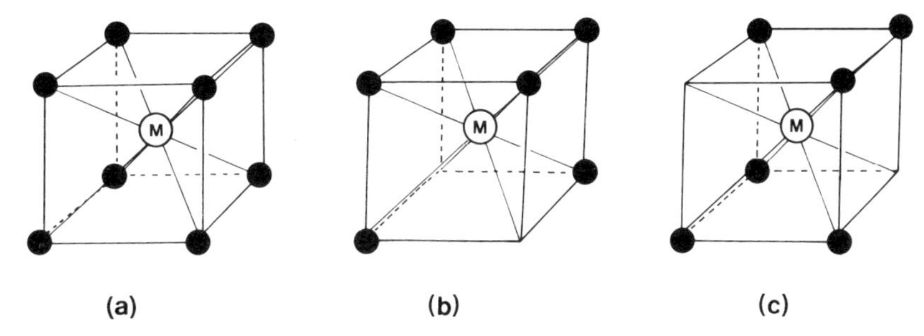

Figure 9-4 (a) A portion of the CaF_2 (fluorite) structure showing cubic coordination around calcium. (b) and (c) show the two different types of coordination around M in the C–M_2O_3 structure, and the relation to fluorite

heavier elements. This is commonly achieved only by acceptance of some unusual coordination arrangements about the metal.

Why is there random distribution of metal atoms in γ-Ga_2S_3 and related compounds? The reason may be associated with the method of preparation of the material. What is clearer is that a higher entropy is associated with β-Ga_2S_3 than with the ordered α-phase but that this is not sufficient to destabilize the structure. The enthalpy term, dominated by the energy of formation of strong covalent bonds, is of sufficient magnitude to keep the free energy at an acceptable value.

Further series of compounds are based upon zinc blende with $\frac{3}{4}$ or $\frac{7}{8}$ of the cation sites occupied either in a regular or a random manner. Typical of these is Ag_2HgI_4; the β-phase has a regular distribution of cations resulting in the structure shown in Figure 9-3d. Above 51°C a transformation to the α-form takes place: the iodine atoms remain in their positions but the metal atoms are now randomly distributed. The silver ions can move throughout the lattice via the vacant sites thereby impairing an ionic conductivity some 10^3 times that of a normal ionic compound at similar temperature, though not so great as in the superionics.

Finally, we note that simultaneous randomization is also possible on the fully occupied anion sublattice as in Zn_3PI_3 and Zn_3AsI_3.

9.2 NON-STOICHIOMETRIC COMPOUNDS

The early concept of a crystal as a perfectly ordered periodic array gave way to a more accurate and comprehensive view with the classic contributions of Schottky and Wagner, of Fowler and others, whose work on the statistical mechanics of crystals showed that there is an inherent statistical disorder in all crystals which perturbs the periodicity of the structure. The classic defects described in Section 1.4, namely vacancies and interstitial atoms between the normal sites are known as *point defects*. Point-defect theory is perfectly adequate to explain many optical, conduction and other transport properties of effectively stoichiometric crystals, such as those which are highly ionic and the tetrahedral structures typical of many semiconductors which have nearly constant composition.

The modern concept of the perfectly *stoichiometric* crystal is that it does *not* represent a condition of perfect order but that it is a system in which there are exactly balanced, extremely low, concentrations of intrinsic point defects. This is the condition of the overwhelming majority of solid compounds. However, there are a few systems which are clearly distinguished from normal stoichiometric ones by two objective operational criteria:

(*a*) They exist over a chemically significant *range* of composition and are thermodynamically bivariant.

(*b*) Their unit cell size as determined by X-ray diffraction apparently varies smoothly with composition.

Such systems are said to be '*non-stoichiometric*'; in comparison with the vast array of known stoichiometric solids they are few in number but they are of great significance, not only for the industrial importance which some have, but also for the extraordinary insights into the structure, stability and dynamics of solids which are emerging from current research into them. It should be clearly appreciated that non-stoichiometry is essentially a high-temperature phenomenon: this point is developed below.

For many years it seemed that non-stoichiometric systems of wide compositional range could be understood in terms of a parent lattice with a high concentration of randomly distributed point defects; this idea is no longer tenable. The current understanding of non-stoichiometric compounds is moving away from explanations based upon classical 'point' defects, although their existence as *inherent* defects is beyond dispute. It is too early to enunciate general principles, but the fact remains that for each non-stoichiometric system that has been investigated in detail, at significant 'point'-defect concentrations interaction is found to occur with formation of a new, ordered, structure element: we discuss examples below.

In the following sections we investigate the results and concepts which form the basis of the modern approach to non-stoichiometry. It should be emphasized that we are dealing with a complex area of current research in which both facts and theories are far from settled, although great advances have been made. We attempt to give a consensus view.

9.2.1 Operational Criteria of Non-stoichiometric Behaviour

We are chiefly concerned to find out the nature of the 'defects' in non-stoichiometric systems, and to learn how a basic structure type is progressively modified to accommodate them. But first we must relate the experimental criteria which distinguish this class of materials.

The first operational criterion of a non-stoichiometric system is a structural one. Typically it is found that the X-ray diffraction pattern throughout the compositional range shows spots characteristic of the parent lattice, but that the unit cell size varies smoothly with composition. By combining this information with the density of the sample, the atomic composition of the *average* unit cell can be established. Thus, for example, the data for so-called 'super-stoichiometric' $UO_{2+\chi}$ show that both oxygen and uranium sublattices are complete but that additional oxygen atoms are also present.

The second operational criterion is thermodynamic. For an equilibrium between two coexistent phases of definite composition (i.e. stoichiometric) the chemical potentials are functions of temperature only. In contrast, for a non-stoichiometric system, equilibrium in the phase diagram depends upon *both* temperature *and* composition of the solid phase. Such systems are said to be thermodynamically bivariant.

At any given temperature, pressure and composition, the stable phase of a system is that of lowest Gibbs free energy, G. Neglecting the pressure variable, we may represent the stability of a phase by means of a surface in G—T—χ space. For a normal compound of fixed composition (χ is a constant) such a surface degenerates to an infinitely thin sheet. However, Schottky and Frenkel defects must be present in low concentration above 0 K and there is the concommitant possibility of slight deviations from stoichiometry: such deviation occurs only at the expense of a larger increase in G. The term 'line phase' is often used in referring to such (normal) compounds.

In contrast, χ enters as a variable for a non-stoichiometric compound of detectable compositional range. The associated free-energy surface is much broader and is commonly quite asymmetric about the stoichiometric composition, reflecting the differing nature of the defects which give rise to metal-rich or metal-poor composition. Figure 9-5 summarizes the situation.

$G = f(T)$ alone when the equilibrium condition of a system corresponds to a phase of definite composition (i.e. univariant behaviour), but a non-stoichiometric system is thermodynamically bivariant since $G = f(T, \chi)$.

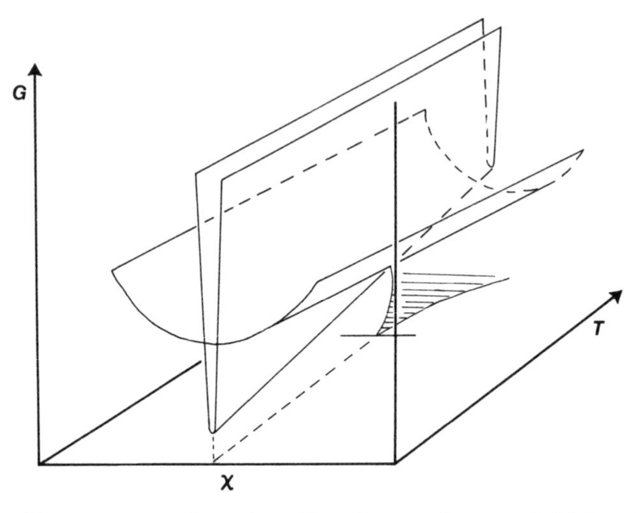

Figure 9-5 Free-energy surfaces for a line phase and a non-stoichiometric phase intersecting in G–T–χ space, showing stabilization of the latter at high temperature. (Anderson (1970))

9.2.2 Non-stoichiometry, Temperature and Composition

Figure 9-5 makes another important point, best seen in the projection on the T–χ plane. It is that *above* some particular temperature the non-stoichiometric phase is more stable than the line phase with which it is in equilibrium; the free-energy surface of the non-stoichiometric phase is seen to be the lower. It is also commonly found that the non-stoichiometric phase changes reversibly, on cooling, into a mixture of two or more compounds of classical composition: such a situation is represented diagrammatically in Figure 9-6. Thus, non-stoichiometric ferrous oxide, given the delightful name wüstite, is thermodynamically stable only above 570°C with iron-deficient composition $Fe_{1-\chi}O$; below this temperature it disproportionates to Fe_3O_4 and α-iron as the stable phase pair.

$$(1 - 4\chi)Fe + Fe_3O_4 \rightleftharpoons 4Fe_{1-\chi}O$$

The phase diagram of this system is shown in Figure 9-7, from which it is seen that the compositional range of wüstite widens progressively with increase of temperature. Note especially that the ideal composition FeO is *not included* in the field of stability: it is thermodynamically impossible for stoichiometric FeO to exist in a stable form, although it has been claimed in metastable form. Outside the wüstite phase limits (above 570°), solid solutions of wüstite with iron (α- or γ-iron according to the temperature) or Fe_3O_4 are formed depending upon whether one is working at the iron-rich or iron-poor end of the range.

It is possible to obtain metastable wüstite at room temperature by quenching, because the decomposition into α-iron and Fe_3O_4 which takes place upon annealing is a diffusion-controlled process of relatively high activation energy.

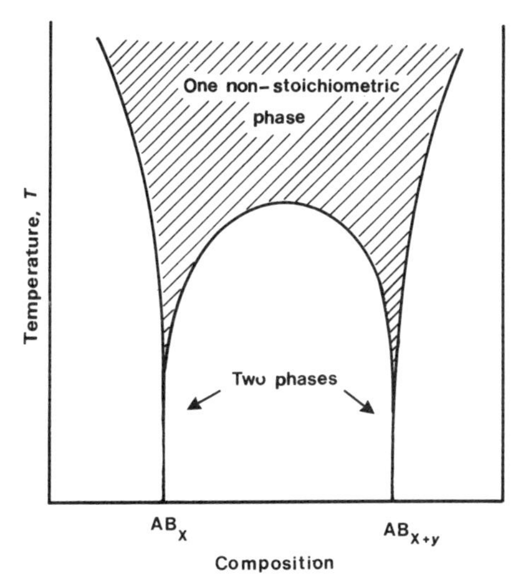

Figure 9-6 Schematic phase diagram showing the relationship between a broad compositional range non-stoichiometric phase at high temperature, and the two compounds into which it separates at lower temperatures

That *non-stoichiometry is essentially a high-temperature phenomenon* also follows from 'third law' considerations. In a non-stoichiometric system the arrangement of defects contributes a 'configurational' term to the entropy in addition to the usual thermal component. If defects are distributed randomly, the configurational term is

$$S_{\text{config}} = k\{\chi \ln \chi + (1 - \chi) \ln (1 - \chi)\}$$

where 'defects' are randomly isothermally mixed with $(1 - \chi)$ mole fraction of the parent lattice atoms. In general, defects are not randomly distributed and other complications appear, but appropriate formulae for S_{config} can sometimes be deduced. Whatever the precise form of S_{config}, the third law of thermodynamics requires entropy to tend to zero as $T \rightarrow 0$ K. Dispersal of the configurational entropy term on cooling therefore implies either that complete ordering will take place (forming a new structure related to the high-temperature phase) or that there will be disproportionation into a pair of phases of zero configurational entropy. However, both these processes are diffusion controlled and high activation barriers are commonly involved. It is therefore possible to study high-temperature structures by quenching, thereby freezing in the disordered configuration. Disproportionation can, and does, take place provided that the change from the high- to the low-temperature structure is at high enough temperature to permit diffusion.

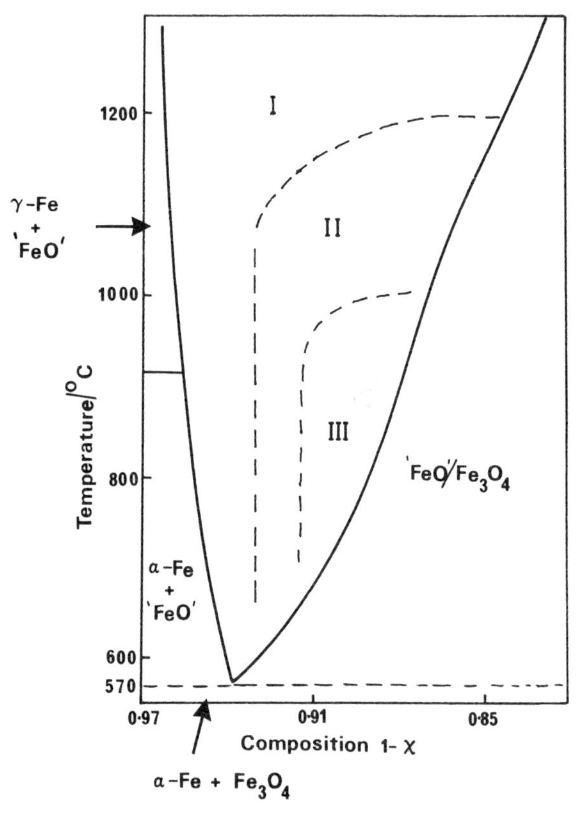

Figure 9-7 Phase diagram of the Fe–O system. I, II and III, together comprise the wüstite region. (Anderson (1970))

It may be that given *any* proportions of the combining atoms it should be possible to form an ordered periodic structure provided that a sufficiently large unit cell is used. Very recently it has been shown that something like this probably does happen in a few cases ('infinitely adaptive' structures, J. S. Anderson, 1973). But most non-stoichiometric compounds, when cooled, either separate into phases of classical composition (as does $Fe_{1-x}O$) or form an ordered compound which is commonly one of a structurally related series (e.g. the PrO_{2-x} system discussed below.)

The phase diagram of the praesodymium–oxygen system is shown in Figure 9-8. At low temperatures, a series of oxides Pr_nO_{2n-2} exists, each of extremely narrow range: they are referred to as a 'homologous series' as they have a basic structural relationship (the term was introduced by Magneli).

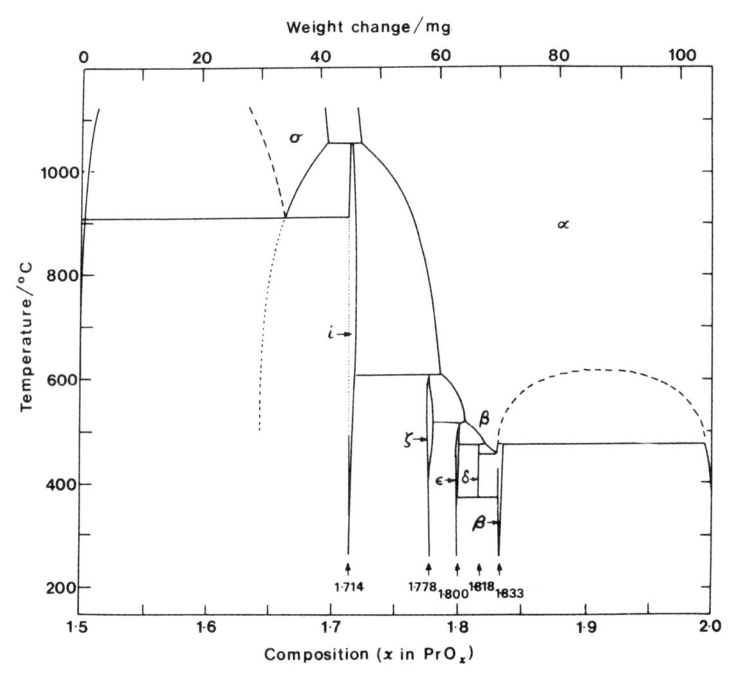

Figure 9-8 Phase diagram of the praesodymium–oxygen system (Eyring and O'Keeffe (1970))

Heated above the temperatures shown, they yield one or other of two non-stoichiometric phases, α and σ (separated by a narrow miscibility gap), overlaying the entire composition range. From the phase diagram the following information may be read.

Oxide	n	χ in PrO_x	Range of Stability at T°C	T°C
Pr_4O_6 (= Pr_2O_3)	4	1·500	1·500–1·503	1000
Pr_7O_{12}(i)	7	1·714	1·713–1·719	700
Pr_9O_{16}(ζ)	9	1·778	1·776–1·778	500
Pr_5O_9(ε)	10	1·800	1·799–1·801	450
$Pr_{11}O_{20}$ (δ)	11	1·818	1·817–1·820	430
Pr_6O_{11}(β)	12	1·833	1·831–1·836	400
PrO_2	∞	2·000	1·750–2·000 1·999–2·000	1000 400

Heating any of the compounds with $n \geqslant 9$ above the temperatures shown yields broad-range non-stoichiometric α-praesodymium oxide.

9.2.3 The Occurrence of Non-stoichiometric Phases, and their Compositional Ranges

In seeking to understand why some materials but not others are non-stoichiometric, and the apparent ranges of phase stability, it is helpful to return to the question of 'point' defects.

Point defects are certainly more complicated than the simple descriptions in Section 1.4 imply. Removal of an atom will cause some relaxation around the vacancy whilst injection of one into an interstitial position causes similar changes in the opposite sense. In a metallic or semi-metallic crystal the effects will probably be fairly localized and largely confined to the shell of nearest neighbours. In such structures the classical concept of a point defect probably comes nearest to realization, although evidence is now accumulating which suggests that vacancies tend to be ordered in these materials as well. For more ionic materials, effects will be felt over considerably greater distances. The interactions described by the Madelung constant of a crystal clearly provide a physical mechanism by which the disturbance of a 'point' defect can be transmitted throughout a relatively large volume. Even with concentrations of point defects as low as 10^{-4} molar there is evidence for their interaction; in general such interaction is cooperative and leads to a new structure related to the parent lattice. In forming this new structure one of two processes occurs: *either* the vacancies are ordered *or* they are eliminated. Examples of these processes are given below.

The ease of creation of defects, and their type, is highly structure-dependent. Since structure-type adoption is intimately related to the electronic properties of the constituent elements (see Chapter 5), explanation of the occurrence of non-stoichiometric phases must reside in the relative energy levels of different atoms.

Consider creation of a cation vacancy in a *metal-deficient* material. Since electrical neutrality must be maintained, the deficit will be balanced by oxidation either of cations or anions nearest the vacancy. A clear example is provided by wüstite. X-ray work indicates that this material has the rock salt lattice. However, preparation at various oxygen pressures yields samples which all appear to have the rock salt structure but the unit cell size is found to vary smoothly with oxygen pressure, indicating change of composition. From the unit cell size, and the measured density of each sample, an average composition results. In this way it was shown that wüstite has an essentially complete oxygen lattice but is deficient in iron, $Fe_{1-x}O$. To balance each Fe^{2+} vacancy, *two* neighbouring cations must be oxidized to Fe^{3+}. This in turn creates further disturbance of the lattice (along the lines outlined above) so that creation of a single cation vacancy in $Fe_{1-\chi}O$ can only be considered as a 'point' defect in a very formal sense.

If a higher cation oxidation state is not energetically accessible, deviations from stoichiometry are usually negligible, although we must still explore the alternative mechanism of anion oxidation by processes such as $O^{2-} \rightarrow O^{-}$ and $X^{-} \rightarrow X$. Perusal of tables of atomic energy levels suggests that changes in cation oxidation state are the more likely, although there are exceptions. One such is the creation of very slightly iodine-rich potassium iodide (a few parts per million) by the incorporation of iodine *atoms* in the crystal. The possible charge-balancing processes are:

$$K^{+} \rightarrow K^{2+} + 3060 \text{ kJ mol}^{-1}$$
$$I^{-} \rightarrow I + 316 \text{ kJ mol}^{-1}$$

From the ionization potentials there is no doubt about the preferred mechanism. Relaxation of the lattice effectively delocalizes the positive hole, giving essentially an I_2^- molecular ion.

A *metal excess* can be accommodated, in principle, by reduction of either component of a lattice by such processes as $Fe^{3+} \rightarrow Fe^{2+}$, $Na^{+} \rightarrow Na$ or $O^{2-} \rightarrow O^{3-}$, $Cl^{-} \rightarrow Cl^{2-}$. Again, change of cation valence is overwhelmingly favoured.

It therefore emerges that the most probable defect-creation processes involve changes in cation oxidation states and that non-stoichiometric behaviour is associated primarily with compounds of those elements for which such changes are facile. *Broad-compositional range non-stoichiometry is therefore a property of transition-metal compounds* (*d*-block, lanthanides and actinides). However, it should be clearly appreciated that by no means all lattice compounds of transition metals are non-stoichiometric. For example, PtS is highly stoichiometric, as are carbides such as Cr_3C_2 and Cr_7C_3. The implication is that creation of defects is energetically expensive in such compounds, which is not unexpected as they are highly covalent.

A second class, of more limited compositional range, is formed by the classic semiconductor compounds of B-metals. Typical members are chalcogenides such as CdS, CdTe and PbTe. This group also exhibits variable cation valency: Pb^{II}, Pb^{IV}; Cd^{I}, Cd^{II}.

The compositional ranges or widths of non-stoichiometric phases vary considerably. Thus, CdO, typical of the second group mentioned above (it has the rock salt structure) has very few defects. The non-stoichiometric nickel-telluride phase covers the range $NiTe_{1\cdot20}$ to $NiTe_{2\cdot00}$, while the closely related trio MnO, FeO and CoO, all have different ranges of stability. $Mn_{1-\chi}O$ has $1 - \chi$ values 1·00 to 0·956; $1 - \chi$ for $Fe_{1-\chi}O$ is 0·95 to 0·88 whilst CoO has a narrow field of variability. As already indicated, the wide stability ranges suggested by these figures in reality mask a fascinating variety of nearly stoichiometric structures interrelated by processes which, for the most part, are not yet fully comprehended. However, the existence of a *range* of compounds in any one metal–non-metal system is clearly related to the ease of oxidation of the metal and to the energetics of its incorporation in various coordination environments. The existence of a compositional range may also be understood in thermodynamic terms; this is considered below.

The accessible compositional range of a non-stoichiometric phase is defined thermodynamically by the familiar 'tangent' method, illustrated in Figure 9-9 for the case of a non-stoichiometric phase with its upper and lower composition limits set by coexistence with two line phases AB_2 and A_2B. For the line

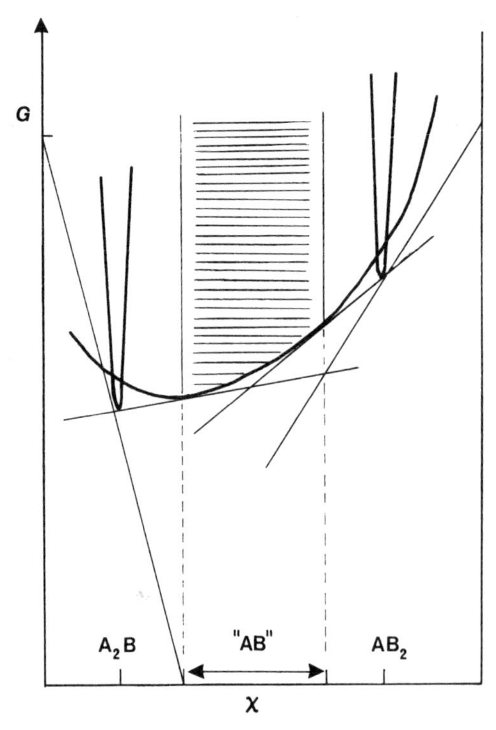

Figure 9-9 Phase limits for a broad-range non-stoichiometric compound 'AB' set by coexistence with line phases A_2B and AB_2. (After Anderson (1970))

phases, the tangent point varies hardly at all on the composition axis even for large changes in G, but for the non-stoichiometric phase the tangent point covers a significant range: the stable range of the latter phase depends upon the thermodynamic properties of the adjacent phases. It is possible to obtain some non-stoichiometric phases in metastable condition outside their thermodynamically predicted stability ranges if no nucleus of the line phase, with which it is in supposed equilibrium, is formed.

Consider now a situation, such as we met in the series Pr_nO_{2n-2}, in which there is a succession of closely related line phases which are formed from the non-stoichiometric phase which spans the same compositional range. The existence of these line phases means, thermodynamically, that below the critical temperature for the change to bivariant behaviour, the free-energy surface of the bivariant phase lies everywhere above the free-energy minima of the line phases. The energy differences between such line phases are very small but apparently sufficient to require formation of ordered structures. Experimentally, unless the determination of free energy versus composition curves is very precise, it is entirely possible that a succession of line phases be mistaken for a bivariant phase since the phase coexistence tangents of the line phases can merge into an envelope (Figure 9-10).

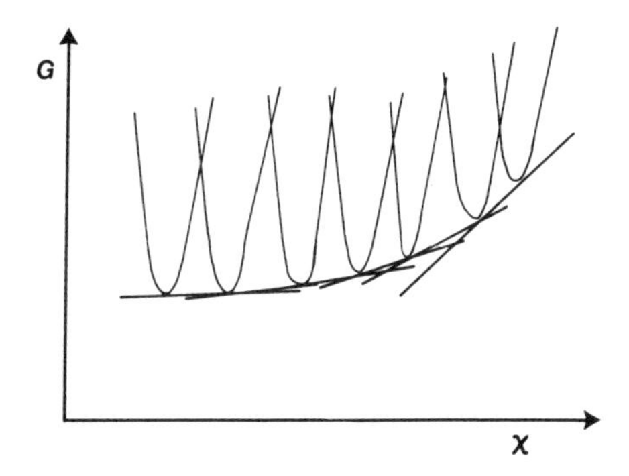

Figure 9-10 Illustration of quasicontinuous envelope formed by drawing coexistence tangents for a close succession of phases

9.2.4 The Microdomain Hypothesis

The broad-range non-stoichiometric phases formed at high temperatures are probably not homogeneous: it is almost certainly incorrect to regard them as having randomly distributed point defects. It is a reasonable assumption that they have at least some short-range order with structural elements closely related to those of their low-temperature precursors, a general view proposed by Ariya. On his hypothesis, a broad-range non-stoichiometric phase is not regarded as homogenous. It contains *microdomains* of structure and composition closely related to those of the low-temperature compounds in the phase diagram; they change shape and size continuously and intergrow coherently with other microdomains within a single particle. These dynamically fluctuating regions are very small, with an upper limit to their dimensions perhaps only three or four times those of the unit cell of a parent structure (i.e. not more than about 100 unit cells, since volume $\propto a^3$); they are sufficiently small not to act as nuclei for growth of another phase and are not evident in X-ray photographs of the material because of their random distribution. Due to this difficulty of observation, Ariya's hypothesis remains a hypothesis, although an attractive one, but it is by no means universally accepted.

Recent evidence suggests that the broad high-temperature α-phase in the Pr—O system is in fact composed of four distinct regions, each composed of different combinations of microdomains with structures related to those of the discrete low-temperature compounds. Thus:

Phase region	Microdomain structure elements
α_1	$Pr_7O_{12} + Pr_5O_9$
α_2	$Pr_7O_{12} + Pr_{11}O_{20}$
α_3	$Pr_7O_{12} + Pr_6O_{11}$
α_4	$Pr_7O_{12} + PrO_2$

Similarly, the wüstite 'phase' consists of three distinct regions (I, II and III, in Figure 9-7): however, the transitions between these regions are certainly not of the normal type (first or second order); their significance and interpretation remain obscure, especially as it has recently been shown by Fender and Cheetham that the same type of defect cluster (Koch type: see Section 9.2.6) is present throughout the wüstite region.

9.2.5 Reconstructive Processes in Defected Solids

We now proceed to a discussion of the various ordering and reconstructive processes which have been observed in non-stoichiometric solids. This is approached from the premise that only a very low concentration of point defects can be tolerated by an initially stoichiometric crystal before they are rearranged into some form having lower energy. With further change of composition, successive reconstructive processes occur, the final stage being that these ordered defects are *either* assimilated, becoming an integral part of a new structure related to that of the parent phase, *or* they are eliminated. The following chart, due to Anderson (1972), summarizes the possibilities, which are now introduced mainly in terms of case histories.

9.2.6 Defect Assimilation: Clusters and Multiclusters

$Fe_{1-x}O$

This material has an essentially complete f.c.c. oxygen sublattice. To maintain electrical neutrality in the defected material there are necessarily two Fe^{3+} ions per vacancy. The difference in magnetic behaviour between ferrous and ferric iron allowed Roth to show that a proportion of the Fe^{3+} was on *tetrahedral* sites. (We recall that since Fe^{3+} is d^5 and will be high-spin in an oxygen lattice, it has no preference for octahedral as opposed to

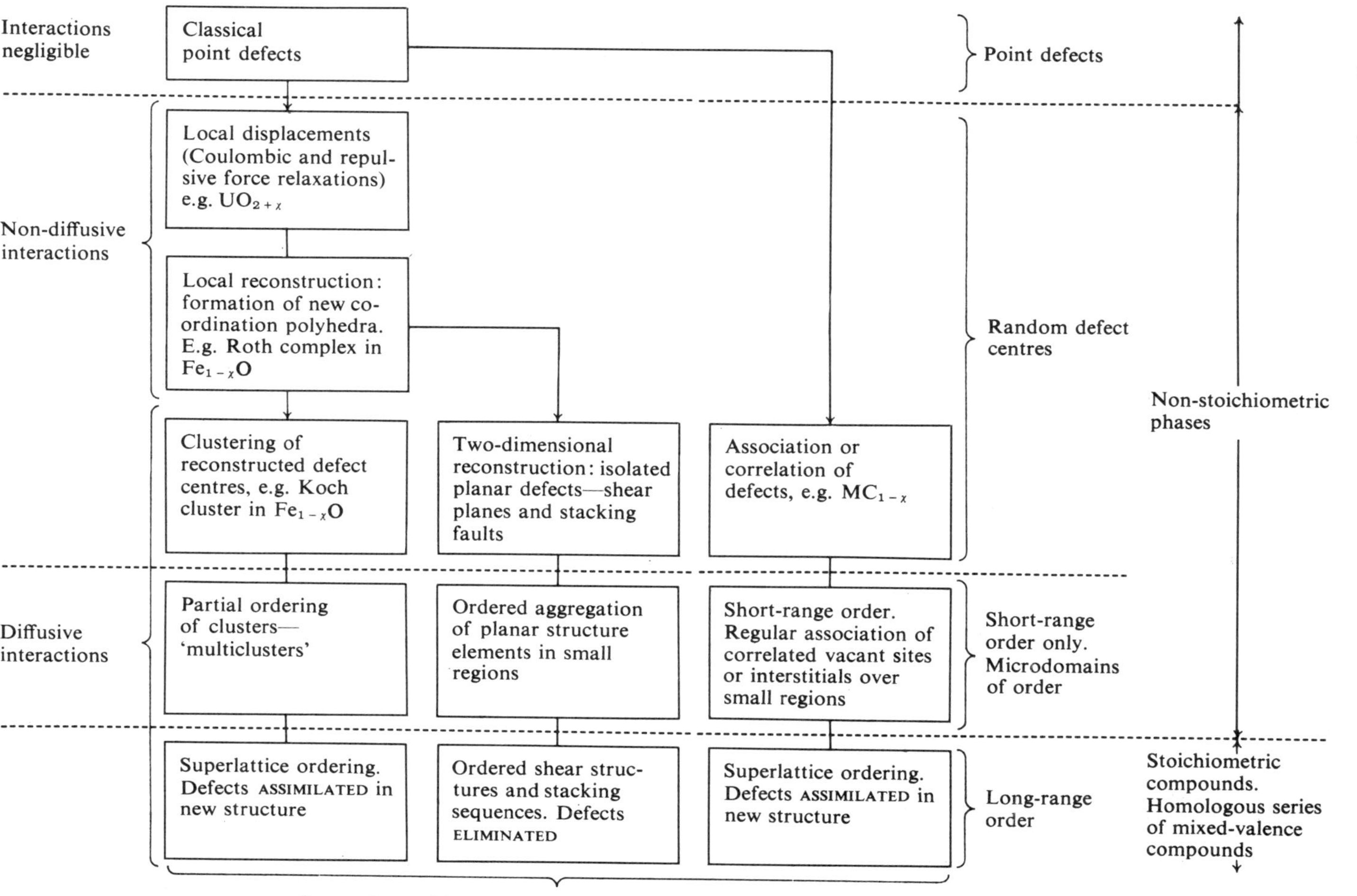

Interactions negligible
Classical point defects
Point defects
Non-diffusive interactions
Local displacements (Coulombic and repulsive force relaxations) e.g. $UO_{2+\chi}$
Local reconstruction: formation of new co-ordination polyhedra. E.g. Roth complex in $Fe_{1-\chi}O$
Random defect centres
Non-stoichiometric phases
Clustering of reconstructed defect centres, e.g. Koch cluster in $Fe_{1-\chi}O$
Two-dimensional reconstruction: isolated planar defects—shear planes and stacking faults
Association or correlation of defects, e.g. $MC_{1-\chi}$
Diffusive interactions
Partial ordering of clusters—'multiclusters'
Ordered aggregation of planar structure elements in small regions
Short-range order. Regular association of correlated vacant sites or interstitials over small regions
Short-range order only. Microdomains of order
Superlattice ordering. Defects ASSIMILATED in new structure
Ordered shear structures and stacking sequences. Defects ELIMINATED
Superlattice ordering. Defects ASSIMILATED in new structure
Long-range order
Stoichiometric compounds. Homologous series of mixed-valence compounds
Successions of intermediate phases (homologous series)

tetrahedral sites on the basis of crystal field stabilization energy (Table 14, Chapter 5), although size considerations may affect the distribution). As a minimum disturbance to the structure, Roth proposed that complexes existed consisting of one Fe^{3+} on a tetrahedral site together with *two* adjacent Fe^{3+} vacancies (i.e. one due to the original 'point' defect, the other formed by moving a neighbouring Fe^{3+} from its normal octahedral site to a tetrahedral one.) Thus:

$$(Fe^{3+})_{oct}\,\square_{oct} \longrightarrow (Fe^{3+})_{tet}\,(\square_{oct})_2$$

the new complex, set in its environment of perfectly regular rock salt structure being viewed as an element of the spinel structure of Fe_3O_4, the next highest oxide of iron.

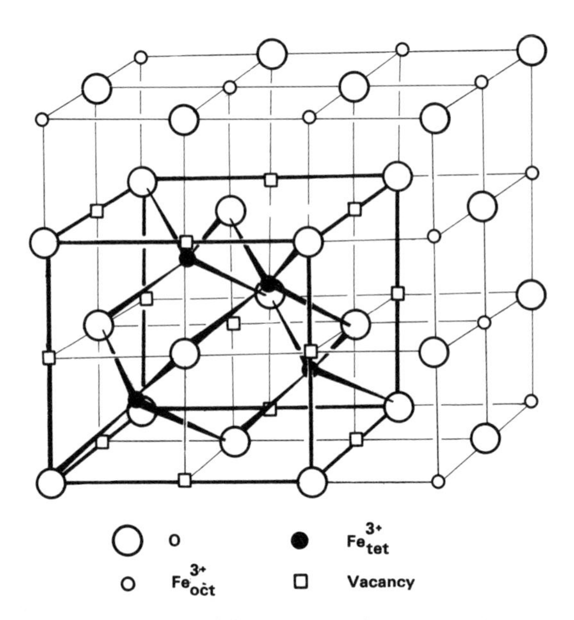

Figure 9-11 The Koch cluster of four tetrahedrally coordinate iron atoms and thirteen cation vacancies in $Fe_{1-x}O$, shown with parts of its surrounding NaCl lattice

Although complexes of the Roth type may exist at low cation vacancy concentrations and high temperature, it is now reasonably certain that a *cluster* of these constitutes the basic defect complex of the wüstite phase field. These so-called Koch clusters can be viewed as four Roth complexes plus five other vacancies; thus, overall they consist of four iron atoms in tetrahedral sites and thirteen octahedral vacancies. The net negative charge which they would otherwise bear is compensated by a sheath of Fe^{3+} ions on octahedral sites. The initial defects are therefore seen to have been *assimilated* into the new structure unit (the Koch cluster) which intergrows coherently with portions of undisturbed rock salt structure, see Figure 9-11. Each Koch complex (including the compensating Fe^{3+} ions) has a cell of size 2 × 2 × 2 times that of the parent rock salt lattice. A structure composed of an ordered array of them would have the composition $Fe_{23}O_{32}$ (i.e. almost Fe_3O_4); although its existence has not been finally established, it is quite probable that it is formed by annealing wüstite in the metastable range below 570°C. Certainly annealing leads to formation of groups of Koch complexes (i.e. 'multiclusters'), partially ordered with respect to each other, which intergrow coherently with portions of essentially unperturbed rock salt structure. The full details have still to be worked out but the general structural features of this classic bivariant, wide-range, non-stoichiometric material are now clear.

The Oxygen-rich UO_2 System

$UO_{2\cdot00}$ and U_4O_9 ($\equiv UO_{2\cdot25}$) exist in equilibrium as line phases at low temperatures. With increase of temperature their compositional widths increase until at 1127°C a single phase covers the range $UO_{2\cdot00}$ to $UO_{2\cdot25}$. In this so-called 'superstoichiometric' phase, additional oxygen has been

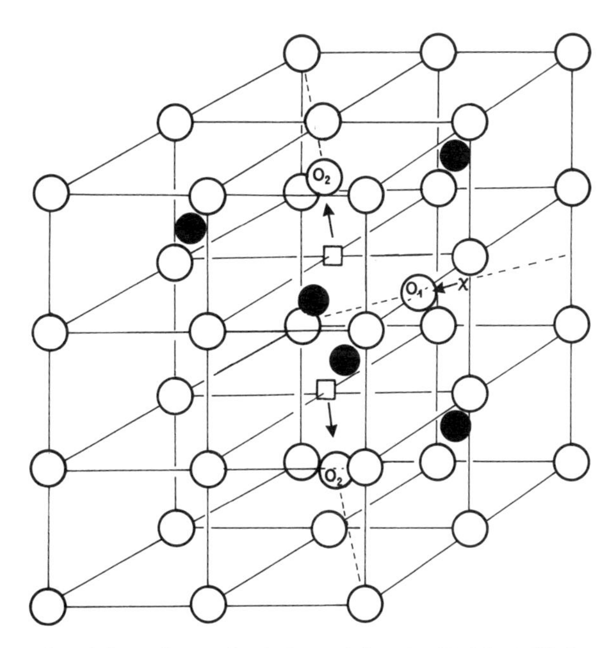

Figure 9-12 The defect cluster in UO_{2+x} (after B. T. M. Willis), consisting of two vacancies, □; one interstitial supernumerary oxygen, O_1; and two interstitial oxygens, O_2, shifted from their normal lattice positions. O_1 is displaced from the cube centre

incorporated into the fluorite lattice of UO_2. Contrary to reasonable expectation, the extra oxygen is not accommodated in vacant octahedral sites but occupies an interstitial position of low symmetry accompanied by two further oxygens which have been shifted from their normal lattice sites; see Figure 9-12. Quite probably some or all of the defect complexes are larger than this (for example, two supernumerary oxygens associated with three anion vacancy pairs) and there may be some further local relaxation. One of the compensating charges on the $O^{2-}(O^{2-}_{int.})_2 \square_2$ cluster is localized on a neighbouring cation which changes from U^{6+} to U^{5+}, the other is distributed statistically over neighbouring cation sites. At the oxygen-rich limit ($UO_{2 \cdot 25} \equiv U_4O_9$) the same type of vacancy–interstitial cluster is present but here they are ordered into a superstructure. As in wüstite there is a degree of correlation between the defect cluster in $UO_{2+\chi}$ depending upon composition, increasing with χ. $UO_{2+\chi}$ may therefore be regarded as containing microdomains of the U_4O_9 structure.

An analogous process is believed to take place in some mixed crystals also based upon the fluorite structure. Thus, in $CaF_2 . \chi yF_3$ mixed crystals, supernumerary fluorines are associated with a minimum of one $(F^-)\square$ pair created by removal of one anion from its normal site. These simple clusters also associate into multiclusters but are then largely prevented from further correlation and ordering by the large y^{3+} ions.

Defect Clusters in Other Systems

There can be little doubt that the pattern of defect formation and clustering described above for $Fe_{1-\chi}O$ and $UO_{2+\chi}$ will be found repeated in many other systems when more detailed investigations have been completed. There is good evidence that clusters are formed in $Mn_{1-\chi}O$ and probably also in $Co_{1-\chi}O$ although their structures may well differ from the Koch complexes of wüstite. Charge balancing in $Mn_{1-\chi}O$ impies creation of $Mn^{3+} d^4$ ions: these are most unlikely to occur in tetrahedral sites due to their highly unfavourable excess octahedral stabilization energy of 95·2 kJ mol^{-1}. Further, due to the necessity of accommodating Mn^{3+} in a tetragonal environment (Jahn–Teller distortion) a further entropy contribution arises from the possibility of orienting the tetragonal environment with respect to neighbours. It is to be expected, therefore, that the defect clusters in $Mn_{1-\chi}O$ will consist of Mn^{3+} in tetragonally distorted octahedral sites with, possibly, some of the Mn^{2+} shifted to tetrahedral sites (allowed since it is high-spin d^5). Since this process requires more disturbance of the oxygen sublattice than in wüstite, it may account for the narrower stability region of the $Mn_{1-\chi}O$ phase.

Clusters of seven vacancies are believed to occur in $V_{0 \cdot 79}O$. The data can be interpreted either in terms of two tetrahedra sharing a vertex, or platelets of vacancies lying parallel to the close-packed planes of the NaCl structure.

Coherent Intergrowth

The microstructure of materials such as those discussed above consists of defect complexes or multiclusters dispersed, and sometimes partially ordered, within a matrix of parent structure. For the coherent intergrowth of the new structure element with that of the parent there must clearly be close matching of their dimensions; this is a prerequisite for the transfer of atoms from one region of structure to another, the mechanism by which ordering of multiclusters takes place.

The general condition for coherence is that at least one sublattice of sites should be common to the hybrid structure. Reordering of atoms is then accomplished via the other sublattice. This is seen to be the case in both wüstite and UO_{2+x}: in each there is a common anion sublattice which runs coherently through defect clusters and intervening portions of parent structure alike. The same condition can be satisfied even with a partly defected sublattice as in superstructures based upon TiO, in which both cation and anion sublattices are incomplete; see next section.

9.2.7 Defect Assimilation: Superstructure Ordering of Vacancies

Vacancies may be *assimilated* into a structure by ordering them. The new structure is generally based upon the lattice of the parent but since, in a formal sense, it is now a ternary system (anion, cation, vacancy) with a more elaborated arrangement of the constituents in space, the repeat unit of the new structure is larger than that of the parent and of lower symmetry. It is referred to as a 'superstructure' of it. Because of the development of our understanding of non-stoichiometric phases, they are usually considered as having defected structures. More logically they should be regarded as the particular structures of materials with a proportion of atoms in each of two oxidation states. Just as the Ga_2S_3 structures (p. 289) are *incorrectly* described as having 'incomplete' or 'defected' ZnS lattices, so the superstructures of transition-metal compounds based upon other lattices should now be regarded as entities in their own right.

Superlattice ordering of vacancies has been demonstrated in a considerable variety of materials of which we consider two as representative. Of course, there is no clear-cut division between the processes considered in this section and the multicluster kind of structure above. In wüstite, for example, it seems likely that at the oxygen-rich limit the Koch clusters are ordered into a superlattice, $Fe_{23}O_{32}$.

The Titanium Monoxide Phase, TiO_x

Titanium and vanadium both have oxide phases $MO_{1\pm x}$ which exist over a range of composition centred about the stoichiometric value. In this they are distinguished from the other first-transition series monoxides (M = Mn, Fe, Co, Ni; CrO appears not to exist) which form phases only on the oxygen-rich side. This observation has been generalized by Reed: for an oxide MO_p oxygen-rich compositions MO_χ with $\chi > p$ are formed only if a higher oxide MO_q ($q > p$) exists. Similarly, metal-rich compositions exist only if a lower oxide can be formed. Thus, titanium and vanadium monoxides have a compositional range on both sides of MO since both the lower (Ti_2O; VO_χ, $\chi < 0{\cdot}5$) and higher (Ti_2O_3, V_2O_3) oxides exist. In contrast, Fe_2O does not exist and the stable range of 'ferrous oxide' ceases at the composition $Fe_{0{\cdot}95}O$. Although the existence of a higher or lower oxide is a prerequisite for existence of a non-stoichiometric range in one or other direction, the *actual* existence of such a phase may also depend upon other factors as well. For example, $Ti_2O_3 \equiv TiO_{1{\cdot}5}$ does not exist for $\chi < 1{\cdot}5$ because although the structure will tolerate cation defects, it cannot accommodate extra cations.

The monoxide phases of titanium and vanadium are both based upon the rock salt lattice and exhibit metallic conductivity due to formation of a conduction band from their spatially extensive t_{2g} orbitals (see Section 5.5.4). For titanium the compositional range is:

$$TiO_{0{\cdot}7} \text{ to } TiO_{1{\cdot}25} \text{ at } 1400°C$$
$$TiO_{0{\cdot}9} \text{ to } TiO_{1{\cdot}25} \text{ at } 900°C$$

while for vanadium the stable range is $VO_{0{\cdot}86}$ to $VO_{1{\cdot}27}$.

The feature of note in the TiO_χ phase is the concurrent existence of defects in *both* sublattices. In $TiO_{0{\cdot}7}$ the cation lattice is almost perfect but it is highly oxygen deficient (ca 33%), whereas by the composition $TiO_{1{\cdot}25}$ the oxygen lattice is nearly perfect but about 25% of the cation sites are vacant. Stoichiometric TiO is *not* a defect-free material; it has some 15% of both cation and anion vacancies. Above 900° these vacancies appear to be randomly distributed in the rock salt lattice, but below this temperature a succession of vacancy-ordered structures develops unless the material has not been sufficiently annealed, in which case the randomized structure appears also at low temperature in metastable condition. The details of the ordered structures depend upon the composition and are summarized as follows:

(*a*) TiO (range $TiO_{0{\cdot}9}$ to $TiO_{1{\cdot}1}$). Half the titanium and half the oxygen atoms are missing alternately in every third (110) plane.

(*b*) $TiO_{1{\cdot}25}$ (a line phase of very narrow range). The f.c.c. oxygen sublattice is complete, but one in five titanium atoms are missing; the arrangement of the vacancies is shown in Figure 9-13.

(*c*) $TiO_{1{\cdot}19}$. This is an intermediate phase in which portions of the TiO and $TiO_{1{\cdot}25}$ sructures are coherently intergrown: this is possible since both structures are based upon the same f.c.c. oxygen sublattice.

It is a reasonable inference that the spatially extensive *d*-orbitals on titanium are basic to the mechanism by which vacancies are ordered in these oxide structures.

Phases with Structures Based upon NiAs and Related Lattices

The NiAs (B8) and cadmium iodide (C6) structures are both based upon an h.c.p. lattice of non-metal with cations in octahedral holes. They differ in that in the C6 structure only alternate layers of octahedral sites are occupied whereas all are filled in NiAs. Between the limits represented by these two structures there exists a considerable range of materials, mostly chalcogenides of transition metals, of composition $AB_{1-\chi}$ where $0 < \chi < 1$. At first it was thought that these materials with intermediate values of χ were formed by the gradual random filling of the vacant layer in C6 or the creation of vacancies in B8. More recent studies have shown that, provided the specimens are properly annealed, these materials usually exist at low temperatures as well-defined compounds whose structures arise from the finite number of ways in which vacancies can be distributed in an ordered fashion in alternate layers of the C6 lattice. Figure 9-14 illustrates five such possibilities, each of which has been found in practice. Consider arrangement (a). It represents the structure of the octahedral site layer that would be vacant in CdI_2: XMX □ XMX □X . . . The overall composition of the material is therefore given by the repeat unit MX_2□ where □ represents the entire *layer* (a). In (a) one third of the cation sites are occupied. The composition of the crystal is therefore $MX_2M_{1/3} \equiv M_2X_3$. The relation of these structures to composition is summarized below.

Layer type (Figure 9-13)	Formula	Composition
NiAs (B8)	MX	MX
(d)	M_7X_8	$MX_{1.143}$ or $M_{0.875}X$
(b)	M_5X_6	$MX_{1.200}$ or $M_{0.833}X$
(c)	M_3X_4	$MX_{1.333}$ or $M_{0.750}X$
(a)	M_2X_3	$MX_{1.500}$ or $M_{0.666}X$
(e)	M_5X_8	$MX_{1.600}$ or $M_{0.625}X$
CdI_2 (C6)	MX_2	MX_2

These vacancy-ordered structures are the low-temperature forms of the materials they represent. Upon heating, the general pattern of behaviour is progressive change from line phase to broad-range non-stoichiometric phase via an intermediate condition of measurable compositional width, just as shown in Figure 9-6 for the Pr—O system. There is, however, good reason to believe that the broad-range high-temperature region is *not* to be described in terms of layers of randomized cation defects. Energetically, it is always more favourable to have layers in which defects are ordered rather than randomly distributed. The true situation is most probably one in which the *stacking sequences* are disturbed, but that the individual layers retain their ordered forms.

Series of phases of the above type have been reported for sulphides, selenides and tellurides of all the first-row transition metals, Ti to Ni, although the

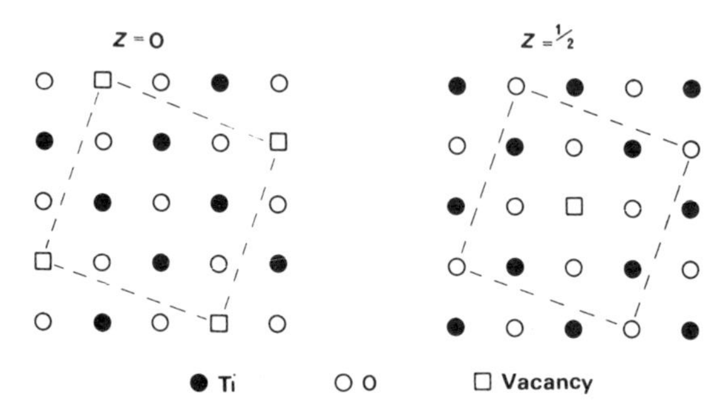

Figure 9-13 The structure of ordered $TiO_{1\cdot25}$, showing two successive layers of the (outlined) tetragonal cell, and its relation to the f.c.c. lattice. (Watanabe and coworkers, in Eyring and O'Keeffe (1970))

series is not necessarily complete in each case. One of the most thoroughly studied is the Cr–S system (Jellinek, 1957). Various proportions of the elements were heated under vacuum at 1000°C and then annealed at 300°C. The results were as follows:

Phase		Homogeneity range
	CrS	
	Cr_7S_8	$Cr_{0\cdot87-0\cdot88}S$
	Cr_5S_6	$Cr_{0\cdot85}S$
	Cr_3S_4	$Cr_{0\cdot76-0\cdot79}S$
Trigonal	Cr_2S_3	$Cr_{0\cdot69}S$
Rhombohedral	Cr_2S_3	$Cr_{0\cdot67}S$

The range of homogeneity apparently does not contain the ideal composition for Cr_3S_4, although further studies are desirable to check this. Their structures are those indicated above, with the exception of Cr_7S_8. In this material the vacancies are still confined to alternate layers of octahedral sites but are randomly distributed within the layers. On the other hand, Fe_7S_8 *does* have the ideal ordered structure, although on heating it above 360°C it undergoes a transformation to the same structure as Cr_7S_8 ,while above 400°C the cations are completely disordered over *all* cation sites.

Since Fe_7S_8 represents a stage of oxidation intermediate between ferrous and ferric, it must be described formally in terms of a mixture of oxidation states, as was $Fe_{1-x}O$ for example. Formally, we can write $Fe_5^{2+}Fe_2^{3+} \square S_8$ since creation of one vacancy by (hypothetical) removal of Fe^{2+} from FeS must be balanced by oxidation of two neighbouring cations. However, it is misleading in this case because an important element of the bonding in NiAs and related structures (of which this is one) is due to *d–d* overlap. We therefore prefer to write $Fe_7^{2+} \square S_8h_2^+$ where h_2^+ stands for two electron holes, the location of which is not specified (Ward, 1970). The Mössbauer evidence

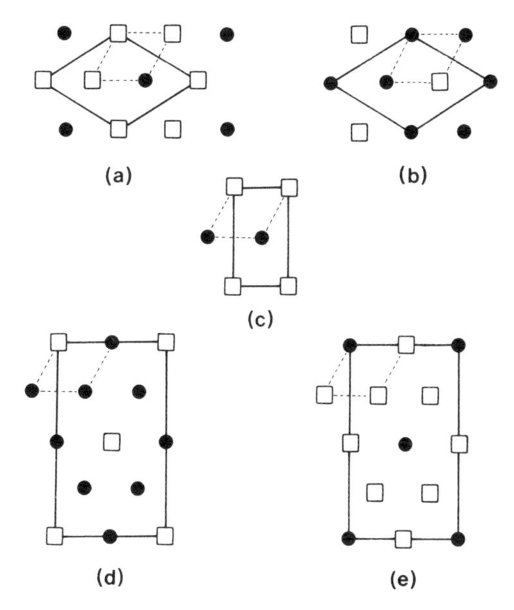

Figure 9-14 Five ways of ordering cation vacancies in the NiAs/CdI_2 series of transition-metal compounds. (Anderson, in Rao (1970)) ● = filled cation site; □ = cation vacancy

supports the view that all the iron is of the same type, but, taken together with other evidence on current carrier mobility, leads to the remarkable conclusion that the holes are present *not* as Fe^{3+} but associated with the *sulphur* sublattice. In terms of band theory, this means that the valence band (mainly associated with sulphur) is only partly filled, consistent with the known high electrical conductivity of the material.

Although the pattern of vacancy ordering and assimilation outlined above is commonly encountered, it by no means follows that related materials of the same formula have the same structure (see also Section 1.1). Thus the Cr_3S_4 structure is basically that of CdI_2 with the 'vacant' layers consisting of ordered vacancies and chromium atoms in the pattern of Figure 9-14c. In contrast, Fe_3S_4 (in the form of the mineral Smythite) has an elaborate 'sandwich' structure in which sheets of four close-packed sulphur atoms enclosing three (full) iron layers are stacked in the sequence ABCABC. Cations are totally absent from the layers on either side of the complex sheet (cf. CdI_2). Thus: □SFeSFeSFeS□SFeSFeSFeS□ . . .

Compounds in the B8/C6 sequence that have been studied in detail by a variety of physical techniques commonly exhibit a number of phase transitions. Some of these are associated with magnetic ordering, a topic which we will not discuss here; the interested reader is referred to Goodenough (1963). Others are more readily understood in structural terms as follows, examples having been introduced above.

(*a*) There may be an order–disorder transition within the partially filled sheets of the CdI_2-type structure which alternate with filled sheets.

(*b*) While retaining ordered configuration in the filled and part-filled cation layers, the stacking sequence may be disturbed.

(*c*) Defects may be randomly distributed over all cation sites.

The general structural theme upon which the B8/C6 series of defected materials is based (i.e. essentially stoichiometric slabs XMX interleaved with partially filled and ordered layers of cations plus vacancies) is capable of much further variation. Without going into details, it is clear that materials can be envisaged in which XMX slabs are linked by metal cations other than M; e.g. $FeCr_2Se_4$. Permutation of two metals and one non-metal yields a very large number of possible materials which should provide a regular showcase full of physical properties.

9.2.8 NiAs and Related Structures

In view of its central position as the parent structure to which many non-stoichiometric phases are closely related, it follows that a clear understanding of the nickel arsenide structure is basic to more complete appreciation of factors influencing the defected materials. The NiAs structure is certainly the most enigmatic of the simple AB_n structures and the reasons for its adoption are by no means self-evident. It is an almost uniquely flexible structure. It has the ability to absorb internal stresses by changes in axial ratio c/a (as indicated on p. 56) which retain the symmetry of the structure, rather than by a phase change. A wide range of composition can also be accommodated, *either* on the metal-deficient side of MX by the processes described above for the B8/C6 hierarchy of structures, *or* by accommodating extra metal atoms up to the limit M_2X. In the latter case (the structure is referred to as that of Ni_2In) the additional atoms are fed into the trigonal–bipyramidal holes; since there is *one* of these per packing atom the limit is clearly M_2X. This is a typically metallic structure and has c/a values near 1·22.

The octahedral and trigonal–bipyramidal sites share a common face, making movement from one type of site to the other especially easy and accounting (at least in part) for some of the experimental difficulties encountered in trying to form ordered materials. Commonly it proves impossible to obtain a compound in the NiAs structure with the ideal composition.

With the exception of some alloy phases, *the NiAs structure is essentially restricted to high temperatures*. It is avoided in the equilibrium state at low temperatures by one of three means (Kjekshus and Pearson, 1964):

(*a*) A change of stacking sequence, e.g. TiP.
(*b*) A structural deformation, e.g. MnP.
(*c*) The absence of any homogeneous phase at the ideal composition.

Titanium Phosphide, TiP

This structure differs from NiAs only in its layer-stacking sequence, ABAC, which is illustrated in Figure 9-15. As a geometrical consequence of this more elaborate stacking, two crystallographically different environments

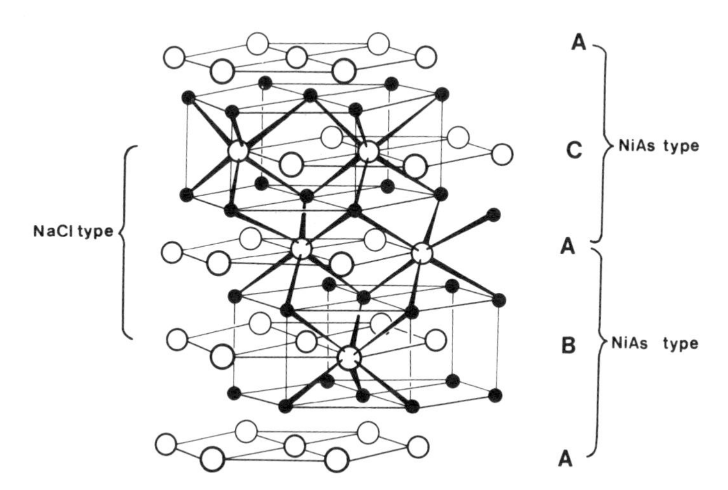

Figure 9-15 The TiP structure, showing the two kinds of coordination of the phosphorus atoms (trigonal–prismatic and octahedral). (Reprinted with permission from H. Reiss (Ed.), *Progress in Solid State Chemistry*, Vol. 1, Pergamon Press Ltd., Oxford, 1964)

exist for the packing atoms (P), some being in trigonal–prismatic coordination with six metal atoms (as in NiAs), the others having octahedral coordination to metal. In a formal sense the structure can be described as slabs of NiAs structure interleaved with others of the rock salt type. TiAs, ZrP and ZrAs, also adopt this structure.

Manganese Phosphide, MnP

This structure is almost as common as NiAs, being adopted principally by phosphides, arsenides, silicides and germanides, but apparently *not* by chalcogenides (other than $TiSe_{0\cdot95}$). It is characterized principally by a shift of the metal atoms from the octahedral centres, resulting in somewhat irregular coordination about each metal atom: in MnP itself the six Mn—P distances vary from 2·29 to 2·39 Å.

Reasons for Adoption of the NiAs, TiP and MnP Structures (Pearson, 1972; Kjekshus and Pearson, 1964)

In a sense NiAs is three structures in one: there are three distinct sets of values of the axial ratio c/a which must be accounted for. (a) Those with the ideal value 1·63, corresponding to undistorted closest packing, (b) values $>1\cdot63$, (c) values $<1\cdot63$. This flexibility of axial ratio implies a pronounced lack of rigidity in the bonding indicative of some ionic and/or metallic nature. In contrast, for the hexagonal wurtzite structure (only adopted by non-transition metal compounds) the axial ratio never varies by more than 3% from 1·63. We recall that the NiAs structure is adopted by transition-metal compounds only.

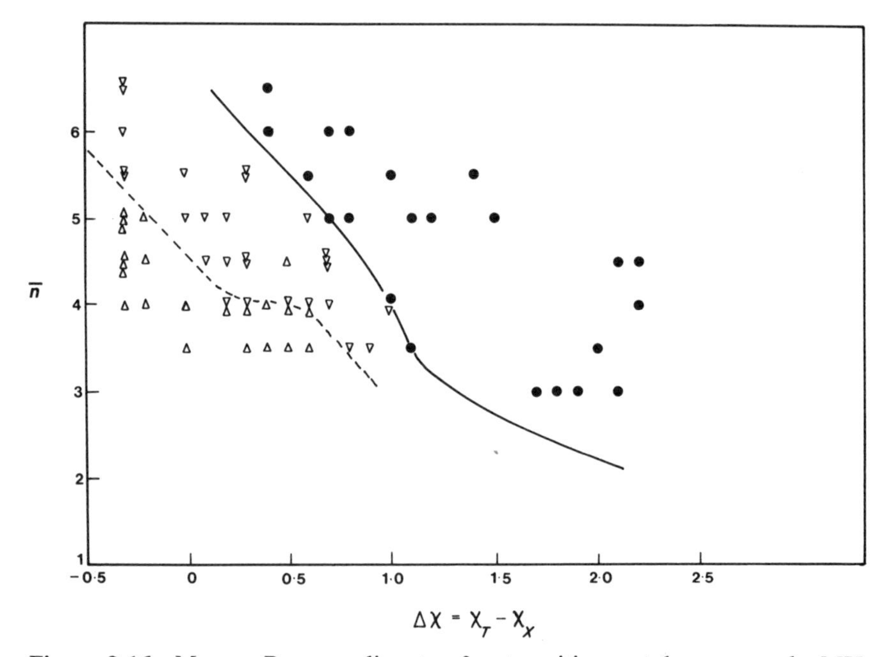

Figure 9-16 Mooser–Pearson diagram for transition-metal compounds MX (interstitial carbides and nitrides are not included). ● = NaCl, ▽ = NiAs, △ = MnP structure. (Kjekshus and Pearson (1964))

Compare the Mooser–Pearson diagrams of Figures 9-16 and 5-8. The boundary between directionally bonded (wurtzite) compounds and those with less than the critical degree of directional bonding (NaCl structure) for non-transition metals (Figure 5-8) is *identical* with that on the transition-metal diagram (heavy line). For transition-metal compounds, the progression of structures, rock salt → NiAs → MnP, represents increasingly directional bonding. Adoption of the NiAs structure with c/a close to 1·63 is undoubtedly due to the highly directional character of the bonding; the avoidance of a tetrahedral structure is probably due to *d*-electron stabilization associated with many d^n configurations. Since materials with the MnP structure lie in the extreme directional bonding domain of the $\bar{n}$ vs $\Delta\chi$ plot we must conclude that this particular distortion of NiAs occurs to allow better covalent overlap.

Phases with *low* c/a ratios (e.g. compounds of Rh, Pd and Pt) are largely metallic in bond type. Lowering c/a brings two next-nearest metal neighbours close to any one metal in an octahedral site thereby giving it a c.n. of (effectively) eight. At the same time this close approach allows *d–d* overlap along the *c*-axis.

Ratios $c/a > 1{\cdot}63$ are believed to be due to cation–cation repulsion in the most ionic of the compounds that adopt the NiAs structure. Compounds of this group appear to have relative atomic sizes incompatible with the NaCl lattice, but which can be accommodated in NiAs whilst relieving cation–cation repulsion by increasing c/a. Note, however, that transition metal *oxides* MO never adopt the NiAs structure, partly because of the different atomic sizes involved, partly due to their higher ionicity.

9.2.9 Defect Elimination: Crystallographic Shear

In Sections 9.2.5 and 9.2.6 it was observed that a non-stoichiometric system is intolerant of a high random defect concentration. Defects can become *assimilated* into a new structure by incorporating them as a new structure element. The most notable examples of this behaviour were found in the monoxides $M_{1-x}O$, where M = Mn, Fe, Co, Ni. In contrast, in an earlier region of the Periodic Table, notably among the higher oxides of titanium, molybdenum and tungsten, the reaction of the system to point defect creation is entirely different and results in their *elimination* by a process known as 'crystallographic shear' (c.s.). The term c.s. was coined by Wadsley, although structures in which the process occurs were first studied by Magneli (1953) and coworkers. We first outline the concept and then show how it assists in rationalizing many experimental observations.

In Section 7.2, reasons were outlined for describing some structures in terms of coordination polyhedra. The ReO_3 structure was specifically considered: it is conveniently described as ReO_6 octahedra joined to others by sharing all vertices (i.e. oxygen atoms). Imagine now the creation of a number of oxygen defects in the ReO_3 structure. There is evidence that in some oxide phases the locus of the reduction reaction resulting in c.s. is concentrated at a small cluster of sites, since imperfections can be seen (under the electron microscope) at exceedingly small departures from perfect stoichiometry. Figure 9-17 shows how one specific configuration of anion vacancies in ReO_3

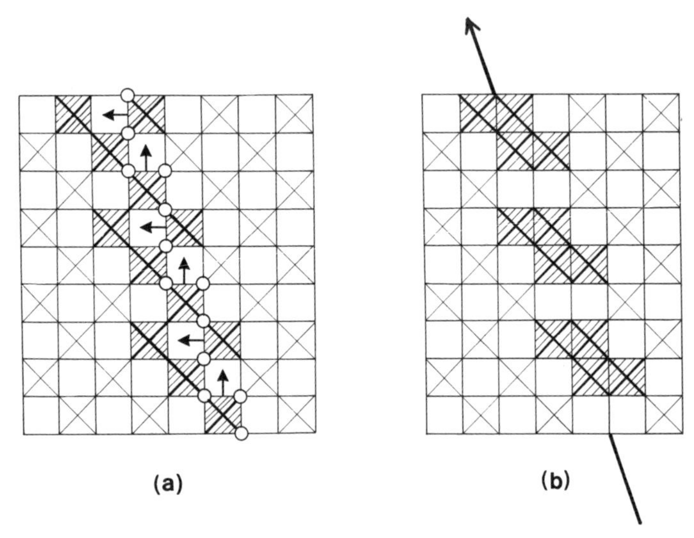

Figure 9-17 Schematic representation of defect elimination by crystallographic shear in the ReO_3 structure. (a) Shows one layer of the structure, viewing the vertex-shared octahedra from above. If oxygen vacancies (O) are introduced as shown, the CS movement indicated by arrows results in (b), in which the vacancies have been removed by edge sharing. (After Wadsley, in Mandelcorn (1964))

can be *eliminated by a change in anion coordination* along a specific crystallographic plane, cation coordination remaining octahedral. The process may perhaps be made clearer by considering the case of two pairs of octahedra as shown below.

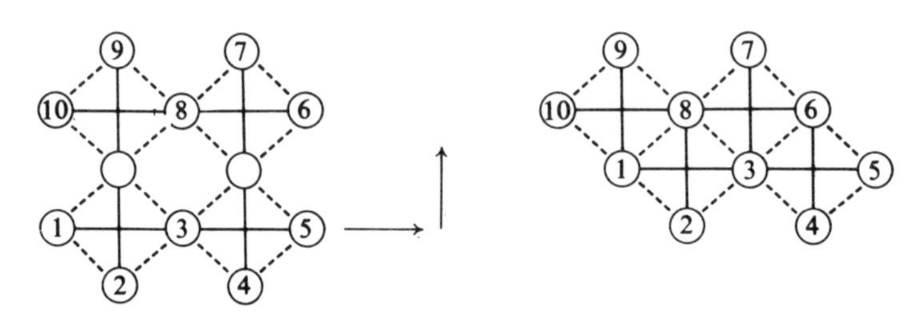

Two oxygens have been eliminated (corresponding to vacancies in the lattice): atoms 1, 3, 6 and 8, have suffered a change from linear two-coordination to

90° two-coordination (atoms 1 and 6) while 3 and 8 are now three-coordinate. Note especially that one part of the original structure has moved both along *and* normal to the fault or shear line: it is the movement *normal* to the shear plane that results in elimination of the defects and change of stoichiometry. For this reason it is essential to use the specific term 'c.s.' rather than just 'shear' since the latter implies no movement normal to the mechanical-shear plane. In a portion of ReO_3 structure in which this kind of crystallographic shear has taken place, slabs of essentially perfect ReO_3 structure are separated by a narrow slice of different structure and composition.

Remarkable though this process is, the most extraordinary feature is that not only do c.s. planes run right across entire microcrystals, but they are also generally found to be *regularly spaced.* Slabs of effectively unperturbed parent structure are separated by narrow c.s. planes, as described above, the slabs being of the same width. Clearly, a cooperative mechanism is implied. The *overall stoichiometry* of the resulting crystals depends upon the width of the unperturbed slabs, and the particular crystallographic plane in which the c.s. occurs. Commonly, a homologous series of structures is formed, having definite composition (i.e. line phases), a definite ratio of mixed-valence cations of formula MO_{an-m} where MO_a represents the parent structure, m is the number of anion sites eliminated by c.s., and n is the width (in octahedra) of the slabs between the new planes. Typical of these is Ti_nO_{2n-1}.

Among binary systems, c.s. has been observed on the rutile, MoO_3, α-PbO_2, and ReO_3 structures. Ternary and other systems give rise to still greater complexities and variations. Many c.s. phases are based upon the ReO_3 lattice as this is particularly suited to several different types of reconstructive process. Figure 9-17 illustrates c.s. along a {120} plane which results in groups of four ReO_6 octahedra joined by common edges: c.s along {130} planes yields groups of six octahedra (Figure 9-18). Oxides of the M_nO_{3n-1} series such as Mo_8O_{23} and Mo_9O_{26} are of this type. They are both formed by {120} c.s. of the ReO_3 lattice and differ only in the widths of the slabs between successive c.s. planes, viz. 8, 9 octahedra respectively. $W_{20}O_{58}$ of the M_nO_{3n-2} series, is sheared on {130} with ten octahedra between successive c.s. planes.

The series Ti_nO_{2n-1} and V_nO_{2n-1} are both based upon the rutile structure. Depending upon which set of planes are used for c.s., different families of structures are formed. Thus, in the titanium series, the family $n = 4$–10 involves c.s. along {121} planes, whilst use of {132} planes yields $n = 16$–36. Along the c.s. planes, octahedra are joined to each other by sharing *faces*; this is readily understood by reference to Figure 3–32.

The sensitivity of these c.s. systems to change of oxygen tension is remarkable and rather well pointed up by simply filling in the value of n in formulae such as those above. Thus, the *entire* family $n = 16$–36 falls inside the range $TiO_{1\cdot9375}$ to $TiO_{1\cdot9730}$, and the compositional difference between any two members is exceedingly small. For example, $n = 20$ $TiO_{1\cdot9500}$, $n = 21$ $TiO_{1\cdot9524}$.

It should be emphasized that for any one oxide system the various line phases are not necessarily related by the same structural theme. The point is well made by the oxides intermediate between MoO_2 and MoO_3 (Table 1), which show a variety of ways of reacting to change of oxygen tension.

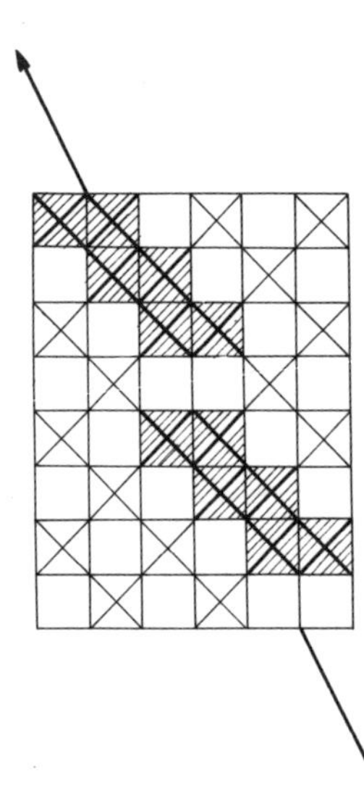

Figure 9-18 Crystallographic shear along a {130} plane of the ReO_3 structure. Compare with Figure 9-16 which shows the equivalent process along {120}

The **mechanism** by which vacancies are cooperatively ordered to give rise to c.s. is currently the object of much speculation and research. It seems probable that the process starts at a crystal edge, probably at a dislocation, and propagates into the body of the material. The succeeding processes are illustrated by reference to the various structures produced by reduction of rutile, as this system has been the object of intensive study.

At exceedingly small defect concentrations it now seems likely that point defects cluster together in platelets along what will eventually become c.s. planes, if reduction continues. With marginally higher defect concentrations randomly, but very widely, spaced c.s. planes appear along specific crystal planes, the {132} set in nearly stoichiometric rutile in the compositional range $TiO_{1\cdot999}$ to TiO_{2-x} with $x \ll 0\cdot001$.

In the approximate range $TiO_{1\cdot9370}$ to $TiO_{1\cdot999}$ {132} c.s. planes group together in lamellae in which the c.s. planes are regularly spaced defining high members of the Ti_nO_{2n-1} series. These lamellae coexist with nearly stoichiometric rutile saturated with platelets of the type mentioned above. Further reduction then leads into the $n = 16$ to 36 family of c.s. phases.

Between $TiO_{1\cdot937}$ ($n = 16$ for the {132}-plane group of c.s. structures) and $TiO_{1\cdot900}$ at which a new homologous series based upon {121} c.s. planes is

Table 1 Molybdenum oxides and their structural relationships (after Greenwood (1968))

χ in MoO_χ	Formula	Family	Description
2·000	MoO_2		Rutile.
2·750	Mo_4O_{11}	M_nO_{3n-1}	Slabs of octahedra joined by tetrahedra.
2·765	$Mo_{17}O_{47}$	$W_{18}O_{49}$	Complicated linking patterns of octahedra involving seven-co-
2·800	Mo_5O_{14}	$W_{18}O_{49}$	ordinate pentagonal-bipyramidal Mo, Figure 9-19.
2·875	Mo_8O_{23}	M_nO_{3n-1}	Slabs of ReO_3 structure eight octahedra thick.
2·889	Mo_9O_{26}	M_nO_{3n-1}	lSabs of ReO_3 structure nine octahedra thick.
	$Mo_{18}O_{52}$		MoO_3-type strips connected by tetrahedra, with increased edge-sharing at boundaries.
3·000	MoO_3		Figure 7-15c.

established (i.e. $n = 10$), there is a remarkable transition. The details were worked out for a mixed Cr/Ti oxide system of the rutile type, but it is almost certainly the same in the pure Ti—O system. As the compositional range is traversed from $TiO_{1\cdot937}$ the {132} c.s. planes pivot around, passing through every possible intermediate orientation, until they stabilize on {121} for $TiO_{1\cdot900}$. For *any* composition within this range the structure is well ordered: put another way, continuous changes in composition are accommodated within a continuous series of *ordered* structures which may have c.s. planes with apparently irrational indices. Systems which react in this manner are said to be 'infinitely adaptive' (Anderson, 1973), and have only been recognized very recently.

Block Structures: Double Crystallographic Shear

An exceedingly subtle and flexible form of defect elimination is found in the so-called 'block' structures, which are based upon the ReO_3 lattice. Block structures are found in the Nb—O system and extensively in ternary systems M—Nb—O where M = Ti, Mo, W or P, and in some related oxide fluorides. They are formed by c.s. in two orthogonal directions in the parent oxide, thereby dividing it into columns or blocks of essentially infinite length and of cross-section $m \times n$. The block size in cross-section, expressed as the number of octahedra sharing vertices, (i.e. the mode of linking characteristic of the parent structure), is always between 3×3 and 5×5. Up to three different block sizes have been found in a given phase.

At the shear planes there is a displacement of half an octahedron diagonal along the column axes; thus, the ReO_3 oxygen sublattice is left unaltered but

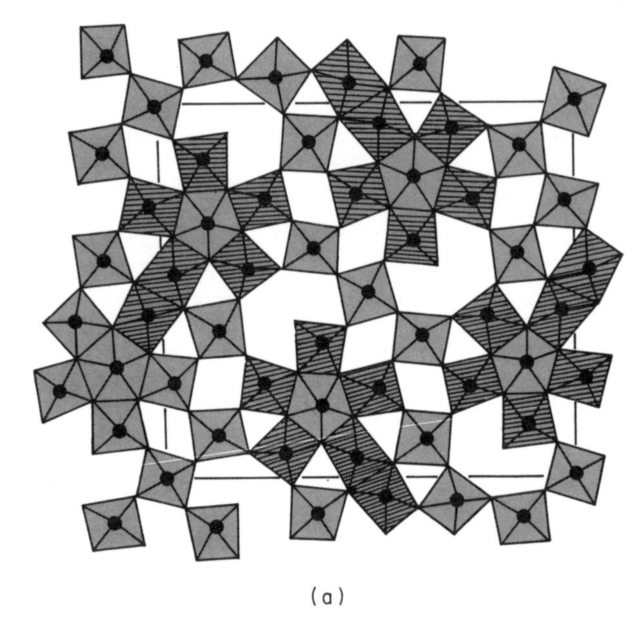

(a)

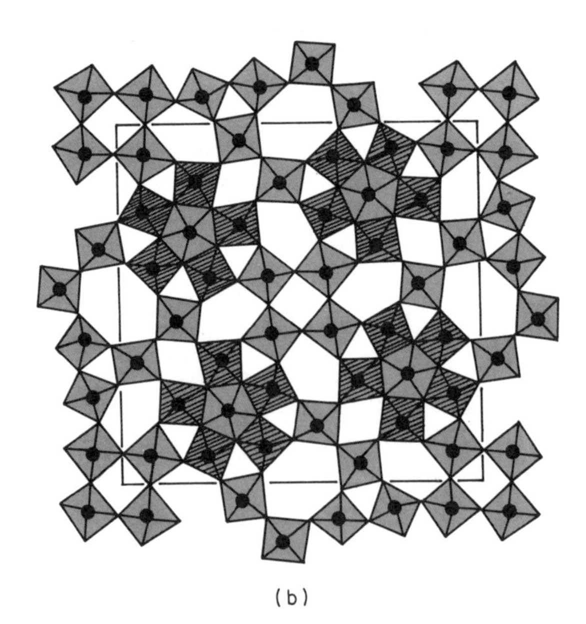

(b)

Figure 9-19 The structures of (a) $Mo_{17}O_{47}$ and (b) Mo_5O_{14}. (Parts of each diagram are shaded to assist appreciation of the different polyhedral connections. (Modified from L. Kihlborg, in *Adv. Chem. Series*, No. 39 (1963) by courtesy of the American Chemical Society)

the metal atoms are now at two different levels. The c.n. of oxygen at the outside of a block is 1 at unshared block corners, 3 at the edges, and 4 at shared corner sites. (Of course, it remains linear two-coordinate in the interior of each block of ReO_3 structure.) The two or more sets of blocks in any one compound form a space-filling rectangular array in which individual blocks are found either singly, in pairs, or joined together to form infinite arrays.

The formulae of block structures may be represented in terms of the blocks $(m \times n)_p$ they contain, where p = number of blocks joined together (usually 1, 2 or ∞). If p is finite, tetrahedral sites are created at the free corners. The overall composition of a given block type is given by $M_{mnp+1}O_{3mnp-p(m+n)+4}$ which, in the limit $p \rightarrow \infty$, becomes $M_{mn}O_{3mn-(m+n)}$. When more than one block type is present, the contributions from each set must be summed. Thus:

(*a*) Blocks of type $(3 \times 4)_2$ are of formula $M_{25}O_{62}$.
(*b*) Blocks of type $(3 \times 4)_\infty$ are of formula $M_{12}O_{29}$.
(*c*) A structure composed of two types of block $(3 \times 4)_2 + (3 \times 4)_\infty$ has formula $M_{25}O_{62} + M_{12}O_{29} = M_{37}O_{91}$.

This remarkable process of double c.s. provides an extraordinarily flexible means of accommodating exceedingly small changes of M:O ratio whilst maintaining a high degree of order. The sensitivity to change is indicated by the following series of eight distinct block structures which lie between $\chi = 2{\cdot}4100$ and $2{\cdot}5000$ in NbO_χ. It should, however, be emphasized that coherent intergrowth of adjacent phases is common.

χ	Block structure	Formula
2·4167	$(3 \times 4)_\infty$	$Nb_{12}O_{29}$
2·4545	$(3 \times 4)_\infty + (3 \times 3)_1$	$Nb_{22}O_{54}$
2·4681	$(3 \times 4)_\infty + (3 \times 3)_1 + (3 \times 4)_2$	$Nb_{47}O_{116}$
2·4800	$(3 \times 4)_2$	$Nb_{25}O_{62}$
2·4872	$2(3 \times 4)_2 + (3 \times 4)_1 + (3 \times 5)_\infty$	$Nb_{39}O_{97}$
2·4906	$(3 \times 4)_2 + (3 \times 4)_1 + (3 \times 5)_\infty$	$Nb_{53}O_{132}$
2·5000	$(3 \times 4)_1 + (3 \times 5)_\infty$	H–Nb_2O_5
	$(4 \times 4)_\infty$	M–Nb_2O_5
	$(4 \times 4)_\infty$	N–Nb_2O_5

There are a sufficient number of parameters involved such that the same formula may be given by different block structures, which may therefore be regarded as polymorphs. For example, $M_{47}O_{116}$ can be given by either $(3 \times 5)_\infty + (3 \times 4)_1 + (3 \times 3)_2$ or $(3 \times 4)_\infty + (3 \times 3)_1 + (3 \times 4)_2$. Nb_2O_5 exhibits no less than fourteen modifications of which ten are probably block structures; in M– and N–Nb_2O_5 the $(4 \times 4)_\infty$ columns are linked in different ways. Several of the salient features of block structures are illustrated by Figure 9-20 which represents the structure of H–Nb_2O_5.

Figure 9-20 The block structure of H–Nb_2O_5, $(3 \times 4)_1 + (3 \times 5)_\infty$. The structure consists of isolated blocks of ReO_3 structure of size (3×4) octahedra (heavy lines), together with blocks of size (3×5) which are linked to each other by common edges thereby forming strips $(3 \times 5)_\infty$

9.2.10 Non-stoichiometry and Chemical Specificity

Our brief, and necessarily oversimplified, survey of non-stoichiometric phases has revealed a remarkable degree of chemical specificity in the preferred form of ultra microstructure. This is so marked as to imply that the manner of dealing with defects in a compound of a given metal is directly related to its electronic structure, and hence to the relative ease of accessibility of lower oxidation states. The full details of these relationships have not yet been worked out, but we can note some more or less obvious features.

The rock salt type oxides MO of the first transition series (Ti to Ni), show two kinds of behaviour. Those of Ti and V are metallic in type, whilst those of Mn to Ni are semiconductors: non-stoichiometry is handled differently in the two types. In 'TiO' and 'VO', high concentrations of vacancies are ordered in a way which must depend upon the cooperative nature of the bonding (see p. 157); 'MnO', 'FeO', etc., handle defects by assimilating them via clusters of various types.

With higher oxides such as MO_2, M_2O_5 and MO_3, a wonderful variety of processes for accommodating departures from stoichiometry is observed. However, it should perhaps, be emphasized that all these ways of assimilating defects in shear, block and infinitely adaptive structures, operate within the framework of the ideal lattice type of the perfectly stoichiometric end member, and that this end member structure is determined by the principles outlined earlier in this book.

Crystallographic-shear and double c.s. (i.e. block) structures are exhibited by Ti, V, Nb, Mo, W and (most probably) Re, elements which form a diagonal band across the *d*-block series. Beyond this band (i.e. to higher atomic number in each *d* series) shear structures are not known. By the extreme end of the 4*d* and 5*d* series the corresponding materials show negligible departures from stoichiometry, reflecting the relatively large energy intervals between oxidation states. The tantalum oxides do not exhibit c.s. but have infinitely adaptive structures in which the relative proportions of six-and seven-coordinate cations vary with M:O ratio, and some U—O systems (U_3O_8 especially) behave similarly. Finally, the Zr and Hf oxides are based upon the fluorite lattice.

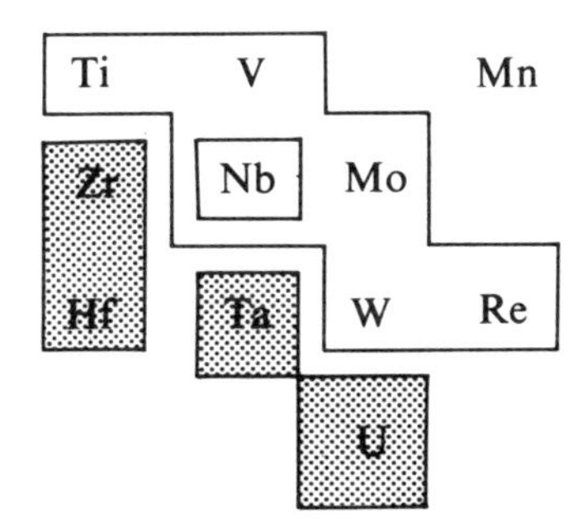

This area of structural chemistry still poses major problems and challenges and there is clearly a great deal of experimental work yet to be done. Whatever the final picture, it is sure to have surprises in store.

BIBLIOGRAPHY

Anderson, J. S., *Bull. Soc. Chim. France*, **7**, 2203 (1969).

Anderson, J. S., in *Surface and Defect Properties of Solids*, Vol. 1 (Ed., W. M. Roberts and J. M. Thomas), The Chemical Society, London, 1972.

Anderson, J. S., *J. Chem. Soc.* (*Dalton*), **1973**, 1107.

Burch, R. and F. A. Lewis, in *Annual Reports*, **67A**, 231 (1970). The Chemical Society, London.

Bursill, L. A. and B. G. Hyde, Chapter 6 in *Progress in Solid State Chemistry*, **7** (Ed., H. Reiss and J. O. McCaldin), Pergamon, Oxford, 1972.

Eyring, Le Roy and M. O'Keeffe (Ed.), *The Chemistry of Extended Defects in Non-Metallic Compounds*, North Holland, Amsterdam, 1970.

Goodenough, J. B., *Czech. J. Phys.*, **B17**, 304 (1967).

Goryunova, N. A., *The Chemistry of Diamond-like Semiconductors*, Chapman and Hall, London, 1965

Jellinek, F., *Arkiv Kemi*, **20**, 447 (1962).

Kihlborg, L., in *Adv. Chem. Series*. **39**, American Chemical Society (1963).

Kjekshus, A. and W. B. Pearson, in *Progress in Solid State Chemistry*, **1** (Ed., H. Reiss), Pergamon, Oxford, 1964.

Koch, F. and J. B. Cohen, *Acta Cryst. B.*, **25**, 275 (1969).

Magneli, A., *Acta Cryst.*, **6**, 495 (1953).

Mandelcorn, E. (Ed.), *Nonstoichiometric Compounds*, Academic Press, New York, 1964.

Pearson, W. B., *Crystal Chemistry and Physics of Metals and Alloys*, Wiley, London, 1972.

Rabenau, A. (Ed.), *Problems of Nonstoichiometry*, North Holland, Amsterdam, 1970.

Rao, C. N. R. (Ed.), *Modern Aspects of Solid State Chemistry*, Plenum Press, New York, 1970.

Roth, W. L., *Acta Cryst.*, **13**, 140 (1960).

Suchet, J. P., *Crystal Chemistry and Semiconduction in Transition Metal Binary Compounds*, Academic Press, London, 1971.

Ward, J. C., *Rev. Pure Appl. Chem.*, **20**, 175 (1970).

Formula Index

Subject Index